Reflective
Liquid Crystal Displays

Wiley-SID Series in Display Technology

Editor:
Anthony C. Lowe

Display Systems:
Design and Applications
Lindsay W. MacDonald and **Anthony C. Lowe** (Eds)

Electronic Display Measurement:
Concepts, Techniques, and Instrumentation
Peter A. Keller

Projection Displays
Edward H. Stupp and **Matthew S. Brennesholz**

Liquid Crystal Displays: Addressing Schemes & Electro-Optical Effects
Ernst Lueder

Reflective Liquid Crystal Displays
Shin-Tson Wu and **Deng-Ke Yang**

Published in Association with the *Society for Information Display* SID

Reflective Liquid Crystal Displays

Shin-Tson Wu
HRL Laboratories, Malibu, California, USA

Deng-Ke Yang
Kent State University, Ohio, USA

JOHN WILEY & SONS, LTD
Chichester • New York • Weinheim • Brisbane • Singapore • Toronto

Other Wiley Editorial Offices

John Wiley & Sons, Inc., 605 Third Avenue,
New York, NY 10158-0012, USA

Wiley-VCH Verlag GmbH
Pappelallee 3, D-69469 Weinheim, Germany

John Wiley, Australia, Ltd, 33 Park Road, Milton,
Queensland 4064, Australia

John Wiley & Sons (Canada) Ltd, 22 Worcester Road
Rexdale, Ontario, M9W 1L1, Canada

John Wiley & Sons (Asia) Pte Ltd, 2 Clementi Loop #02-01,
Jin Xing Distripark, Singapore 129809

Library of Congress Cataloguing in Publication Data

Wu, Shin-Tson
  Reflective liquid crystal displays/Shin-Tson Wu, Deng-Ke Yang.
    p. cm.
  Includes bibliographical references and index.
  ISBN 0-471-49611-1
  1. Liquid Crystal Displays.   I. Yang, Deng-Ke.   II. Title.
  TK7872.L56 W82 2001
  621.3815'422 - dc21   00-054572

British Library Cataloguing in Publication Data

A catalogue record for this book is available from the British Library

ISBN **0 471 49611 1**

Typeset in 10/12pt Times by Thomson Press (India) Ltd., Chennai
Printed and bound in Great Britain by Antony Rowe Ltd., Chipppenham, Wilts
This book is printed on acid-free paper responsibly manufactured from sustainable forestry,
in which at least two trees are planted for each one used for paper production.

Contents

5   Super-twisted Nematic Displays   113

6   Guest–Host Displays   125

Series Editor's Foreword

Reflective Liquid Crystal Displays by Shin-Tson Wu and Deng-Ke Yang is the fifth volume in the Display Technology series published by John Wiley & Sons in association with the Society for Information Display.

Although transmissive backlit displays have captured the majority of the flat panel display market, the rapid growth of wireless communications and the availability of light weight, low power storage technology have created an increasing demand for light weight communication devices in markets that include telephony, web-based communications and electronic readers. Recently, much effort has been devoted to improving the power consumption and visual performance of reflective displays to the extent that they can be used as real alternatives to the printed page. Yet, despite these efforts, a display that offers a combination of high reflectivity, high contrast, colour, low power and low cost remains an elusive target. The difficulty of developing such a display is such that reflective displays present one of the greatest intellectual challenges in the entire field of display science and technology.

Thus this book addresses an aspect of display technology that is of great technical interest and of significant economic importance. It has been written by two authors of international repute who have both made leading contributions to their respective fields, as the reader will deduce from the very extensive bibliography provided with each chapter. Although the object of this book is to discuss liquid crystal displays, a comprehensive review of all reflective display technologies is provided in the first chapter and this will assist the reader to place the different reflective display technologies in context. There follows a chapter on the properties of liquid crystals, which is of sufficient detail to satisfy the needs of all but the most advanced reader. The next six chapters comprehensively describe all known reflective LCD effects and the final three chapters discuss the optimisation of important display attributes; fast response, low voltage operation and wide viewing angle.

This is a very complete work. Those readers who wish to understand the derivation of the physical relationships that govern display effects will find what they require. But this book offers far more. It contains a wealth of practical detail and information that the practicing engineer will find of great value and which hitherto has been unavailable in a single volume.

Anthony C Lowe
Braishfield, UK, 2000

Preface

In a chilly winter at Kobe, Japan, Shin-Tson Wu met with Dr Anthony Lowe of IBM while attending the 1998 International Display Workshops. Shin-Tson Wu warily accepted his challenge to write this book, knowing the diversities in this technical area. Later, Professor Denk-Ke Yang was persuaded to partake in this challenge.

There are not many experts in the liquid crystal display (LCD) area because it covers such broad technology spectra. On the material side, many types of material ranging from solid state, liquid to polymers, are involved. For example, a LCD panel may consist of glass or plastic substrates, polarisers, indium-tin-oxide electrodes, pigment color filters, silicon thin-film transistors, liquid crystal mixture, polymeric phase retardation films, optics and lighting system. How to design and formulate liquid crystal mixtures with proper physical properties is a challenging task. To optimise one parameter is probably not too difficult. However, to optimise half a dozen parameters simultaneously, e.g. wide temperature range (from $-40°C$ to $90°C$), birefringence, rotational viscosity, elastic constants, dielectric anisotropy, photo-stability and resistivity, is an art. On the device side, questions of how to achieve a wide viewing angle without sacrificing light throughput, high brightness at low power consumption, high contrast ratio at low voltage, and fast response time at low temperature, remain to be answered. On the addressing electronics side, the selection and processing of amorphous silicon, low or high temperature poly-silicon and single-crystal silicon, affect the device resolution, light throughput and manufacturing yield. On the optics side, liquid crystals are not self-emissive so that backlighs are required for direct-view displays and lamps for projection displays. How to convert unpolarised light to linearly polarised light requires a deep understanding of optics. Liquid crystal display is truly an interdisciplinary field; it requires tremendous teamwork. Every expert stands a chance to contribute. A breakthrough in one area could lead to a chain reaction for the whole industry.

It is known that technologies advance rapidly, but the underlying operation principles remain. Therefore, in this book we focus on the basic operation principles and some exemplary device structures. Owing to the vast amount of published literature, it is impossible to cover all the details. We can only select a few landmark approaches and analyse their merits and demerits.

Although the key emphasis of this book is on reflective displays, some common aspects relevant to transmissive displays are also discussed. Three types of reflective display have been developed and have found widespread application. These are direct-view, projection and microdisplays. Reflective direct-view liquid crystal displays offer advantages in low power consumption, sunlight readability, light weight and compact size, and are particularly attractive for mobile displays and personal entertainment. Reflective projection displays are targeted at projection monitors, multi-media projectors and electronic cinema. Finally, microdisplays (virtual projection) open new applications for mobile Internet and near-eye displays.

Several promising liquid crystal technologies have been developed, depending on whether two, one or zero polarisers are employed. Chapter 1 briefly highlights the state-of-the-art reflective display technologies. Chapter 2 describes the basic properties of nematic liquid crystal materials that are the dominant electro-optic media for display applications. Chapter 3 describes the phase retardation effect and associated display devices. The most commonly employed twisted nematic cells are analysed in detail in Chapter 4. Computer simulation results of each operation mode are compared and the competitive advantages discussed. Chapter 5 summarises the super twisted nematic cells using one or two polarisers. A trans-reflective display that is intended to accommodate a dark ambient is presented. In the following three chapters, a guest-host display, a polymer-dispersed liquid crystal and a cholesteric liquid crystal display that does not require any polariser are introduced. In the absence of a polariser, the reflectivity can reach as high as 60%. However, the trade-off is in the reduced contrast ratio. Chapter 9 presents the operation principles and device structures of some bistable nematic displays. Three most critical display issues, fast response time, low operation voltage and wide viewing angle, are treated separately in Chapters 10, 11 and 12, respectively.

We are greatly indebted to Drs Anthony Lowe, Allan Kmetz and Ernst Lueder for the critical reviews of our outlines and for providing useful suggestions. Ongoing collaborations with Dr C.S. Wu of Hughes Research Laboratories (now at Compaq Computers), Drs F.C. Luo and D.L. Ting of Unipac Optoelectronics, Drs C.L. Kuo and C.K. Wei of Chi-Mei Optoelectronics, and the ERSO display group are greatly appreciated. Deng-Ke Yang would like to thank Ms E. Landry of KSU for patiently proof-reading his manuscripts. He also would like to thank Professors J. Kelly and P. Bos for many useful discussions. Many colleagues kindly provided us with drawings, as acknowledged with the figures. We are also grateful for the long-term financial support of AFOSR (through Dr. Charles Lee) HRL Laboratories and NSF ALCOM centre. The earnest prayers and spiritual encouragement from our family members and friends gave us strength during the course of manuscript preparation.

Shin-Tson Wu
Deng-Ke Yang

Foreword

About 30 years ago RCA demonstrated the first LCDs, based on the dynamic scattering effect which involves a transition between scattering and clear states. Off-axis illumination from the rear was used initially to obtain good contrast and visibility; light from a hidden lamp passed straight through the clear background away from the observer, but where the display was energised the light was forward-scattered into a wide angle that included the observer's eye, resulting in a bright image on a dark background. A four-function calculator the size of a hardcover book with a pair of D-cells to power the backlight appeared on the market, and wall-mounted digital clocks were installed in many airports, but backlit LCDs had a hard time competing with emissive displays. Replacing the backlight by a mirror created a reflective-mode LCD whose power consumption was three orders of magnitude less than competing displays, a unique performance advantage.

Television was then the only mass market for electronic displays, and it was locked up by the cathode-ray tube. There remained only niche markets for products like digital scientific instruments and elevator indicators, where one could find examples of LED, plasma, electroluminescent, vacuum fluorescent and incandescent displays, not to mention edge-lighted numerics and mechanical flaps and drums. The situation changed dramatically when integrated circuits appeared and IC manufacturers started looking for new markets. The Sharp Corporation in Japan found some success with a sleek pocket calculator whose pop-up lid shielded the reflective LCD to prevent the glare of bright light sources from reaching the eye of the user. Digital wristwatches could not use that stratagem so LEDs took an early lead despite the need to push a button for momentary activation of the power-hungry display. A breakthrough came with the discovery of a different liquid-crystal electro-optic effect: the twisted nematic cell between crossed polarisers produced black characters on a clear background, so the troublesome mirror could be replaced by a white diffusing reflector. The reflective TN-LCD was clearly visible under many viewing conditions including full sunlight, with power consumption so low that a wristwatch could display the time continuously for more than a year. A new consumer mass market swept the world. Multiplexing was developed to reduce the cost of addressing more characters, and batch fabrication brought down manufacturing costs. Soon the reflective TN-LCD, often powered by solar cells, came to dominate another burgeoning market for pocket calculators.

When the personal computer was introduced in the early 1980s, it quickly became clear that there was a need for a portable version, but no flat-panel technology existed that could display a full page of information with good battery life at low cost. The multiplexed TN-LCD evolved over years toward a somewhat wider viewing angle, better contrast and higher information content, in part by adopting 'first minimum' designs that demanded critical manufacturing control of cell thickness. The LCD industry in Japan despaired of reaching the required performance and undertook a massive effort to develop active-matrix addressing, replacing multiplexing with direct control of each pixel by its own thin-film transistor. A breakthrough came again from an inventive modification of the electro-optic effect: the SuperTwisted Nematic was conceived through improved analytical tools and fabricated using precise cell thickness to achieve the desired full-page capacity. Very soon portable PCs with reflective monochrome STN displays were everywhere, confirming yet another huge consumer market.

By then the desktop PC had converted to color and consumers demanded the same performance in the flat panels. Designers installed red-green-blue microfilters in the LCD but only 1% of the light came through. They were forced to develop increasingly clever lightguides behind the LCD to distribute white light from bright fluorescent lamps; they sacrificed the advantages of reflective operation to gain color. The challenge of manufacturing millions of TFTs on large glass panels with virtually perfect yield and low cost was eventually accomplished, and backlit active-matrix LCDs now dominate the huge market for laptop computer displays. Through sophisticated analyses of LCD physics and optics, compensator films and new LCD effects such as in-plane switching have been developed that improve the field of view, contrast and speed enough for AMLCDs to encroach on the traditional CRT domain of desktop monitors and television.

Recently other important trends have become apparent. The Internet, cellular telephones and digital coding of images are revolutionising our society, leading to a demand for multimedia connectivity wherever we go. The Palm Pilot reminded people that a reflective display makes possible a truly portable 'personal digital assistant' with long battery life. Portable DVD players and 'electronic books' are testing the waters, but again the need for color is a problem. Researchers are now producing a plethora of electro-optic effects in liquid crystals and other materials to devise a reflective display comparable to a color print in a magazine. They seek to improve brightness by eliminating one or both polarisers and to avoid parallax by putting the reflector inside the display cell. Thanks to the availability of active-matrix technology, they are no longer restricted to effects with an inherent threshold for multiplexing. They are helped by improved simulations, newly developed materials and fabrication techniques. Many see opportunities in making the internal reflector itself a silicon integrated circuit containing drivers and interface functions; a miniature display can be combined with optics to make a compact full-page viewer for cell phones or an inexpensive projection monitor. The search is on for the next breakthrough in a reflective display!

Allan R. Kmetz

1

Overview of Reflective Displays

1 Introduction

The reflective display is a non-emissive device from which the readout light is reflected rather than transmitted to the viewer. Two types of reflective display are frequently encountered: direct-view and projection. In most direct-view displays, ambient light is used to read the displayed information; front lighting is needed only in a dark environment. In projection the light source could be a lamp for video projectors or light-emitting diodes for near-eye virtual displays.

Reflective direct-view displays offer three advantages over transmissive ones: low power consumption, sunlight readability and film-like image quality. First, reflective direct-view panels do not require a backlight so that their power consumption and panel weight are both reduced. Second, reflective displays have superb sunlight readability. In a bright outdoor environment the images of a transmissive display could be washed out by sunlight. Since reflective displays utilise ambient light as the reading source, the brighter the ambient light, the more vivid the displayed images. Third, the active-matrix addressed reflective displays maintain a large (90%) aperture ratio for high-resolution devices. The 'screen-door' effect, as observed in low-aperture ratio transmissive active-matrix liquid crystal displays, is eliminated.

Several reflective display technologies have been actively pursued, including liquid crystal display [1], digital micromirror device, grating light valve, interferometric modulation, electrophoretic display and rotating ball display [2]. Each technology has its own merits and unique applications. Some devices are more suitable for direct-view and others are preferred for projection displays.

Within the liquid crystal display category, several mechanisms for modulating light have been developed and implemented in flat panel devices. These mechanisms include phase retardation, polarisation rotation, absorption, light scattering and Bragg reflection. Phase retardation and polarisation effects need to incorporate at least one polariser. A reflective direct-view display employing a polariser can achieve high contrast ratio and good color saturation, except that its light throughput is sacrificed. On the other hand, display mechanisms involving absorption, light scattering and Bragg reflection do not require a polariser. High brightness and a wide viewing angle are the inherent advantages. However, their contrast ratio is usually limited to the 5–10 : 1 range.

Although this book is mainly devoted to reflective liquid crystal displays, some common issues that are also essential to transmissive displays, such as LC materials, operation modes, fast response time, low operation voltage and wide viewing angle are addressed as well.

In this overview chapter, we start with the operation principles of reflective direct-view, projection and virtual liquid crystal displays, then move on to digital micromirror devices, grating light valves, interferometric modulation and electrophoretic displays.

2 Reflective Direct-view LCDs

Thin-film transistor (TFT) based transmissive liquid crystal displays offer some competitive advantages over cathode ray tubes (CRT) in thin profile, light weight, low power consumption, high resolution and high contrast ratio. They have dominated the portable displays and have gradually penetrated the desktop monitor and video projector markets. In a transmissive direct-view LCD, the viewer and light source are on the opposite sides of the display panel, as depicted in Figure 1.1(a). Since liquid crystal is non-emissive, a backlight is required to illuminate the display device. The scattered ambient light from the front LCD surface is negligible compared with the backlight intensity. As a result, the display exhibits a high contrast and saturated colors. However, the employed backlight consumes a few watts of power and the display brightness is insufficient in sunlight. Therefore, such transmissive

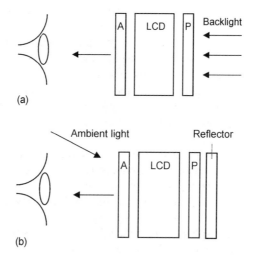

Figure 1.1 A schematic drawing of (a) a transmissive LCD using a backlight and two polarisers, and (b) a reflective LCD using ambient light for displaying images

displays are more suitable for high-end products, such as desktop monitors, notebook computers and large-screen projectors.

By contrast, the reflective display sketched in Figure 1.1(b) utilises ambient light to read out the displayed images. A reflector is preferably embedded in the inner side of the rear LC substrate. Such a display does not require a backlight. Thus, its power consumption and weight are greatly reduced. Since it is a reflective device, a brighter ambient light would lead to a more vivid image. However, the background ambient light could be reflected from the front surface to the viewer, overlap with the displayed images and degrade the contrast ratio. A typical contrast ratio for a reflective LCD is in the range 5–30. A low-contrast display will cause unsaturated colors. Thus, reflective LCDs are more suitable for handheld and outdoor applications, such as a personal digital assistant.

Besides reflective direct-view displays, reflective LCD panels are good candidates for large-screen display and virtual microdisplay applications. Liquid crystal light valves using a silicon photoconductor (both crystalline [3] and amorphous [4] silicon) and liquid crystal on silicon (LCOS) have been demonstrated for film-like electronic cinemas and high-definition television for home theaters. For such reflective displays the light from a lamp is reflected, instead of being transmitted by the light valve and then projected on to the screen. Thus, the light valve can tolerate the high luminous flux required for large screen display. For virtual displays, in order to save space and cost, a single LCOS chip and three light-emitting diodes (LEDs) with primary colors are often used. A key challenge here is to get a fast response LC (less than 4 ms) for color sequential displays.

Several types of reflective direct-view liquid crystal display have been developed. They can be roughly categorised by the number of polarisers employed: 2, 1 or 0. The benchmark for these types of reflective display is white paper, as shown in Figure 1.2. White paper and newspaper, respectively, have reflectivities of 80 and 55% and contrast ratios of 12 : 1 and 6 : 1. All the electronic displays developed so far still cannot match the performance of white paper in reflectivity and viewing angle.

In the following sections, we will review the operation principles of each display.

2.1 Two-polariser R-LCD

A common example of such a display is a wristwatch. Its configuration is shown in Figure 1.3. A LC cell is sandwiched between two crossed linear polarising plates. The

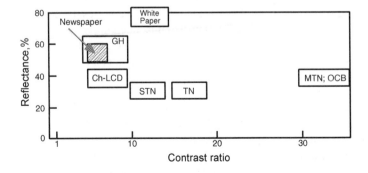

Figure 1.2 A qualitative comparison of various reflective LCDs in brightness and contrast ratio

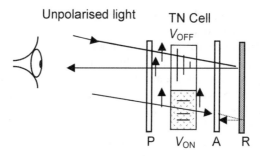

Figure 1.3   A reflective LCD which uses a 90° TN cell and two crossed polarisers

90° twisted nematic (TN) cell [5] is used as a light switch. The 90° TN cell has been widely used for transmissive displays due to its high efficiency and weak color dispersion, high contrast ratio, and low operation voltage. However, when a 90° TN cell is used for a reflective display, two crossed polarisers are required to achieve a high contrast ratio. The second polariser (around 200 μm thick) sitting between the back substrate and the reflector not only reduces brightness by around 28% but also causes parallax (double image) which limits the device resolution. Removing the second polariser would eliminate the parallax; nevertheless, its light modulation efficiency is greatly reduced and there is no common dark state for the RGB colors even in the crossed-polariser configuration, i.e. adding a quarter-wavelength film. Details are given in Chapter 4.

2.2   Single-polariser R-LCD

A linear polariser transmits about 45% of unpolarised light and 85% of linearly polarised light. For a reflective LCD using one polariser, the light traverses the polariser twice so that the maximum light throughput is limited to 38%. However, as compared to the two-polariser LCD, the single-polariser LCD offers two advantages: the removal of the second polariser sitting between the back glass substrate and the reflector eliminates parallax; and the elimination of the second polariser enhances the reflectivity by around 28%. On the other hand, the single-polariser LCD exhibits a better contrast ratio than a non-polariser LCD, although its reflectivity is lower. A high contrast ratio is essential for color displays. If the contrast ratio is 5–10, light leakage in a color pixel is relatively large, resulting in a mixed color. If the contrast ratio exceeds 30:1, the color saturation is much improved, although it is still far from perfect. In a reflective display, due to the front surface reflections, it is not easy to obtain a contrast ratio higher than 50:1 under ambient light conditions. The single-polariser LCD offers a compromise between display brightness and contrast ratio, and has gradually become the mainstream approach.

Several cell configurations have been developed for single-polariser reflective LCDs. For example, the mixed-mode TN (called MTN) cell, super-twist TN (STN) cell, homeotropic cell, film-compensated homogeneous cell and π-cell all have their own merits. In the single-polariser reflective LCD, the polariser behaves like two parallel polarisers for the incident light. In such a parallel-polariser configuration (i.e. no $\frac{1}{4}\lambda$ film is used), it is more difficult to find a LC operation mode that possesses a good common dark state for the red (R), green (G) and blue (B) light employed for color display. Details are presented in Chapter 4. Light leakage that results from a low contrast causes desaturation and unsatisfactory color purity. To obtain the crossed-polariser configuration, a quarter-wave film is usually inserted beneath the polariser, i.e. outside the LC cell for convenient assembly.

The viewing angle of a reflective cell is equivalent to a two-domain transmissive cell. This is because of the mirror image effect experienced by the incident and reflected beams as shown in Figure 1.4. For monitor applications, such a $\pm 45°$ viewing angle is insufficient. However, for personal handheld applications, the panel can be conveniently adjusted to be normal to the viewer. Therefore, most reflective LCDs provide an adequate viewing angle.

2.3 Zero-polariser R-LCD

An LCD that does not require a polariser has the advantage of high reflectivity and a wide viewing angle, although the contrast ratio is often limited to 10 : 1 or less. In some applications where high brightness is more critical than contrast ratio, these LCDs are attractive. Three reflective LCDs that do not require a polariser have been developed. These are the polymer-dispersed liquid crystal display (PDLC), the cholesteric liquid crystal display (Ch-LCD) and the guest–host (GH) display.

PDLC

The operation principle of a PDLC is based on light scattering induced by the refractive index mismatch between micron-sized LC droplets and the surrounding polymer matrices [6]. Both normally black and normally white modes have been demonstrated. In the normally black mode, LC droplets scatter light at $V=0$ due to index mismatch and transmit light when V is greater than the threshold voltage (V_{th}) [7]. Conversely, the reverse mode

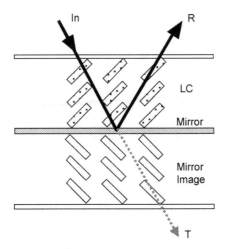

Figure 1.4 Mirror image effect of a reflective LCD

PDLC transmits light at $V=0$ and scatters light at $V > V_{th}$. In a reverse-mode PDLC, the substrates were covered with a homeotropic alignment layer (Nissan polyimide 7511L). An ultraviolet curable monomer was mixed with a negative $\Delta\varepsilon$ LC. The monomer, which contained 5 wt% methyl benzoate as an initiator, was dissolved in the liquid crystal. The liquid crystal and monomer ratio is between 98/2 and 90/10 [8]. After the cell was injected with the LC solution, it was subjected to UV exposure. Standard light strength and time are 1.3 mW/cm^2 and 10 minutes.

Figure 1.5 shows a reverse mode PDLC developed by a Seiko-Epson group, described as internal-reflection inverted-scattering (IRIS) [9]. In the voltage-off state, unlike the normal mode PDLC, the LC droplets are aligned nearly perpendicularly to the substrates. As a result, the refractive index of the extraordinary ray of the LC matches that of the polymer matrix. The device is clear and more than 90% of the incoming light transmits through the PDLC layer and is reflected by the reflector. In a high voltage state, the LC droplets are reoriented by the electric field so that the index mismatch occurs. In addition, substantial light scattering in the forward direction is observed.

The voltage-dependent light reflectance of an IRIS device is shown in Figure 1.6 compared with a normally white TN cell (open circles). Due to the removal of the polariser and to the restricted viewing angle, IRIS has a brightness 2.5 times higher than a TN cell. However, the TN cell still has advantages in lower operating voltage and higher contrast ratio.

Cholesteric LCD

Cholesteric LCD is another display that does not require a polariser. The operation principles are illustrated in Figure 1.7. A cholesteric LC has two stable states: reflective planar and focal conic texture [10]. In the voltage-off state, the planar texture reflects brilliant colored light if the Bragg reflection condition is satisfied, i.e. $\lambda=np$, where λ is the central wavelength of the reflection band, n is the average LC refractive index and p is the pitch length. The bandwidth $\Delta\lambda$ of the reflected light is equal to $p\Delta n$; here Δn is the birefringence of the LC. For a display in the visible region, $\lambda\simeq550$ nm, $n\simeq1.6$ and $\Delta n\simeq0.2$, the pitch length $p\simeq344$ nm and the bandwidth $\Delta\lambda\simeq70$ nm.

In a helical structure, the incident white light is decomposed into right and left circular polarised components with one component reflected and the other transmitted.

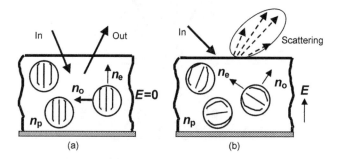

(a)   (b)

Figure 1.5 Reflective display based on the reversed mode PDLC: (a) clear state when voltage is off, (b) scattering state when voltage is on

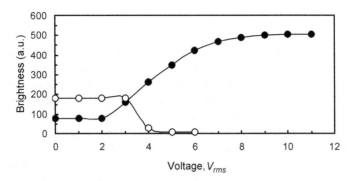

Figure 1.6 The voltage-dependent reflectance of an IRIS (solid circles) and a TN cell (open circles) after [9]

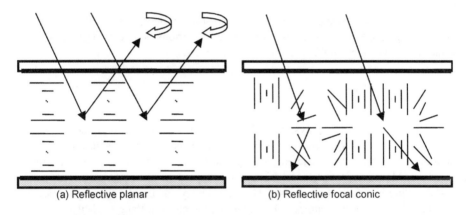

Figure 1.7 A cholesteric LCD at (a) voltage-off and (b) voltage-on state

The transmitted light is absorbed by the black paint coated over the rear substrate, as shown in Figure 1.7(a). In a voltage-on state (Figure 1.7(b)), the periodic helical structures are changed to focal conic. As a result, the Bragg reflection is interrupted. The incoming light is absorbed by the dark paint and a dark state is obtained.

An important advantage of the Ch-LCD is that the reflected light is already colored. Therefore, a color display without using color filters can be achieved. In addition, no polariser is needed. Thus, the brightness of a Ch-LCD is at least three times higher than a reflective color display employing one polariser. Without using a polariser, the viewing cone of a Ch-LCD is wider than 60° except that its contrast ratio is in the 5–10 range. To achieve multiple colors, two or three RGB panels can be stacked together [11]. A potential problem of the stacked approach is parallax, i.e. the incident light and reflected light pass through different pixels. Parallax leads to colour mixing which becomes a serious problem for a high-resolution LCD.

The operating voltage of a Ch-LCD is inversely proportional to the pitch length. Owing to the short pitch involved for visible displays, the operating voltage of a Ch-LCD is in the vicinity of 30 V. Detailed operation mechanisms are described in Chapter 8.

Guest-host display

The operating principle of a guest-host display is to utilise the absorption anisotropy of dichroic dye molecules doped in an aligned LC host [12]. Normally, a guest-host display system consists of 1–5% of absorbing dye molecules dissolved in a liquid crystal host. The host LC is highly transparent in the visible spectral region and the guest dyes strongly absorb one polarisation of the incoming light and transmit the other.

To avoid using any polariser in a guest-host display, special device configurations have been developed. For instance, the Cole–Kashnow cell [13] uses a homogeneous GH cell with a quarter-wave film between the rear substrate and the reflector. The White–Taylor cell [14] uses chiral dopant to form twisted GH layers so that both polarisations of the incident light can be absorbed while traversing through the GH LC layer twice. The double cell [15] uses two orthogonal homogeneous cells; the first GH layer is responsible for one polarisation and the second layer for the other. All the three above-mentioned cells have their own pros and cons.

The typical performance of a GH LCD is 50–60% reflectivity, 5–10 : 1 contrast ratio, $\pm 60°$ viewing angle and 50–80 ms response time. Various device configurations and performances are discussed in Chapter 6.

2.4 Reflectors

For reflective direct-view displays, the reflector design and fabrication is probably the most critical issue. A mirror-type reflector reflects light at a specific angle according to the law of reflection. This mirror has restricted viewing characteristics and is not suitable for display applications. Moreover, the displayed image overlaps with glare reflected from panel substrates, resulting in a poor contrast ratio. On the contrary, the Lambertian-type reflector, such as a sheet of white paper, scatters light almost uniformly to all angles. It is ideal for group viewing, but for a personal display the light scattered to unwanted areas is wasted.

An ideal reflector needs to meet two criteria. It directs images to the viewer without overlapping with glare; if the images mix with glare, the contrast ratio would be reduced substantially. It also provides a sufficiently wide viewing cone and optical gain. A specular mirror has too narrow a viewing angle and cannot be used alone for reflective displays. By contrast, a Lambertian reflector has uniform scattering over all the angles so that the light intensity in the preferred viewing cone is diluted.

Extensive efforts have been made into reflector design and fabrication with acceptable viewing angles while providing optical gain and preserving good contrast ratio. So far, four types of reflector have been developed: light control films, rough surface reflectors, holographic reflectors and cholesteric reflectors. The light control film is laminated to the front surface of the display without altering the internal device structures. On the other hand, the rough reflector approach involves a fundamental structural change in mirror fabrication. The holographic reflector can be either laminated to the reverse of the display or integrated with color filters. Although the former approach is simple, it creates undesirable parallax so that its usefulness is limited to low-resolution displays. The internal holographic reflector approach is challenging, but holds promise for high-brightness, high-contrast and saturated color displays. Lastly, a cholesteric liquid crystal layer that reflects colors without color filters has also been used as a reflector for some birefringent color reflective LCDs.

Light control films

In 1995 a Tohoku research group proposed the idea of a color reflective LCD using a front scattering film [16]. This configuration consists of an external light control film together with an internal flat aluminum reflector, as depicted in Figure 1.8 [17]. In terms of the panel fabrication process, this is the simplest approach because it involves straightforward aluminum mirror deposition on the inner side of the back substrate. The uniform LC alignment and cell gap are maintained. As the mirror functions as a specular reflector, the viewing angle control relies entirely on the light control film. For simplicity, the diffusive film could be laminated together with the $\frac{1}{4}\lambda$ film and polariser. The light control film diffuses the incoming light to a cone of about $\pm 15°$. When the reflected light passes through the diffuse film for the second time, the viewing cone is further increased.

Using a light diffusive film, the ERSO/Hughes team demonstrated a full-color reflective TFT-LCD with contrast ratio greater than $15:1$ [18]. However, the ERSO TFT arrays have only a 60% aperture ratio. The incident light passes through the TFT twice so that the brightness and contrast ratio are reduced dramatically.

3M and Sumitomo have developed several image directing films and brightness enhancement films for both transmissive and reflective displays. In the 3M Vikuiti^{TM} image directing films, microprisms are fabricated using modified acrylic resin on the polyester substrate. The nominal film thickness is 155 μm. The repeating microreplicated prism has a 50 μm pitch. Due to refraction and reflection of the film, the incident light is directed to a controlled angle. Two types of film with 10° and 20° deviation angles are commercially available. As nearly all of the light that falls on the film is transmitted through it, the redirected images are virtually distortion-free.

A NCTU group has reported a light control film using an asymmetric microlens array (AMA) [19]. Figure 1.9 depicts the device structure. The off-axis AMA light control film is implemented with a binary Fresnel microlens structure that can be fabricated with standard VLSI processes. The film is then laminated to the top surface of a reflective LCD. The reflected light is directed near to the normal of the display surface. As a result, an optical gain of 4–10 times the MgO standard white at the 10–20° viewing cone is achieved.

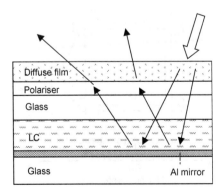

Figure 1.8   A reflective LCD using a light control film and a specular reflector

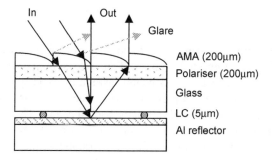

Figure 1.9 Device structure of an asymmetric microlens array-based reflective LCD after [19]

Rough surface reflectors

Several versions of a rough surface reflector have been reported. For example, Sharp has developed a high brightness reflective color LCD incorporating microreflector structures (MRS) [20, 21]. The microrough reflectors as depicted in Figure 1.10, implemented on the inner side of the rear substrate, have three important functions, as follows. The TFT structures are hidden beneath the reflector. Thus, the aperture ratio is greater than 90%. A high aperture ratio not only enhances the display brightness but also makes the images like film. The screen-door effect that results from low-aperture TFT arrays is eliminated. Secondly, the microrough reflectors steer and focus the reflected light to a desired viewing direction. Owing to this focusing effect, the display brightness is about three times greater than the standard Lambertian MgO diffuser within the $\pm 30°$ viewing cone, as shown in Figure 1.11 [22]. It directs images to a different angle ($\simeq 15°$) from reflected glare so that a high-contrast image can be obtained.

NEC also developed microreflectors by controlling the surface morphology [23]. These rough-surface reflectors are used in a guest-host display. An aperture ratio as high as 94% has been achieved. However, the peak reflection angle is not well separated from specular reflection. As a result, the contrast ratio is affected.

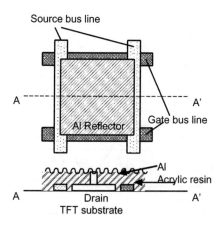

Figure 1.10 The structure of Sharp's microreflectors after [22]

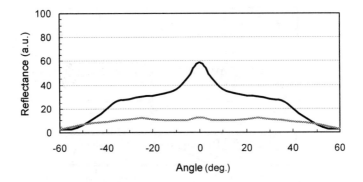

Figure 1.11 The angular-dependent light reflectance of Sharp's MRS after [22]

Hitachi developed a blazed reflector plus scattering materials as a diffuser [24], as shown in Figure 1.12. The inclination angle is 20° and pitch length is 30 μm. The reflectance of this diffuser and blazed reflector is 87%, as compared to 92% for a conventional aluminum mirror. By laminating this blazed reflector to an STN, the display brightness and contrast ratio are both improved.

ERSO also developed a diffusive microslant reflector [25]. The reflectors are integrated with color filters, as shown in Figure 1.13. Due to the asymmetric structure of the slant reflector, the peak of the reflected light is shifted away from glare by $\simeq 12°$. Therefore, a contrast ratio in the range 20–30 : 1 has been achieved [26]. Due to the optical gain, the reflectance normal to the surface is about 42% of a Lambertian white standard. It is about four times brighter than a LC cell with a mirror reflector. The ERSO display structure is apparently not optimised. Its TFT aperture ratio is only 60%. Since the light passes through the TFT arrays twice, the effective aperture ratio may drop to below 40%. If the TFT array is hidden beneath the reflector, as shown in Figure 1.10, a much higher contrast ratio and brightness should be obtainable.

Holographic reflectors

A holographic reflector is a volume hologram that diffracts selected wavelengths of the ambient light to a specific angle, which is different from the specular reflection angle

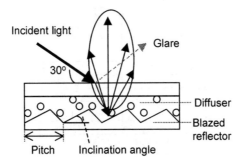

Figure 1.12 The structure of Hitachi's blazed reflector after [24]

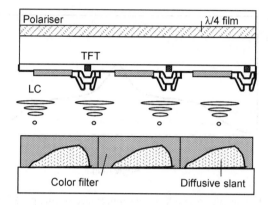

Figure 1.13 ERSO's reflective TFT-LCD structure (courtesy of Dr C. J. Wen of ERSO)

[27, 28]. Holographic components have been commonly used for direct-view and projection displays [29]. Figure 1.14 depicts the device configuration of a direct-view LCD with a holographic reflector. As the diffracted light is separated from the glare, the display contrast is improved substantially. Also, the diffracted light is confined within a finite viewing cone so that the display has an optical gain of two to three in brightness.

Polaroid has developed holographic reflectors for TN and STN displays [30, 31]. A typical STN panel has a contrast ratio of about 7 : 1. With the assistance of a holographic reflector, the contrast ratio is improved to 17 : 1. However, the holographic reflector is placed outside the back glass substrate. This leads to parallax, which limits the resolution of the display devices.

Sharp has elegantly combined the functions of holographic directive reflectors with color filters [32]. The holographic reflectors not only make the reflected light deviate from glare, but also generate colors without using color filters. The targeted Sharp holographic reflectors are implemented in the inner side of the rear substrate. An optical gain of 3.5 $\times$ has been realised. This boosts the total display brightness to 60% of the standard white in a single-polariser LCD. For the single-polariser reflective LCD, a contrast ratio greater than 15 : 1 can be achieved without much difficulty. In addition, the reflected colors have a relatively narrow

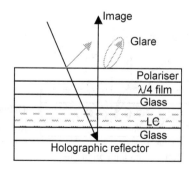

Figure 1.14 A reflective LCD using a holographic reflector outside the LCD panel

bandwidth (± 25 nm) and the color shift is insensitive to the incident angles. This ensures that saturated reflective colors can be obtained.

Cholesteric reflectors

From Figure 1.8, a cholesteric LC layer reflects colors without color filters. The Philips group has applied this concept to demonstrate reflective color displays using a cholesteric LC layer as a broadband reflector [33]. The cholesteric reflector could be implemented outside or inside the LC cell. A similar parallax problem occurs when the reflector is placed behind the rear substrate. While implementing the cholesteric reflector on the inner side of the substrate, a planarisation layer is needed. The ITO electrode is then deposited on top of this buffer layer.

Two problems relating to a cholesteric reflector need to be addressed. First, the cholesteric reflector only reflects one component of circularly polarised light and transmits the other. Thus, the LC switching states need to be designed accordingly. Second, the reflectivity of the cholesteric reflector depends on the pitch length and birefringence of the LC employed. It is not uniform across the whole spectral bandwidth.

2.5 Color filters

How to achieve saturated colors is a critical issue for reflective liquid crystal displays. In a direct-view reflective LCD employing no polariser, the contrast ratio is limited to 5–10:1. In a single-polariser LCD the contrast ratio is boosted to 15–30:1. Even for a 30:1 contrast ratio, the dark state is still not sufficiently black. Therefore, the colors of a reflective display are not so pure as with a transmissive display.

For a direct-view LCD, colors are achieved using pigment color filters [34]. The three primary colors are centered at R$\simeq$630, G$\simeq$550, and B$\simeq$460 nm, and the bandwidth is about ± 40 nm. The single-path transmissions of these color filters are depicted in Figure 1.15. In a reflective LCD, no matter whether color filters are implemented on the front or rear substrate, the incident light passes through the color filter twice. Thus, the transmittance is related to the absorption coefficient α and filter thickness d as $T = \exp(-2\alpha d)$. For a given color pigment used in a reflective display, the filter thickness or pigment concentration is reduced to half in order to obtain the same performance characteristics as a transmissive

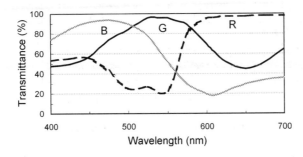

Figure 1.15 Single-path transmittance of reflective-type RGB color filters after [34]

color filter. Reducing the color filter thickness or increasing the bandwidth would increase the transmittance at the expense of color chromaticity or purity. The transmittance, contrast ratio and color purity need to be balanced.

2.6 Trans-reflective displays

Reflective LCDs rely on ambient light to read out the displayed image. When the ambient light is dim, the display is not readable if no built-in light is available. To overcome this problem, a trans-reflective (or transflective) display as shown in Figure 1.16 has been proposed [35]. The device consists of two crossed polarisers, a reflective polariser which reflects P and transmits S polarisation, and a gray film for absorbing unwanted light.

To illustrate the operating principles, the dashed lines divide a LCD panel into the voltage-on (left) and voltage-off (right) states. Figure 1.16(a) shows a reflective display. In a voltage-on state, let us assume the top linear polariser transmits the P polarisation of the incoming light. In the voltage-on state, the LC directors are reoriented along the electric field direction. Thus, the P wave is not phase retarded by the LC layer and is reflected by the imbedded reflective polariser. The reflected P wave transmits through the LC layer and polariser for a second time, resulting in a bright state. In the voltage-off state, the incoming P wave is changed to S by the LC layer. This S wave transmits through the reflective polariser and the bottom polariser, and finally gets absorbed by the gray film. Therefore, the dark state results.

At dark ambient, the backlight is turned on, as sketched in Figure 1.16(b). The bottom polariser transmits the S component and absorbs the P. Subsequently, the S polarisation passes through the reflective polariser. In the voltage-on state (left), the S wave passes through the LC layer with little phase retardation. Thus, it is absorbed by the top polariser, and a black state results. In the voltage-off state, the S wave is converted to P by the LC layer and transmitted by the top polariser. As a result, a white state appears.

Although the trans-reflective mode extends the application of a reflective LCD to dark ambient, it has a few problems. First, the reflective and transmissive modes have inverted

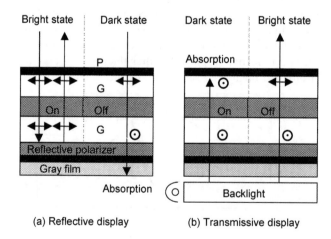

(a) Reflective display   (b) Transmissive display

Figure 1.16 A device configuration of a trans-reflective LCD: (a) reflective display, (b) transmissive display after [35]

images. The bright pixels in a reflective display become dark in transmission, and vice versa. Therefore, if the backlight is turned on at a not-so-dark ambient, the reflective and transmissive images appear simultaneously. This would lead to a poor image contrast. Second, the reflective polariser embedded beneath the bottom substrate will cause parallax in the reflective mode. Unless a thin bottom substrate is used, the parallax would limit the device resolution. Third, the semi-transparent reflective polariser and gray film would undoubtedly cause severe optical losses for the backlight. Thus, the power consumption is increased substantially. For example, a trans-reflective, 4096 color, 640×240 resolution elements, thin-film diode, LCD panel demonstrated by Seiko-Epson consumes 100 mW power in reflective mode. In the transmissive mode when the backlight is used, its power consumption jumps to 750 mW.

LG Philips developed a transflective LCD [36] using divided pixels as shown in Figure 1.17. Each pixel is divided into two reflective (R) and transmissive (T) sub-pixels. The cell gaps for the R and T sub-pixels are d and $2d$, respectively. In a bright ambient, the incident light on the T sub-pixel is absorbed by the lower polariser and the device functions as a reflective display. In a dark ambient the backlight transmits the T sub-pixels and the device works as a transmissive display. A technical challenge of such a transflective display is the color balance between the R and T sub-pixels. For R sub-pixels, the light transverses the color filters twice. However, the light transmits color filters only once in T sub-pixels. To have the same color saturation, the color filters in the T sub-pixels should have double pigment concentration as compared to those of the R sub-pixels.

2.7 Front lighting

To overcome the above-mentioned drawbacks, a front lighting apparatus, as depicted in Figure 1.18, has been introduced [37]. This front lighting system consists of a compact, flat and transparent light pipe designed with microprisms on its front surface, and an optical compensator which has a surface profile complementing the surface of the light pipe. The light from the source passes through a collimating section first. Light injected into the light pipe is reflected to the display panel by total internal reflection from the microprism surface. With properly constructed collimating prisms, light of any incident angle is collimated to have a divergence angle within $\pm 18°$ when entering the LCD panel [38]. The reflected light from the LCD transmits through the light pipe and reaches the

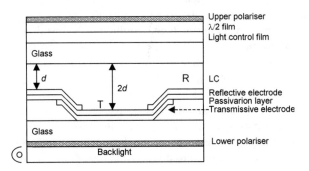

Figure 1.17 Device configuration of a transflective LCD

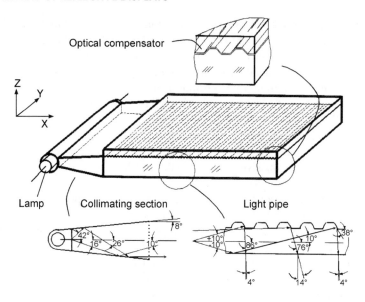

Figure 1.18 A device configuration of a front lighting system for reflective display (courtesy of Dr C. Tai of CLIO Technologies)

observer. The optical compensator shown in Figure 1.18 plays an important role in reducing image distortions. The prism surface has a grooved structure. The light exiting at different parts of the grooves will result in image distortion. The optical compensator corrects this distortion.

Some Sharp reflective LCD panels have implemented a front lighting device, and the results are encouraging. A display contrast ratio greater than 5 : 1 and decent brightness have been achieved.

3 Reflective Projection Displays

For a large display both direct-view and projection can be considered. The most recent technology cannot quite make it economic to make direct-view displays with diagonal sizes larger than 30 inches. To overcome the panel size limitation, projection has been used to achieve wide screen computer monitors, high definition TVs, boardroom presentations and electronic cinemas. Both transmissive and reflective LCDs have been developed for projection displays. LCD projectors using three transmissive thin-film transistor (TFT) panels have been described in great detail [39]. In this section we emphasise the projection and virtual displays using reflective liquid crystal light valves.

Liquid crystal on silicon (LCOS) [40] has been developed as the optical engine for projection and virtual display systems. In the projection display, a real magnified image is projected onto a screen through a series of lenses. Two types of projection display are often configured: front and rear projections. Front projection configurations utilise a distant screen to view the magnified image. Rear projection configurations enclose the magnification optics behind an imaging screen to produce a self-contained system. LCOS has been used in both front and rear projection systems for business presentations where the device is connected to a laptop computer, home entertainment (HDTV), projection monitors, data/video projectors, simulation systems and digital cinemas.

3.1 Liquid crystal on silicon (LCOS)

Almost all electronic components are shrinking in size, except the display part. Many products are now dominated by the display, both physically and from a cost perspective. Liquid crystal on silicon (LCOS) [41, 42], also called microdisplay, is a reflective image transducer that is capable of accepting a video signal and converting it into a high-brightness, high-contrast, large-screen image.

Figure 1.19 shows a cross section of an LCOS pixel developed by JVC [43]. In the silicon backplane, a CMOS transistor and a capacitor are fabricated and connected. The CMOS transistor can support a high data rate owing to the high electron mobility of single-crystal silicon. The electron mobility of c-Si is about two orders of magnitude higher than that of amorphous silicon (a-Si). The reflective aluminum electrode is connected to the drain of the transistor and the capacitor through a light blocking metal layer. After smoothing, the aluminum pixel mirror approaches a reflectivity of 91%. The dimensions of the electrode are $13 \times 13\,\mu m$ and the gap between electrodes is $0.5\,\mu m$. Thus, the aperture ratio is about 93%. For a 1365 (H) $\times$ 1024 (V) imager in a 4 : 3 aspect ratio, the diagonal is 23 mm.

To prevent the leaked light from the electrode gaps activating the transistors and then smearing the device resolution and degrading the image quality, a light blocking layer, as shown in Figure 1.19, is implemented. Based on the light blocking and thermal loading, the light valve is able to produce the 15 000 lumens required for an electronic cinema using a 5 kW Xenon lamp.

3.2 Key performance parameters

From the system viewpoint, the resolution, contrast ratio, projection optics, overall light throughput, color separation method, lamp and screen are the key performance parameters. In the following sections we will briefly discuss each parameter.

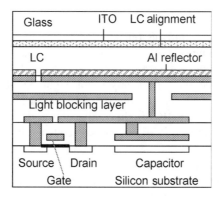

Figure 1.19 Device configuration of liquid crystal on silicon

Resolution

The definition (number of rows and columns of color triads), viewable screen size and dot pitch are interdependent. Computer monitors have a fixed dot pitch, usually about 0.2–0.3 mm. Dot pitch increases with screen size. A typical 50 inch home theater screen has a triad pitch of nearly 1.5 mm. The demand for higher-definition displays is increasing in both the computer and entertainment markets. The standard for the desktop is fast becoming XGA (1024×768) which results in a 15 to 16 inch viewable screen diagonal. Even higher-definition monitors are growing increasingly popular.

Microdisplays can have much higher pixel density than direct-view display devices. Many direct-view LCDs employ a-Si TFTs; because of the low mobility of a-Si the pixel size is normally larger than 80 μm. As a result, display pixel density is around 100 lines/inch. For a polysilicon (p-Si) transmissive TFT-LCD panel, the pixel sizes can be reduced to 20–40 μm because of the increased electron mobility. On the other hand, many reflective-type micro-displays use a CMOS silicon backplane and the pixel size can be as small as 6–12 μm so that the pixel density is greater than 1000 lines/inch. The aperture ratio is greater than 90% and the LCOS provides high resolution film-like images.

Contrast ratio

Contrast for displays is typically quoted for measurements made under laboratory conditions with almost no ambient light. The actual contrast observed by a viewer under office or home lighting conditions is much lower than the specified value. A typical CRT has a contrast ratio greater than 100 : 1.

The contrast ratio of a LCD depends heavily on the LC alignment. A TN-type LCD can have a contrast ratio of 300 : 1 and a homeotropic cell can have an unprecedented contrast ratio of more than 1000 : 1. The response time of a direct-view display is in the 20–40 ms range. However, for a projection display, a typical frame (rise + decay) time of an a-Si LCLV is less than 16 ms for a 3 μm LC layer because of the elevated operating temperature. The viscosity of a LC mixture typically drops by a factor of two per $10°$ rise in temperature.

Projection optics

An optical projection system employing reflective LCDs is more complicated than one using transmissive light valves. Figure 1.20 shows the optical system for a professional large-screen projector employing three reflective LCOS light valves [44]. The cold mirror is used to remove unwanted infrared and reflect only visible light. The condensing and integrator lenses serve to homogenise the beam to achieve $> 85\%$ uniformity over the screen diagonal. The color beam splitter directs the RGB light components to the respective light valves. The X-cube recombines the modulated RGB beams and sends the images to the screen through the projection lens.

To reduce complexity, two compact projection prisms have been developed by Philips and OCLI, as depicted in Figure 1.21(a) and (b), respectively [45]. The OCLI prism, modified from the Philips prism, incorporates two identically shaped prisms and one larger prism. This design offers the advantages of lower system weight and reduced manufacturing cost.

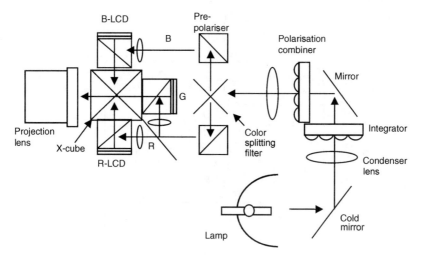

Figure 1.20 Projection display using three-channel reflective LCDs

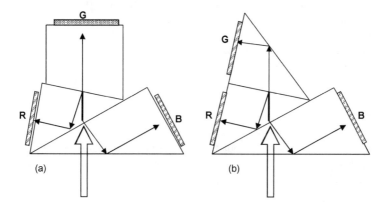

Figure 1.21 Two prism designs for three-panel reflective LCDs: (a) Philips prism, (b) OCLI prism after [45]

In the on-axis projection system, a broad band polarising beam splitter (PBS) is commonly required. Whilst on-axis reflective designs offer the potential to minimise the product package size, the need of PBS severely limits contrast as the f-number of the projection lens is decreased. To overcome these problems, Aurora Systems has developed a relatively compact off-axis projection system [46]. In the off-axis approach, three 45°-TN LCDs are used. The incident angle is 12° from the surface normal. Two crossed sheet polarisers, instead of a PBS, are used. A contrast ratio greater than 300:1 and light efficiency greater than 7 lumens/W (with polarisation conversion) has been demonstrated.

In addition to color beam splitters, the holographic color filter (HCF) is an efficient method for separating colors. A transmission-type HCF is demonstrated to have about three times higher light efficiency than the conventional absorption-type color filters [47]. By using holographic color filter arrays, the RGB light components are directed to each sub-pixel, as illustrated in Figure 1.22. In the reflective device developed by JVC, each sub-pixel

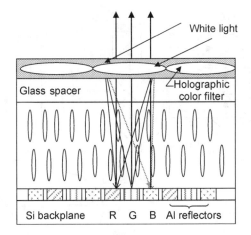

Figure 1.22 A spatial color approach using holographic color filters after [44]

has dimensions of 7.6 μm (H) and 14.8 μm (V). A full-color projector containing 3840 (H) × 1028 (V) sub-pixels using a single LCD light valve has been demonstrated. To generate a hologram, a photosensitive polymer is used to record the laser interference patterns. Diffraction efficiency depends on the wavelength. In the green spectral band, a diffraction efficiency as high as 90% has been routinely obtained [48].

Light throughput

The need for a bright display is obvious to most viewers, especially in brightly illuminated environments. The benchmark of brightness for monitors is 70 to 100 cd/m². Televisions are typically much brighter, running at 300 to 1000 nits, to permit viewing under bright daylight conditions.

The brightness of a projector is determined by several factors, such as lamp efficacy, polarisation conversion efficiency, LCD aperture ratio, and the numerical aperture of the projection lens. The brightness is often quantified as ANSI lumens. However, ANSI lumens are often over-emphasised as a factor in comparing one projector to another. In general, there needs to be around 30% difference in the ANSI lumen rating to produce a noticeable difference when the two projectors are shown side by side. For example, projector brightnesses rated 600 or 700 lumens may look about the same. Most of the products rated 900 to 1100 lumens will also look about the same. However, projectors rated 900 to 1100 lumens will produce noticeably brighter pictures than those of 600 to 700 lumens.

Colors

Three approaches have been commonly used to obtain full-color projection displays: three light valves, each of which modulates a single color [49], a single light valve with embedded color filters, and a single light valve with sequential colors. The first requires splitting the lamp light into three colors, directing each color to a corresponding light valve, and converging the modulated light into a single beam for projection. This approach has advantages in high resolution and high brightness. The disadvantages are in higher cost (more optics and three light valves are used), pixel registration, heavier weight and larger space needed. The second approach divides each pixel into three-color sub-pixels. The problems of color splitting, recombination and pixel alignment are all removed. The burden now falls on the fabrication of a high-resolution light valve itself. Due to the triple pixel numbers, the panel size is increased to 4–6 inches; this means that larger optics are required. The last approach uses a color filter wheel or color switch to sequentially beam each primary color to the light valve for modulation [50]. The challenge of the color sequential approach is in the fast response time of the LCD panel. Inherently, the color filter and color sequential designs have less than one-third of the throughput efficiency of the parallel color approach. The three-light valve system offers the highest efficiency and throughput while the single light valve system offers simplicity and potentially lower cost. Most professional projectors use the three-light valve approach.

Lamps

Light efficacy, color temperature, color rendering index, arc gap and lifetime are important characteristics of a lamp. Table 1.1 compares some metal halide and xenon lamps with different efficacy, arc gap and lifetime. Metal halide lamps offer efficacy of greater than 60 lumens per watt, color temperature of greater than 5000 K, and times to half brightness of more than 5000 hours [51]. On the other hand, the xenon lamp offers a very good color rendering index and high wattage. Thus, it is more suitable for electronic cinema applications.

Reductions in projection lamp arc dimensions have also enhanced projector performance. As the light source gets closer to a point source, it becomes easier to collimate. Of equal importance, lamp lifetime has been dramatically improved. Many presentation and portable projectors use low power (less than 200 W) metal halide lamps to deliver 300 to 600 plus lumens from a three-light valve liquid crystal imaging system [52, 53].

The major reliability issue for a LCD projector is the lamp lifetime. Typical times to half brightness are 3000 to 10 000 hours. For the presentation market where projectors are used for about 10 hours per week, the lamp lifetime does not become an issue. However, for monitor and television applications, the daily operating time is increased. Here projector designs must take convenient lamp changing into account.

Table 1.1 Lamp parameters

Lamp	Ushio	Phillips	Welch Allyn	Ushio
Type	DC MH arc	DC MH arc	DC MH arc	Xenon arc
Product		UHP	Solarc	UXL-16SC
Power (Watt)	125	100	50	1600
Output (Lumen)	7750	6000	3200	60 000
Efficacy (Lm/w)	62	60	64	37
Color temp. (°K)	5600	8500	5000	6200
CRI		60		95
Lifetime (hours)	3000	6000	4000	2000
Arc gap (mm)	2.4	< 1.4	1.2	3.3

Although a higher-wattage lamp is desirable for higher display brightness, it has short-comings in heat dissipation, fan noise and shorter lifetime. For instance, a 270 W lamp will give off double the heat of a 120 W lamp. In a small conference room or meeting room with a number of people, this additional heat may not be welcome. Second, a 270 W lamp requires more cooling than does a 120 W lamp. Thus the fan noise on a projector with a higher-wattage lamp can be more annoying than the noise from projectors with smaller lamps. The third disadvantage is that the lamp lifetime is shorter for a higher-wattage lamp.

Screens

The output brightness of a CRT monitor or television is approximately Lambertian. This means that the brightness is essentially independent of viewing angle. Projection displays with gain screens trade-off the wide viewing angle to achieve increased brightness at normal viewing angles. While the presentation projection market relies on front projection systems with remotely mounted reflective screens, many of the potential markets for microdisplay-based projectors, such as projection monitors, require a rear projection system.

The existing rear projection television screen is a multilayer laminate, which combines lenticular and Fresnel optical elements that assure brightness uniformity, increases image brightness in a preferred viewing cone (gain), and minimises glare and ambient reflections (contrast enhancement and anti-glare). The screen is ideal for microdisplay projection big screen television, but is not suitable for the higher resolution desktop monitor and PCTV applications [54].

High-resolution and high-gain screens that can be used for applications with a pixel pitch of around 0.3 mm (the nominal value for a computer monitor) are under enthusiastic development [55, 56]. Firstly, there are embedded lenticular elements; gain screens have been demonstrated that employ embedded gradient index lenses which offer high gain at high resolution. Dai Nippon has developed a new lenticular screen showing an excellent contrast ratio and fine pitch that eliminates Moiré patterns [57]. Secondly, there is the holographic type [58]; holographic diffraction elements can be embossed or printed on a polymer film by a low-cost process. Thirdly, there is the composite refractive mask. A unique design has been demonstrated that embeds spheres between a black front surface mask and a rear film.

3.3 Virtual projection displays

Virtual projection uses optics to create virtual magnified images. A user looks into a small viewfinder and sees a magnified projected image located in a plane some distance from the viewer. To reduce weight and cost, a single high-resolution and high-speed LCOS is commonly used as the optical engine. Three sequential color LEDs are used to display full-color images. LEDs are known to have a fast response time, light weight, high brightness, low power consumption and a long lifetime. Thus, they are ideal light sources for virtual projection displays.

A critical element for virtual display is lightweight and high-magnification projection optics. If a 12.7 mm diagonal LCOS light valve (800 × 600 pixels) is used as the object, the required magnification is about 15 × in order to view an equivalent 19 inch display

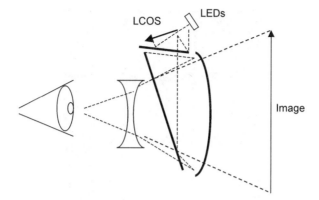

Figure 1.23 A folded compound magnifier for a virtual display after [60]

about 2 m away. Simple optics such as a magnifying glass or an eyepiece are often limited to a maximum magnification of about 10 × . Compound optical systems such as microscopes permit much higher magnification, but are too bulky to be wearable. To achieve a large magnification, a folded compound magnifier as shown in Figure 1.23 has been developed [59]. The eye relief is about 20 mm and the exit pupil is 10 mm.

In addition to optics, a major challenge for a color sequential virtual display is the LC response time. For a frame rate of 85 Hz, the LC frame (rise and decay) time needs to be shorter than 3.9 ms in order to avoid color breakup. To achieve such a fast response, an LC cell gap as thin as 1 μm has been considered. Color sequential operation from − 10°C to 70°C has been demonstrated [60]. Although a thin cell is favorable for speeding up the LC response, the manufacturing yield will be compromised.

4 Digital Micromirror Devices (DMD)

Another reflective display that has been developed for video and data projectors, HDTV, and electronic cinemas is the digital micromirror device (DMD) developed by Texas Instruments.

4.1 Operating principles

A DMD can be described simply as a semiconductor light switch [61]. Thousands of tiny, square, 16 μm × 16 μm mirrors, fabricated on hinges on top of a static random access memory (SRAM) make up a DMD, as illustrated in Figure 1.24. Each mirror is capable of switching a pixel of light. The hinges allow the mirrors to tilt between two states: + 10 degrees for ON or − 10 degrees for OFF [62]. When the mirrors are not operating, they sit in a 'parked' state at 0 degrees.

Applying voltage to one of the address electrodes in conjunction with a bias/reset voltage to the mirror structure creates an electrostatic attraction between the mirror and the addressed side. The mirror tilts until it touches the landing electrode that is held at the same potential. At this point, the mirror is electromechanically latched in place. Placing a binary

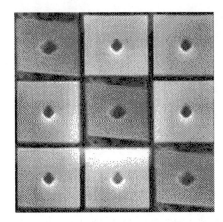

Figure 1.24 A close-up photograph of mirrors on the surface of a DMD. The mirrors can tilt ± 10 degrees (courtesy of Dr L. J. Hornbeck of Texas Instruments)

one in the memory cell causes the mirror to tilt + 10 degrees, while a zero causes the mirror to tilt − 10 degrees. The switching time of the DMD is about 10 μs.

DMDs have been built in arrays as large as 2048 × 1152, yielding roughly 2.3 million mirrors per device. These devices have the ability to show true high-definition television. The first mass-produced DMD was an 848 × 600 device. This DMD is capable of projecting NTSC, phase alternating line (PAL), VGA, and super video graphics adapter (SVGA) graphics, and it is also capable of displaying 16 : 9 aspect ratio sources.

By electrically addressing the memory cell below each mirror with the binary bit plane signal, each mirror on the DMD array is electrostatically tilted to the on or off position. The technique that determines how long each mirror tilts in either direction is called pulse width modulation (PWM). The mirrors are capable of switching on and off more than 1000 times a second. This rapid speed allows digital gray scale and color reproduction.

Figure 1.25 shows a color sequential projection system employing a single DMD. A color wheel is used to illuminate the DMD sequentially with red, green and blue light. In each frame (say, 60 Hz), the encoded information in DMD is sequentially illuminated by each color for about 5.5 ms. The reflected monochromatic images are projected onto a screen. The

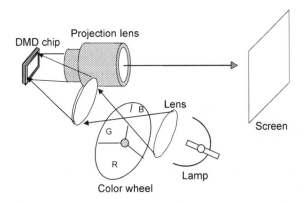

Figure 1.25 A projection system using a single DMD for color sequential display

human eye integrates the three primary colors to perceive a full-color picture. Due to the simple optics involved, this approach is ideal for ultra lightweight (< 1.5 kg) projectors.

4.2 Key DMD features

DMD offers three important features for projection technology:digital nature, high brightness and seamless picture quality.

All digital nature

The inherent digital nature of DMD enables noise-free precise image quality with digital gray scale and color reproduction. The digital nature eliminates the need for analog-to-digital or digital-to-analog conversions so that noise is minimised. This feature also positions DMD to be the final link in the digital video infrastructure [63]. Another digital advantage is DMD's accurate reproduction of gray scale and color levels. Because each video or graphics frame is generated by a digital, 8- to 10-bits-per-color gray scale, the exact digital picture can be recreated time and time again. For example, an 8-bits-per-color gray scale gives 256 different shades of each of the primary colors, which allows for 256^3, or 16.7 million, different color combinations that can be created digitally. The full-color displays have become a common standard for high-end products.

Light efficiency

Because the DMD does not need a polariser, it has a light efficiency of more than 60%, making DMD systems more efficient than most LCD projection displays. This efficiency is the product of reflectivity, fill factor, diffraction efficiency and actual mirror 'on' time.

Most LCDs are polarisation-dependent, so one of the polarised light components is not used. This means that 50% of the lamp light never even gets to the LCD because it is filtered out by a polariser. Other light is blocked by the transistors, gate and source lines in the TFT-LCD panel. The result is that only about 8% of the incident light is transmitted through the LCD panel and onto the screen. There is no doubt that LCD light throughput will be continuously improved. For example, polarisation conversion apparatus has recovered 60–70% of the lost polarisation, and microlens arrays could be implemented to overcome the reduced aperture ratio due to ever-increasing resolution.

Table 1.2 compares the individual factors and overall light efficiency of three microdisplay technologies: p-Si transmissive LCD [64], c-Si reflective LCD and DMD. Obviously,

Table 1.2 Comparison of light efficiency (%) of 3 XGA microdisplay panels: transmissive p-Si LCD, reflective c-Si LCD, and DMD. For both LCD technologies, polarisation conversion is assumed. In the optics category, the f-number is not taken into consideration.

Display	Aperture	Polarisation	Reflectivity	Optics	Efficiency
p-Si T-LCD	60	70	—	90	38
c-Si R-LCD	90	70	90	90	51
DMD	90	—	90	85	69

the results are dependent on the device resolution and advances in technology. In 1996, SVGA (800 × 600) was the mainstream resolution. The aperture ratios of the high-temperature p-Si transmissive LCD, c-Si backplane reflective LCD and DMD are 65, 84 and 77%, respectively. Since the LCD requires linearly polarised light, the light efficiency after passing through two crossed polarisers is cut to 38%. At that time, the polarisation conversion techniques had not yet matured. So the LCD light throughput was only of that of a third DMD. In 1998, XGA (1024 × 760) became the mainstream resolution. The aperture ratio of the transmissive p-Si LCD drops to 45%. For LCOS and DMD, the resolution can increase by enlarging the chip size without sacrificing the aperture ratio. The polarisation conversion method has been implemented in most LCD projectors. As a result, the gap in light throughput between LCD and DMD has been reduced to 20%. As demand for high resolution, such as SXGA (1280 × 1024) or UXGA (1600 × 1200) continues to grow, the transmissive LCD would require microlens arrays [65] to overcome the aperture ratio disadvantage. On a LCOS, an aluminum reflector (90% reflectivity) is normally used. By adding 2–4 layers of dielectric mirror on top of aluminum, the reflectivity can be increased to 97%.

Seamless pictures

The square mirrors on DMDs are 16 μm × 16 μm, separated by 1 μm gaps, giving a fill factor of around 90%. In other words, 90% of the pixel/mirror area can modulate light to create a projected image. Pixel size and gap uniformity are maintained over the entire array and are independent of resolution. A transmissive LCD has, at best, a 70% fill factor for XGA resolution. A low aperture LCD tends to produce the undesirable screen-door effect. The higher DMD fill factor gives a higher perceived resolution and this, combined with the progressive scanning, creates a projected image that is much more natural and lifelike than conventional projection displays.

4.3   Reliability

DLP systems have successfully completed a series of regulatory, environmental, and operational tests, including thermal shock, temperature cycling, moisture resistance, mechanical shock, vibration and acceleration testing [66, 67]. Standard components with proven reliability were chosen to construct the digital electronics used to drive the DMD. No significant reliability degradation has been identified with either the illumination or projection optics. Most of the reliability concerns are focused on the DMD because it relies on moving hinge structures. To test hinge failure, approximately 100 different DMDs were subjected to a simulated one-year operational period. Some devices have been tested for more than 1 trillion cycles, equivalent to 20 years of operation. Inspection of the devices after these tests showed no broken hinges on any of the devices. Hinge failure is not a factor in DMD reliability.

5   Grating Light Valves

The grating light valve (GLV) developed by Silicon Light Machines is a simple micromechanical device [68]. It consists of parallel rows of reflective 'ribbons' suspended in air, as shown in Figure 1.26. The ribbons, made of 100 nm thick silicon nitride, over-coated with 50 nm aluminum, are about 1 μm wide and 110 μm long. Two layers are deposited on to the silicon substrate: the outer layer is 100 nm thick Tungsten and the middle layer is 500 nm thick oxide.

Alternate ribbons can be addressed by applying a voltage across the air gap ($\simeq$ 130 nm). The voltage produces an electrostatic attraction between the nitride ribbon and the substrate, so the ribbon bends towards it. With no voltage applied, reflections from ribbons and substrate add in-phase, and the GLV acts as a mirror. The incoming light is reflected along the same path, resulting in a dark state. When alternate ribbons are deflected by to one-quarter of the incident light wavelength, the outgoing light is diffracted at an angle, as depicted in Figure 1.26(b). The diffracted beams are collected by a Schlieren filter to block the zero order. For the $\pm$ 1 orders, the diffraction efficiency is around 80%. Assuming the ribbon fill factor is around 90% and the aluminum reflectivity is around 90%, the overall light efficiency is about 65%. A contrast ratio of 2000 : 1 at an applied voltage of about 15 V has been demonstrated at the device level. At the system level, a contrast ratio higher than 200 : 1 has routinely been obtained [69].

The GLV switching time is about 20 ns, by far the fastest among all the display technologies developed. Such a fast response time enables a gray scale to be achieved by the pulse width modulation method. The GLV can be addressed in either digital or analog modes. In digital addressing, pixels are switched between on and off states at speeds far faster than the eye can perceive. By varying the amount of time the pixel is on and off, precise gray levels can be obtained. In analog mode, the deflection depth of the ribbons can be precisely controlled by the driving circuitry. When the ribbons are not deflected at all, the pixel is full off. When they are deflected to exactly $\frac{1}{4}\lambda$ of the incident light, the pixel is full on. If the ribbon displacement is between these two limits, various gray scales can be generated.

Unlike a direct-view LCD panel where two-dimensional TFT arrays are used as two-dimensional light switches, the GLV systems use one-dimensional arrays (single columns) that are scanned to produce a two-dimensional image [70]. Using this linear scan, the ribbon fill factor becomes less important. Thus, the ribbons can be made considerably longer, but only the central 18 percent is illuminated. In this way, performance degradation caused by ribbon curvature can be avoided.

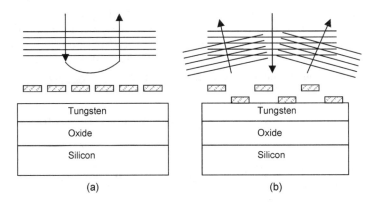

(a)   (b)

Figure 1.26 Operating principles of a grating light valve. (a) When no voltage is applied, the incoming beam is reflected back. (b) When voltage is on, the active ribbons are electrostatically pulled towards the substrate and the device becomes an efficient diffraction grating after [70]

Similar to the DMD, both single and three GLV approaches have been demonstrated for color-sequential and high brightness large-screen projection displays.

6 Interferometric Modulation

A micromechanical device using interferometric modulation (IMod) to create a reflective color display is sketched in Figure 1.27 [71]. The device consists of a self-supporting deformable membrane, a thin-film stack protected by an insulating layer, and a transparent glass substrate. The stack and the metal (Nickel) membrane act as two mirrors in an optically resonant cavity. The colors are determined by the thickness of the insulating layers, as well as the height of the deformable membrane. Where no voltage is applied, the optical distance between the two mirrors is determined by the thickness and refractive index of the insulator, and height of the membrane. When the device is switched, however, the current causes the membrane to be electrostatically attracted to the insulator, stack and substrate. When it makes contact, the optical thickness is reduced and the color changes. By using this method, some of the problems inherent in diffractive systems are overcome. In particular, there is almost no rainbow effect, where the color changes with viewing angle. For a full-color display, three sets of IMods would be fabricated with dimensions selected to allow red-black (gap $\simeq$ 400 nm), green-black (160 nm), and blue-black (300 nm) switching, respectively. White is the addition of the RGB colors and gray scales can be achieved by pulse width modulation. The switching speed of the IMod is faster than 100 kHz. Thus, a digital gray scale can be achieved.

The voltage-dependent reflectance of the micromechanical IMod device exhibits an inherent hysteresis. This is caused by the fact that the restoration force of the membrane is linear whereas the offsetting electrostatic force is nonlinear. If the device is operated between 3 and 5 V, it is bistable. The bistable display is known to have low power consumption and wide viewing angle.

Both alphanumeric and matrix addressed displays using IMod arrays have been demonstrated. A monochrome matrix display with reflectivity more than 50%, contrast ratio of 4 : 1, viewing angle of $\pm 45°$, and drive voltage less than 5 V has been demonstrated [72]. On the other hand, static black and white displays with reflectivity greater than 70%, contrast ratio less than 12 : 1 and viewing angle exceeding $\pm 60°$ have been developed.

A major factor affecting the device lifetime is charge deposition during operation. A device lifetime as high as 7.5×10^{10} actuation cycles has been obtained, which is equivalent to 18 years of operation at 4 hours/day with a refresh rate of 10 Hz and 4 bits of gray.

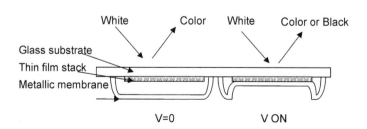

Figure 1.27 The operating principles of an interferometric modulation device after [72]

7 Electrophoretic Displays

The electrophoretic (EP) effect has been actively investigated since the late 1960s, envisaging paper-like flexible displays [73, 74]. A EP display is a non-emissive display based on the transport of charged pigment particles in a collodial suspension [75]. By alternating the polarity of the applied voltage, the charged pigment particles are transported to either the front or the back surface of the display panel. Thus, images with a modest contrast ratio ($\simeq 5:1$) can be displayed.

In the traditional implementation of a reflective EP display [76, 77], white or light colored particles are suspended in a dark dyed fluid between parallel electrodes at the front and back surfaces of the display. When the particles are transported to the front surface by an electric field, the bright state appears. Conversely, when the particles are driven to the rear electrode by reversing the electric field, the viewer sees the color of the dye and the display looks dark. The typical reflectivity of such an EP display is less than 30%.

To improve the reflectivity and contrast ratio, several approaches using absorbing or colored particles in a transparent fluid [78, 79], microencapsulated EP [80], and reverse-emulsion EP [81] have been developed. The electrode design plays a key role in determining the display performance. The in-plane, lines/plate and wall/post structures are shown to have promising features for flexible displays.

7.1 In-plane electrophoretic displays (IP-EPD)

The device structure of the IP-EPD is sketched in Figure 1.28. Two electrodes are embedded in the rear plastic substrate. The top stripe-shaped electrode, spaced at 120 μm, is a titanium thin film coated with a black photoresist in order to obtain a good dark state. The bottom electrode is a continuous aluminum thin film. The cell gap is about 50 μm. The EP medium consists of Isopar (isoparaffinic solvent) dielectric fluid, in which are suspended positively charged black toner particles about 1–2 μm in diameter.

Figure 1.28(a) shows the black state of an IP-EPD. When the D1 electrode has positive and D2 has negative voltage, these positively charged particles are dispersed away from D1.

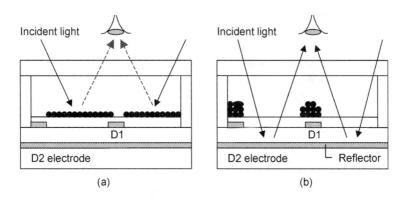

(a)   (b)

Figure 1.28 Device structure of an in-plane electrophoretic display: (a) black state, (b) bright state after [78]

The black toner particles are spread around to cover the bottom area. As the D1 electrodes are coated with black photoresist, the incident light is absorbed everywhere and images look dark. When the polarity of the applied voltage is reversed, the toner particles aggregate around the D1 electrodes, as depicted in Figure 1.28(b). The incident light traverses through the EP medium and is reflected from the middle insulating layer which is dispersed with aluminum fine particles.

The highest contrast ratio of around 8 : 1 is achieved when the area ratio of the black to white electrode is 2.5 : 7.5. The reflectivity is about 50%. The response time of the toner particles depends on the driving voltage. At 100 V, the switching time is 15 ms, and at 40 V it increases to 30 ms. A remarkable feature of IP-EPD is that the image stability is excellent even when the display is mechanically deformed. Also, the images can be retained for a week without any holding voltage. Thus, potential applications of EPD for low power, reflective and paper-like flexible displays are foreseeable.

7.2 Lines/plate EPD

In the lines/plate design, when the voltage is positive, as in the right pixel shown in Figure 1.29, the charged particles are repelled by the top electrode and collected along line electrodes on the bottom surface. The pixel becomes clear. The incoming light is either transmitted or reflected (for reflective mode operation) with high efficiency. When the voltage is reversed (left pixel), the particles are driven off the bottom electrodes and spread across the top plate. The incident light is absorbed by the black particles, resulting in a dark state. A black/white electrophoretic display with 300 dots-per-inch resolution, 61% reflectivity and 9.7 : 1 contrast ratio has been demonstrated.

The lines/plate design has the advantage that the particles have only to travel between the two surfaces, which is about 25 μm. Therefore, the required operation voltage is reduced to around 12 V. The response time of the device depends on several factors, such as the particle size, mobility, cell gap and applied voltage. For the sub-micron particle size

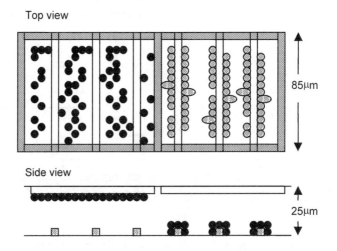

Figure 1.29 Top and side views of the lines/plate electrophoretic display after [79]

Top view

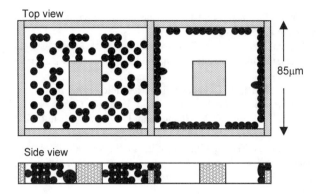

85μm

Side view

Figure 1.30 Top and side views of the wall/post electrophoretic display after [79]

employed, in order to achieve video rate operation, the particle mobility needs to exceed $3500\,\mu m^2/Vs$.

7.3 Wall/post EPD

In the wall/post design, as shown in Figure 1.30, the electrodes surrounding the post electrode are held at ground potential. When the post electrode is positive, the particles are gathered on wall electrodes, resulting in a bright state. When the voltage is reversed, particles are dispersed across the entire pixel. The incoming light is absorbed. This design has the advantage that the entire vertical height of the cell can be used to collect the pixels. The disadvantage is that the electrical field strength is not linear and higher particle mobility ($18\,000\,\mu m^2/Vsec$) is required to achieve video rate. A black/white device with 71% reflectivity and 11.3 : 1 contrast ratio has been demonstrated.

References

1 A. R. Kmetz, A single-polariser twisted nematic display, *Proc. SID* **21**, 63 (1980).

2 N. K. Sheridon, E. A. Richley, J. C. Mikkelsen, D. Tsuda, J. M. Crowley, K. A. Oraha, M. E. Howard, M. A. Rodkin, R. Swidler, and R. Sprague, The Gyricon rotating ball display, *J. SID* **7**, 141 (1999).

3 U. Efron, J. Grinberg, P. O. Brazz, M. J. Little, P. G. Reif, and R. N. Schwartz, The silicon liquid crystal light valves, *J. Appl. Phys.* **57**, 1356 (1985).

4 R. D. Sterling, R. D. TeKoslste, J. M. Haggerty, T. C. Borah, and W. P. Bleha, Video-rate liquid crystal light valve using an amorphous silicon photoconductor, *SID Tech. Digest* **21**, 327 (1990).

5 M. Schadt and W. Helfrich, Voltage-dependent light transmittance of a twisted nematic liquid crystal, *Appl. Phys. Lett.* **18**, 127 (1971).

6 J. L. Fergason, Polymer encapsulated nematic liquid crystals for display and light control applications, *SID Tech. Digest* **16**, 68 (1985).

7 J. W. Doane, N. A. Vaz, B. G. Wu, and S. Žumer, Field controlled light scattering from nematic microdroplets, *Appl. Phys. Lett.* **48**, 269 (1986).

8   H. Murai, T. Gotoh, T. Nakata, and E. Hasegawa, Homeotropic reverse-mode polymer liquid crystal device, *J. Appl. Phys.* **81**, 1962 (1997).

9   T. Sonehara, M. Yazaki, H. Iisaka, Y. Tsuchiya, H. Sakata, J. Amako, and T. Takeuchi, Full-color reflective liquid crystal display using internal-reflection inverted-scattering mode, *J. SID* **7**, 23 (1999).

10 K. Yang, L. C. Chien, and J. W. Doane, Cholesteric liquid crystal polymer gel dispersion bistable at zero field, *Conf. Record of the 11th* Intl Display Research Conf., 49–52 (1991).

11 D. Davis, A. Khan, C. Jones, X. Y. Huang, and J. W. Doane, Multiple color high resolution reflective cholesteric liquid crystal displays, *J. SID* **7**, 43 (1999).

12 G. H. Heilmeier and L. A. Zanoni, Guest-host interactions in nematic liquid crystals, *Appl. Phys. Lett.* **13**, 91 (1968)

13 H. S. Cole and R. A. Kashnow, A new reflective dichroic liquid crystal display device, *Appl. Phys. Lett.* **30**, 619 (1977).

14 D. L. White and G. N. Taylor, New absorptive mode reflective liquid-crystal display device, *Appl. Phys. Lett.* **45**, 4718 (1974).

15 T. Uchida, H. Seki, C. Shishido and M. Wada, Guest-host interactions in nematic liquid crystal cells with twisted and tilted alignments, *Mol. Cryst. Liq. Cryst.* **54**, 161 (1979).

16 T. Uchida, T. Nakayama, T. Miyashita, M. Suzuki, and T. Ishinabe, A novel reflective LCD for high resolution color display, *Asia Display'95*, p. 599 (1995).

17 T. Uchida, Reflective LCDs for low-power systems, *SID Tech. Digest* **27**, 31 (1996).

18 C. L. Kuo, C. L. Chen, D. L. Ting, C. K. Wei, C. K. Hsu, B. J. Liao, B. D. Liu, C. W. Hao, and S. T. Wu, A 10.4$''$ reflective MTN-mode TFT-LCD with video-rate and full-color capabilities, *SID Tech. Digest* **28**, 79 (1997).

19 F. J. Ko and H. P. D. Shieh, Asymmetrical microlens array light control film for reflective LCDs application, *SID Tech. Digest* **31**, 1071 (2000).

20 Y. Ishii and M. Hijikigawa, Development of highly reflective color TFT-LCDs, *Asia Display'98*, p. 119 (1998).

21 Y. Itoh, S. Fujiwara, N. Kimura, S. Mizushima, F. Funada, and M. Hijikigawa, Influence of rough surface on the optical characteristics of reflective LCD with a polariser, *SID Tech. Digest* **29**, 221 (1998).

22 K. Nakamura, H. Nakamura, and N. Kimura, Development of high reflective TFT, *Sharp Technical Journal* **69**, 33 (1997)

23 H. Ikeno, H. Kanoh, N. Ikeda, K. Yanai, H. Hayama, and S. Kaneko, A reflective guest-host LCD with 4096 color display capability, *SID Tech. Digest* **28**, 1015 (1997).

24 I. Hiyama, O. Itoh, K. Kondo, and A. Arimoto, High performance reflective STN-LCD with a blazed reflector, *J. SID* **7**, 49 (1999).

25 C. J. Wen, D. L. Ting, C. Y. Chen, L. S. Chuang, and C. C. Chang, Optical properties of reflective LCD with diffusive micro slant reflectors, *SID Tech. Digest* **31**, 526 (2000).

26 D. L. Ting, W. C. Chang, C. Y. Liu, J. W. Shiu, C. J. Wen, C. H. Chao, L. S. Chuang, and C. C. Chang, A high brightness high contrast reflective LCD with a micro slant reflector, *SID Tech. Digest* **30**, 954 (1999).

27 P. J. Ralli and M. M. Wenyon, Imagix™ holographic diffusers for reflective LCDs, *SID Application Digest* **27**, 3 (1996).

28 D. Kodama, N. Ichikawa and Y. Ohyagi, Holographic light control film for liquid crystal displays, *The 5th Int'l Display Workshops*, p. 319 (1998).

29 T. J. Trout, W. J. Gambogi, K. W. Steijn, and S. R. Mackara, Volume holographic components for display applications, *SID Tech. Digest* **31**, 202 (2000).

30 A. G. Chen, K. W. Jelley, G. T. Valliath, W. J. Molteni, P. J. Ralli, and M. M. Wenyon, Holographic reflective liquid crystal display, *SID Tech. Digest* **26**, 176 (1995).

31 M. Wenyon, W. Molteni, and P. Ralli, White holographic reflectors for LCDs, *SID Tech. Digest* **28**, 691 (1997).

32 Y. Higashigaki, T. Tokumaru, and K. Iwauchi, Holographic directive reflectors for reflective color LCDs, Sharp Technical Journal (May 1999) or http://sharp-world.com/sc/library-e/techn_top/journal2000/4.html

33 R. van Asselt, R. A. W. van Rooij, and D. J. Broer, Birefringent colour reflective liquid crystal displays using broadband cholesteric reflectors, *SID Tech. Digest* **31**, 742 (2000).

34 M. Tani, Color filters for reflective LCDs, *The 5th Int'l display workshops*, p. 287 (1998).

35 T. Maeda, T. Matsushima, E. Okamoto, H. Wada, O. Okumura, and S. Iino, Reflective and transflective color LCDs with double polarisers, *J. SID* **7**, 9 (1999).

36 H. Baek, Y. B. Kim, K. S. Ha, D. G. Kim and S. B. Kwon, New design of transflective LCD with single retardation film, *The 7th Int'l Display Workshops*, p. 41 (2000).

37 C. Y. Tai, H. Zou, and P. K. Tai, Transparent frontlighting system for reflective type displays, *SID Tech. Digest* **26**, 375 (1995).

38 C. Y. Tai, Compact front lighting for reflective displays, *SID Application Digest* **27**, 43 (1996).

39 E. H. Stupp and M. S. Brennesholtz. Projection Displays. (Wiley, New York, 1999).

40 D. Armitage, Silicon-chip liquid-crystal display, *Proc. SPIE* **2407**, 280–290 (1995).

41 F. Sato, Y. Yagi, and K. Hanihara, High resolution and bright LCD projector with reflective LCD panels, *SID Tech. Digest* **28**, 997 (1997).

42 P. M. Alt, Single crystal silicon for high resolution displays, Conference record of the *Intl. Display Research Conf.*, M19–28 (1997).

43 H. Kurogane, K. Doi, T. Nishihata, A. Honma, M. Furuya, S. Nakagaki, and I. Takanashi, Reflective AMLCD for projection displays, *SID Tech. Digest* **29**, 33 (1998).

44 R. D. Sterling and W. P. Bleha, D-ILA technology for electronic cinema, *SID Tech. Digest* **31**, 310 (2000).

45 C. Chinnock, Microdisplays and manufacturing infrastructure mature at *SID 2000, Information display* **16**, 18 (September 2000).

46 M. Bone, "Front project optical system design for reflective LCOS technology," Presented at Microdisplay Conference, Boulder, Colorado (2000).

47 N. Ichikawa, Holographic optical element for liquid crystal projector, *Asia Display'95*, pp. 727–729 (1995).

48 B. A. Loiseaux, C. Joubert, A. Delboulbe, J. P. Huignard, and D. Battarel, Compact spatio-chromatic single-LCD projection architecture, *Asia Display'95*, p. 87 (1995).

49 D. Banas, Ferroelectric liquid-crystal spatial light modulators for projection display, *Proc. SPIE* **2650**, 229 (1996).

50 D. Armitage, Design issues in liquid-crystal projection displays, *Proc. SPIE* **2650**, 41 (1996).

51 T. Higashi and T. Arimoto, Long-life dc metal-halide lamps for LCD projectors, *SID Tech. Digest* **26**, 135 (1995).

52 N. Stewart, D. M. Rutan, and D. J. Savage, High efficiency metal halide lighting systems for compact LCD projectors, *Proc. SPIE* **3013**, 72 (1997).

53 E. Schnedler and H.V. Wijngaarde, Ultrahigh-intensity short-arc long-life lamp system, *SID Digest* **26**, 131 (1995).

54 J. H. Shimizu, Rear-projection screens, *SID Tech. Digest* **26**, 141 (1995).

55 D. W. Vance, A novel high-resolution ambient-light-rejecting rear-projection screen, *SID Digest* **25**, 741 (1994).

56 J. F. Goldberg, Q. Huang, and J. A. Shimizu, Rear projection screens for light valve projection systems, *Proc. SPIE* **3013**, 49 (1997).

57 K. Oda, H. Sekiguchi, and M. Gotoh, Ultra high contrast screen, *SID Tech. Digest* **31**, 198 (2000).

58 R. L. Shie, C. W. Chau, and J. M. Lerner, Surface-relief holography for use in display screens, *Proc. SPIE* **2407**, 177 (1995).

59 N. Bergstrom, C. L. Chuang, M. Curley, A. Hildebrand, and Z. W. Li, Ergonomic wearable personal display, *SID Tech. Digest* **31**, 1138 (2000).

60   D. J. Schott, Reflective LCOS light valves and their application to virtual displays, *Euro Display '99*, p. 485 (1999).

61   L. J. Hornbeck, 128 × 128 deformable mirror device, IEEE Trans. *Electron Devices ED* **30**, 539 (1983).

62   J. B. Sampsell, An overview of the performance envelope of DMD-based projection display systems, *SID Tech. Digest* **25**, 669 (1994).

63   R. J. Gove, DMD display systems: the impact of an all-digital display, *SID Tech. Digest* **25**, 673 (1994).

64   K. Yoneda, State-of-the-art low-temperature-processed poly-Si TFT technology, IDRC, M40–M47 (1997).

65   H. Hamada, A new high definition microlens array built in p-Si TFT-LCD panel, *Asia Display '95*, pp. 887–890 (1995).

66   M. R. Douglass and D. M. Kozuch, DMD reliability assessment for large-area displays, *SID Tech. Digest* (Application Session) **26**, 49 (1995).

67   M. R. Douglass and C. G. Malemes, Reliability of displays using digital light processing, *SID Tech. Digest* **27**, 774 (1996).

68   R. B. Apte, F. S. A. Sandejas, W. C. Banyai, and D. M. Bloom, Deformable grating light valve for high resolution displays, *SID Tech. Digest* **24**, 807 (1993).

69   D. M. Bloom, The Grating Light Valve: revolutionizing display technology, *Proc. SPIE* **3013** (1998).

70   D. T. Amm and R. W. Corrigan, Grating Light Valve technology: update and novel applications, *SID Tech. Digest* **29**, 29 (1998).

71   M. W. Miles, A new reflective FPD technology using interferometric modulation, *SID Tech. Digest* **28**, 71 (1997).

72   M. W. Miles, Digital paper: reflective displays using interferometric modulation, *SID Tech. Digest* **31**, 32 (2000).

73   P. F. Evans, H. D. Lees, M. S. Maltz, and J. L. Daily, Color display device, U. S. Patent 3,612,758 (1971).

74   I. Ota, Electrophoretic display device, U. S. Patent 3,666,106 (1972).

75   A. L. Dalisa, Electrophoretic display technology, IEEE *Trans. Electron Device ED* **24**, 827 (1977).

76   I. Ota, M. Tsukamoto, and T. Ohtsuka, Development in electrophoretic displays, *Proc. SID* **18**, 243 (1977).

77   B. Singer and A. L. Dalisa, An X-Y addressable electrophoretic display, *Proc. SID* **18**, 255 (1977).

78   E. Kishi, Y. Matsuda, Y. Uno, A. Ogawa, T. Goden, N. Ukigaya, M. Nakanishi, T. Ikeda, H. Matsuda, and K. Eguchi, Development of In-Plane EPD, *SID Tech. Digest* **31**, 24 (2000).

79   S. A. Swanson, M. W. Hart, and J. G. Gordon, II, High performance electrophoretic displays, *SID Tech. Digest* **31**, 29 (2000).

80   P. Drzaic, B. Comiskey, J. D. Albert, L. Zhang, A. Loxley, R. Feeney, and J. Jacobson, A printed and rollable bistable electronic display, *SID Tech. Digest* **29**, 1131 (1998).

81   M. Bryning and R. Cromer, Reverse-emulsion electrophoretic display, *SID Tech. Digest* **29**, 1018 (1998).

2

Liquid Crystal Materials

1 Introduction

Color reflective displays have been developed using nematic, ferroelectric and cholesteric liquid crystals. Gray scale is an important requirement for color reflective displays. For an analog nematic device, gray scale can be achieved through voltage control [1]. Thus, the major reflective displays developed so far have been dominated by nematics. Some handheld displays using bistable cholesteric have also been developed [2, 3]. These bistable displays exhibit very low power consumption and high brightness, and are particularly suitable for outdoor applications. Ferroelectric liquid crystals, on the other hand, have been demonstrated on a silicon backplane for virtual and projection displays [4]. Due to its fast response time, the ferroelectric LCD can perform color sequential and time-domain modulation to obtain gray scales.

Almost all the light modulation mechanisms, except the guest-host display (see Chapter 6), utilise a voltage-induced LC refractive index change caused by molecular reorientation. The phase retardation effect of a homogeneous cell [5] or a homeotropic cell [6], and the polarisation rotation effect [7] of a twisted nematic cell all depend on the birefringence (Δn) of the LC employed.

Solely from the birefringence standpoint, no matter what birefringence value an LC has, it can always have a useful application for displays. LCs with low birefringence ($\Delta n < 0.1$) are particularly attractive for reflective displays. In a reflective display, the required $d\Delta n$ value (where d is the cell gap) is about half that of a corresponding transmissive display because the beam traverses the LC layer twice. Two methods can be used to obtain a small $d\Delta n$: reduce the cell gap or use a low-birefringence LC. Reducing the cell gap has the benefit of obtaining a fast response time [8], but the manufacturing yield could suffer. Thus, using a low-birefringence liquid crystal mixture is the preferred approach.

LCs with modest birefringence ($0.1 < \Delta n < 0.2$) are attractive for reflective super-twisted nematic (STN) displays [9] where the required $d\Delta n$ is in the 0.6–0.8 μm range. To maintain a 4 μm cell gap, the proper LC birefringence is about 0.15–0.2.

High-birefringence liquid crystals (with $\Delta n > 0.2$) are particularly attractive for cholesteric displays [10], polymer-dispersed LCs (PDLCs) [11], holographic PDLCs [12], infrared [13] and microwave [14,15] light modulators. For a cholesteric display using Bragg reflection, a high Δn widens the reflection spectral bandwidth and improves the display brightness. For polymer-dispersed LCs, a high Δn would enhance the light scattering efficiency and consequently improve the contrast ratio. For LC-based infrared spatial light modulators or a microwave phase shifter [16], high Δn helps to shorten the response time because of the thinner cell gap requirement.

Birefringence is not the only important parameter for display applications. Other parameters, such as phase transition temperatures, dielectric constants, elastic constants and viscosity all play crucial roles. Dielectric and elastic constants jointly determine the operating voltage, whereas the rotational viscosity and the elastic constant determine the response time. From the molecular structure viewpoint, the macroscopic properties of phase transition behavior, birefringence, dielectric constants, elastic constants and viscosity are interrelated. To optimise one parameter may not be too difficult, but to optimise all the parameters simultaneously is an art.

In this chapter the physical origins of LC birefringence, dielectric constants, elastic constants, viscosity and absorption that are relevant to display applications are described.

2 Birefringence

Based on the single-band [17] and three-band [18] birefringence dispersion models, the LC birefringence is mainly determined by the molecular conjugation, differential oscillator strength and order parameter. Thus, a more linearly conjugated rod-like LC would exhibit a larger optical anisotropy. A general problem of these highly conjugated molecules is that their melting temperature is usually quite high. From the Schröder–Van Laar equation [19, 20], an LC compound with a high melting point T_{mp} and large enthalpy of fusion ΔH would result in a poor solubility while forming eutectic mixtures and make an insignificant contribution to lowering the melting point.

2.1 Wavelength-dependent birefringence

Using the single-band model, the wavelength-dependent birefringence of an LC can be expressed as follows [16]:

$$\Delta n = G \frac{\lambda^2 \lambda^{*2}}{\lambda^2 - \lambda^{*2}} \tag{2.1}$$

In Equation 2.1 the parameter $G = gNZS(f_{\parallel}^* - f_{\perp}^*)$, where g is a proportionality constant, N is the molecular packaging density, Z is the effective number of participating electrons (σ and π), S is the order parameter, $(f_{\parallel}^* - f_{\perp}^*)$ is the differential oscillator strength, and λ^* is the mean electronic transition wavelength. For an LC substance, the electron transition bands are fairly well defined. The three-band model provides a good estimate on each band's contribution to the refractive index or birefringence. However, for a eutectic mixture containing multiple components with different structures, the electron transition bands could overlap and be broadened. In this circumstance, the single-band model is more convenient than the three-band model.

Equation 2.1 involves two adjustable parameters G and λ^*. Thus, by measuring birefringence at two laser wavelengths, these two fitting parameters can be determined. Once G and λ^* have been found, the entire birefringence dispersion curve is determined. From Equation 2.1, as the wavelength increases, Δn decreases gradually and saturates in the near infrared region. In the IR or microwave region where $\lambda \gg \lambda^*$, the LC birefringence is basically independent of wavelength ($\Delta n \simeq G\lambda^{*2}$) except in the vicinities of some local molecular vibration bands.

Figure 2.1 plots the wavelength-dependent birefringence using Equation 2.1 with $G = 3 \times 10^{-6}\,\mathrm{nm}^{-2}$ and $\lambda^* = 200$ and $270\,\mathrm{nm}$. The open circles represent the calculated birefringence for $\lambda^* = 200\,\mathrm{nm}$ and solid circles are for $\lambda^* = 270\,\mathrm{nm}$. The larger λ^* implies a longer molecular conjugation. From Figure 2.1, a highly conjugated LC possesses a higher birefringence and its wavelength dependency is also more pronounced.

2.2 Temperature-dependent birefringence

Temperature has great influence on the performance of a reflective LCD. For handheld displays, the ambient temperature could vary significantly, depending on the geographic locations. For a projection display, the lamp-induced thermal effect could heat up the LCD panel to 50°C or higher. As the temperature increases, the birefringence, dielectric anisotropy, elastic constants and viscosity all drop, but at different rates.

The temperature-dependent birefringence is mainly governed by the order parameter as:

$$\Delta n = (\Delta n)_0 S \tag{2.2}$$

$$S = [1 - T/T_c]^\beta \tag{2.3}$$

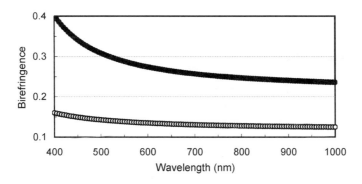

Figure 2.1 The calculated wavelength-dependent birefringence using Equation 2.1. Open circles are results using $\lambda^* = 200\,\mathrm{nm}$ and solid circles are for $\lambda^* = 270\,\mathrm{nm}$. The parameter G is kept unchanged at $G = 3.0 \times 10^{-6}\,\mathrm{nm}^{-2}$

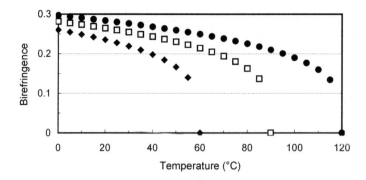

Figure 2.2 The temperature-dependent birefringence of three LC homologues with $T_c = 60$ (diamonds), 90 (squares) and 120°C (circles) using Equation 2.2. $\Delta n_o = 0.4$ and $\beta = 0.25$

In Equation 2.2, Δn_o is the birefringence when the order parameter $S = 1$ (a perfect crystalline state). Equation 2.3 is Haller's approximation for the order parameter where β is a material parameter [21]. This approximation holds reasonably well when $T_c - T > 2$ degrees. From Equation 2.1, Δn_o decreases as wavelength increases. For many of the LC compounds studied $\beta \simeq 0.25$, insensitive to materials [22]. By contrast, β is only dependent on the LC structure and is independent of the wavelength. Therefore, in order to make a fair comparison among different molecular structures, their birefringence has to be compared at the same wavelength and the same reduced temperature.

Figure 2.2 plots the calculated temperature-dependent birefringence using Equation 2.2 with $\Delta n_o = 0.4$ and $\beta = 0.25$. In reality, the assumption of having the same Δn_o but different clearing point can be fulfilled by LC homologues, i.e. the same core but a slightly different side chain length. Three T_c values (60, 90 and 120°C) are used here for comparison. From Figure 2.2, three general trends are found, as follows. As the temperature increases, the birefringence decreases gradually; this trend is more pronounced as the temperature approaches T_c. For the high-T_c homologue, its temperature effect is less sensitive in the vicinity of room temperature; this is because the clearing point is further away. A high-T_c homologue exhibits a slightly higher birefringence than its low-T_c counterpart at a given temperature owing to its higher order parameter.

3   Dielectric Constants

The dielectric constants of a liquid crystal make direct contributions to the operating voltage and resistivity, and have an indirect impact on the viscosity. The threshold voltage V_{th} of a homogeneous cell is related to the dielectric anisotropy ($\Delta \varepsilon = \varepsilon_\| - \varepsilon_\perp$) and splay elastic constant K_{11} as [23]:

$$V_{th} = \pi(K_{11}/\varepsilon_o \Delta \varepsilon)^{1/2} \tag{2.4}$$

Thus, low threshold voltage can be obtained by either enhancing the dielectric anisotropy, reducing the elastic constant, or a combination of both. Before we get into a more detailed discussion about dielectric constants, let us briefly describe the significance of the LC resistivity and charge holding ratio.

3.1 LC resistivity

The LC resistivity ρ affects the charge holding ratio, current density and power consumption of the display device. In a TFT-LCD, if the applied voltage to the image pixels cannot be held constant for about 16 ms, flicker will be noticeable. Thus, a stringent requirement for high resistivity ($10^{13}\Omega$cm) even after thermal cycling and UV radiation is imposed on the chosen LC material. The bulk resistivity of a LC mixture depends on the purity of the individual components and what polar groups are employed. For instance, the cyano group possesses a larger dipole moment than the fluoro group. However, the cyano compound is found more easily to capture impurity ions [24]. These trapped ions tend to accumulate near the alignment layer interfaces. As a result, the LC cell resistivity reduces rapidly as the temperature or ultraviolet radiation increases.

Figure 2.3 shows the temperature effect on the resistivity of three polar (cyano, difluoro and trifluoro) bicyclohexane-phenyl LC compounds [25]. At $T = 0°$C, all compounds have a very high resistivity $\rho \simeq 10^{14}\Omega$cm. As the temperature increases, the resistivity falls at different rates for the three compounds. The trifluoro shows the lowest rate, followed by difluoro and then finally cyano.

The voltage holding ratio (VHR) of these three LC compounds is plotted in Figure 2.4. Both trifluoro and difluoro compounds possess an impressive VHR (greater than 99%) all the way up to 80°C. By contrast, the cyano compound has a 98% VHR at low temperatures, but drops almost linearly to 92% at $T = 80°$C. Low VHR causes the voltage of a pixel image to decay with time and gives rise to undesirable image flickering for active-matrix liquid crystal displays [26]. For color sequential displays, the voltage holding time is so short (less than 2 ms) that the LC is actually in its dynamic switching state. A small amount of flickering is tolerable. Therefore, the high resistivity requirement is not so stringent. In a passive-matrix display, each pixel is addressed momentarily and the LC resistivity is not so demanding as in an active-matrix display.

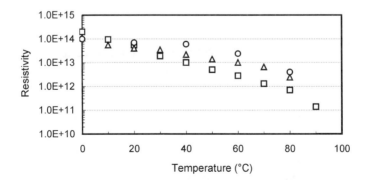

Figure 2.3 Temperature effect on the resistivity (Ωcm) of cyano (squares), difluoro (triangles) and trifluoro (circles) bicyclohexane-phenyl compounds (redrawn from [25])

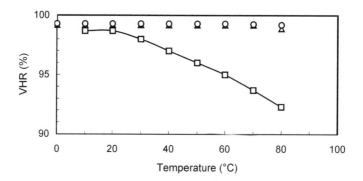

Figure 2.4 Temperature effect on the voltage-holding ratio of cyano (squares), difluoro (triangles) and trifluoro (circles) bicyclohexane-phenyl compounds (redrawn from [25])

3.2 Mean field theory

The dielectric constants of a liquid crystal are mainly determined by the dipole moment, μ, its angle, θ, with respect to the principal molecular axis, and order parameter, S, as described by the Maier and Meier mean field theory [27]:

$$\varepsilon_{//} = NhF\{\langle \alpha_{//} \rangle + (F\mu^2/3kT)[1 - (1 - 3\cos^2\theta)S]\} \qquad (2.5)$$

$$\varepsilon_{\perp} = NhF\{\langle \alpha_{\perp} \rangle + (F\mu^2/3kT)[1 + (1 - 3\cos^2\theta)S/2]\} \qquad (2.6)$$

$$\Delta\varepsilon = NhF\{(\langle \alpha_{//} \rangle - \langle \alpha_{\perp} \rangle) - (F\mu^2/2kT)(1 - 3\cos^2\theta)S\} \qquad (2.7)$$

Here, N stands for the molecular packing density, $h = 3\varepsilon/(2\varepsilon + 1)$ is the cavity field factor, $\varepsilon = (\varepsilon_{//} + \varepsilon_{\perp})/3$ is the averaged dielectric constant, F is the Onsager reaction field, and $\langle \alpha_{//} \rangle$ and $\langle \alpha_{\perp} \rangle$ are the principal elements of the molecular polarisability tensor.

From Equation 2.7, for a non-polar compound, $\mu \simeq 0$ and its dielectric anisotropy is very small. In this case, $\Delta\varepsilon$ is determined entirely by the differential molecular polarisability, i.e. the first term in Equation 2.7. The molecules with a larger differential polarisability in the low-frequency (kHz) regime often exhibit a larger birefringence in the optical frequency regime. Thus, the dielectric anisotropy of a non-polar biphenyl compound is expected to be slightly larger than its phenyl-cyclohexane counterpart with the same chain length. Within the same homologous series, one with a shorter side chain will generally exhibit a slightly larger $\Delta\varepsilon$ due to the denser molecular packing effect, N. As indicated in Equation 2.7, $\Delta\varepsilon$ is linearly proportional to N.

For a polar compound, the dielectric anisotropy depends on the dipole moment, angle θ, temperature T and applied frequency. For a polar compound with its effective dipole at $\theta < 55°$, $\Delta\varepsilon$ is positive. On the other hand, $\Delta\varepsilon$ becomes negative if $\theta > 55°$. As temperature increases, $\Delta\varepsilon$ decreases in proportion to S/T. In addition, at a sufficiently high frequency (around 10 MHz for the cyano-biphenyls), dielectric relaxation occurs and the dielectric anisotropy changes sign. The frequency at which dielectric anisotropy changes sign is called the crossover frequency [28].

Figure 2.5 depicts the simulated temperature-dependent dielectric constants and dielectric anisotropy of a LC mixture using Equations 2.5 and 2.6. The following parameters are assumed: $\theta = 0°$, $T_c \simeq 90°C$, the two $\langle \alpha \rangle$ terms are both ignored, $NhF^2\mu^2 = 270$ and $\beta = 0.25$. As temperature increases, $\varepsilon_\parallel$ decreases and $\varepsilon_\perp$ increases gradually, resulting in a decreased $\Delta\varepsilon$.

When a molecule has two dipole groups (with dipole moment μ_1 and μ_2), its effective dipole moment can be calculated by the vector addition method, as illustrated in Figure 2.6. In Figure 2.6 μ_1 is along the principal molecular axis and μ_2 is at an angle ϕ with respect to μ_1. The resultant dipole moment μ_r can be calculated from the following equation:

$$\mu_r = \left(\mu_1^2 + \mu_2^2 + 2\mu_1\mu_2 \cos\phi\right)^{1/2} \tag{2.8}$$

To help design a LC molecule with proper dielectric constants, the following examples are useful. The position of the dipole in a phenyl ring is described with reference to the following diagram:

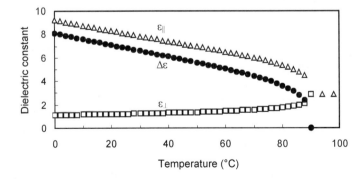

Single polar group (μ_1)

If μ_1 is in the axial (4) position, then $\theta \simeq 0°$ and $\Delta\varepsilon$ is strongly positive. If μ_1 is in the 3 or 5 position and if we assume that the axial group has a negligible dipole moment, then $\theta = 60°$ and $\Delta\varepsilon$ is weakly negative. If μ_1 is in the 2 or 6 position and if we assume that the axial group has a negligible dipole moment, then $\theta = 120°$ and $\Delta\varepsilon$ is weakly negative.

Figure 2.5 Simulated dielectric constants and dielectric anisotropy (solid circles) using Equation 2.7 by neglecting the $\langle \alpha \rangle$ terms and assuming $\theta = 0°$, $T_c = 90°C$, $\beta = 0.25$ and $NhF^2\mu^2 = 270$

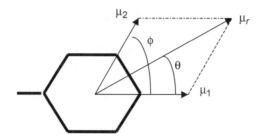

Figure 2.6   Vector sum method for calculating effective dipole moment

Two polar groups (let us assume $\mu_1 = \mu_2$ for simplicity)

If μ_1 and μ_2 are in the 3 and 4 positions, then $\phi = 60°$, $\mu_r = \sqrt{3}\mu_1$, and $\theta = 30°$ so that $\Delta\varepsilon$ is large and positive. If μ_1 and μ_2 are in the 2 and 4 positions, then $\phi = 120°$, $\mu_r = \mu_1$, and $\theta = 60°$ so that $\Delta\varepsilon$ is small and negative. If μ_1 and μ_2 are in the 2 and 3 positions, then $\phi = 60°$, $\mu_r = \sqrt{3}\mu_1$, and $\theta = 90°$ so that $\Delta\varepsilon$ is large and negative. This type of compound is commonly used in homeotropic cells.

Three polar groups (let us assume $\mu_1 = \mu_2 = \mu_3$ for simplicity)

The widely used tri-fluoro groups are in the 3, 4 and 5 positions. In this case, $\mu_r = 2\mu_1$ and $\theta = 0°$. Its effective dielectric anisotropy is about four times as large as the single fluoro group in the axial position.

Due to the smaller dipole moment of the fluoro group, one or more dipole groups may be considered to enhance dielectric anisotropy. In addition, unlike cyano, which is an electron acceptor, the fluoro group is an electron donor. The cyano group tends to increase electron conjugation and subsequently enhance birefringence. The fluoro group tends to have the opposite effect [21]. Increasing the number of dipole groups will also increase the moment of inertia and will lead to a higher viscosity. Thus, dielectric anisotropy, viscosity and birefringence are interrelated. How to optimise one parameter without sacrificing too much on the others is a challenging task.

3.3 Molecular simulations

The rapid advance of simulation software has helped the molecular designs that are made before these compounds are actually synthesised. For example, using the Molecular Orbital Package (MOPAC) program, we are able to calculate the dipole moment and orientation angle of an LC molecule. Some simulation results are listed in Table 2.1 for comparison [29].

From Table 2.1, the cyano-biphenyl (second row, 5CB) has a larger dipole ($\mu = 4.1$ Debye) and smaller orientation angle ($\theta = 7.5°$) than the phenyl-cyclohexane (first line) compound. The slightly larger dipole for 5CB results from its longer conjugation, as

Table 2.1 Molecular orbital packaging (MOPAC) simulations of the dipole μ and orientation angle θ of some LC structures (after [29])

LCs	θ (deg)	μ (Debye)
R₅ —⬡—⬡—CN	13.3	3.9
R₅ —⬡—⬡—CN	7.5	4.1
R₃ —⬡—⬡—⬡—CN	9.9	3.9
R₃ —⬡—⬡—⬡—CN	6.0	4.2
R₅ —⬡—⬡(F)—CN	24.8	4.8
R₅ —⬡—⬡(F,F)—CN	12.6	5.4
R₃ —⬡—⬡—⬡—F	10.8	2.0
R₃ —⬡—⬡—⬡(F,F,F)	10.6	3.8

described by the first term in Equation 2.7. The smaller θ, however, is because the phenyl-phenyl rings are more co-planar than the twisted cyclohexane-phenyl rings. By comparing the third and the seventh compounds, we find that the dipole moment of a single axial fluoro is only half that of a cyano. By contrast, the (3,4,5) tri-fluoro group is equivalent to cyano in terms of effective dipole moment. The differences are in their resistivity, phase transition temperature and viscosity.

4 Viscosity

Viscosity, especially rotational viscosity γ_1, plays a crucial role in the LCD response time. The response time of a nematic LC device is linearly proportional to γ_1. The rotational viscosity of an aligned liquid crystal depends on the detailed molecular constituents, structure, intermolecular association and temperature. As the temperature increases, viscosity decreases rapidly. Several theories, rigorous or semi-empirical, have been developed in an attempt to account for the origin of LC viscosity. However, owing to the complicated anisotropic attractive and steric repulsive interactions among LC molecules, these theoretical results are not completely satisfactory.

A general form for γ_1 is expressed as follows:

$$\gamma_1 = (\alpha_o + \alpha_1 S + \alpha_2 S^2) \exp \frac{ES^m}{k(T - T_o)} \tag{2.9}$$

where α_i are proportionality constants, E is the activation energy of diffusion, m is an exponent, k is the Boltzmann constant, and T_o is the melting point of the LC. From Equation 2.9, several variations have been employed to fit experimental data.

$$\gamma_1 = \alpha_o \exp(E/kT) \tag{2.10}$$

$$\gamma_1 = \alpha_1 S \exp(E/kT) \tag{2.11}$$

$$\gamma_1 = \alpha_2 S^2 \exp(E/kT) \tag{2.12}$$

$$\gamma_1 = \alpha_2 S^2 \exp(ES/kT) \tag{2.13}$$

$$\gamma_1 = \alpha_2 S^2 \exp[E/k(T - T_o)] \tag{2.14}$$

$$\gamma_1 = \alpha_2 S^2 \exp[ES^2/k(T - T_o)] \tag{2.15}$$

$$\gamma_1 = (\alpha_1 S + \alpha_2 S^2) \exp(E/kT) \tag{2.16}$$

$$\gamma_1 = \alpha_1 S \exp(E_1/kT) + \alpha_2 S^2 \exp(E_2/kT) \tag{2.17}$$

The first expression is the well-known Arrhenius law for isotropic fluids. The second one has been used by de Jeu [30] to fit his results on the N-4 LC mixture (the mixture of the two isomers of p-methoxy-p'-butylazoxybenzene, Merck). The third one was derived by Hess [31] and the fifth one was used by Kneppe et al. [32] to fit their results on MBBA and N-4 LC mixture. The fourth one was derived by Martins and Diogo [33] and the sixth one by Diogo and Martins [34] based on their free volume theory. The seventh one is an empirical formula used by Wu and Wu [35] which is modified from Imura and Okano's original expression ($\gamma_1 = C_1 S + C_2 S^2$, where C_1 and C_2 are coefficients) [36] except for the additional exponential term. The eighth is deduced by Belyaev et al. [37] from fitting their experimental results for some LCs studied. These authors found that for nematic LCs containing a

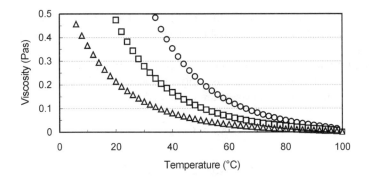

Figure 2.7 The simulated rotational viscosity using Equation 2.11 for $E = 400$ (circles), 380 (squares) and 360 meV (triangles) assuming $T_c = 100°$C and $\beta = 0.25$

long conjugation chain, $\alpha_2 = 0$, and for those containing a fragment with saturated bonds $\alpha_1 = 0$.

Overall, rotational viscosity is a complicated function of molecular shape, moment of inertia, activation energy and temperature. Even in the temperature category, rotational viscosity depends on the absolute temperature T, reduced temperature T_r, through the order parameter S, and clearing temperature T_c. Among these factors, activation energy and temperature are the most crucial ones [38]. Let us illustrate the activation energy and temperature effects using $\gamma_1 = \alpha_1 S \exp(E/kT)$ as an example.

Figure 2.7 depicts the temperature-dependent rotational viscosity of an LC as a function of activation energy using Equations 2.11 and 2.3. The parameters used for the calculations are $T_c = 100°$C, $\beta = 0.25$ and $\alpha_1 = 0.02$ Pascal-second. The circles, squares and triangles in Figure 2.7 represent the simulation results with $E = 400$, 380 and 360 meV, respectively. As the temperature increases, the rotational viscosity decreases drastically. Lower activation energy certainly leads to a lower rotational viscosity. Figure 2.7 shows the temperature-dependent rotational viscosity of an LC with clearing point of $T_c = 100°$C. Circles, squares and triangles represent the calculated data with $E = 400$, 380 and 360 meV.

From the molecular structure standpoint, a linear LC molecule is more likely to possess a lower viscosity [39]. However, all other properties need to be taken into account too. For instance, a linear structure may lack flexibility and lead to a higher melting point. Within the same homologues, a longer side chain will in general (except for the even-odd effect) have a lower melting temperature. However, its moment of inertia is also larger. As a result, the homologue with a longer chain length is likely to exhibit larger viscosity.

5 Elastic Constants

There are three basic elastic constants involved in the electro-optics of an LC cell depending on the molecular alignment: the splay (K_{11}), twist (K_{22}) and bend (K_{33}) elastic constants [40]. Elastic constants affect the display in two respects: the threshold voltage and the response time. The threshold voltage of a homogeneous cell is expressed in Equation 2.4. A smaller elastic constant will result in a lower threshold voltage. However, the response time of an LC device is proportional to the visco-elastic coefficient, the ratio of γ_1/K_{ii}. This means that a small elastic constant is unfavorable from the response time viewpoint. Therefore, a proper balance between threshold voltage and response time should be taken into consideration.

Several molecular theories have been developed to correlate the Frank elastic constants with molecular constituents. Here we use only the mean-field theory [41, 42]. In the mean-field theory, the three elastic constants are expressed as:

$$K_{ii} = C_{ii} V_n^{-7/3} S^2 \qquad (2.18)$$

$$C_{ii} = (3A/2)(Lm^{-1}\chi_{ii}^{-2})^{1/3} \qquad (2.19)$$

where C_{ii} is called the reduced elastic constant, V_n is the molar volume, L is the length of the molecule, m is the number of molecules in a steric unit in order to reduce the steric hindrance, $\chi_{11} = \chi_{22} = z/x$ and $\chi_{33} = (x/z)^2$, where x, $y(=x)$ and z are the molecular width and length, respectively, and $A = 1.3 \times 10^{-8}$ erg cm^6.

Mean-field theory predicts that the ratio of $K_{11} : K_{22} : K_{33}$ is equal to $1 : 1 : (z/x)^2$ and the temperature dependence of elastic constants is basically proportional to S^2. The S^2 dependence has been observed experimentally in many LC compounds. However, the prediction for the relative magnitude of K_{ii} is correct only to the first order.

Figure 2.8 depicts the measured temperature-dependent elastic constants K_{ii} of a difluoro-bicyclohexane-phenyl compound

and a trifluoro bicyclohexane-phenyl

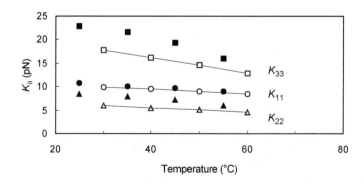

Figure 2.8 Elastic constants of a di-fluorinated (solid) and a tri-fluorinated (open) bicyclohexane-phenyl LC compounds. Squares, circles and triangles are for K_{33}, K_{11} and K_{22}, respectively (redrawn from [25])

compound [43]. At a given temperature, the following trend is observed: $K_{33}>K_{11}>K_{22}$. As the temperature increases, K_{33}, K_{11} and K_{22} decrease, as predicted by Equation 2.18. From the molecular structure viewpoint, the compound with a larger length-to-breadth ratio (z/x), i.e. a more linear structure, should exhibit larger elastic constants. From the molecular packaging simulations [42], the length-to-breadth ratios of the difluoro and trifluoro bicyclohexane-phenyl are calculated to be 3.76 and 3.62, respectively. Therefore, the difluoro compound should have larger elastic constants than the trifluoro homologue. This trend is confirmed in Figure 2.8.

Gelbart and Ben-Shaul generalised the van der Waals theory to explain the detailed relationship between the elastic constants and the molecular dimensions, polarisability and temperature [44]. The following formula for nematic liquid crystals has been derived:

$$K_{ii} = a_i\langle P_2\rangle\langle P_2\rangle + b_i\langle P_2\rangle\langle P_4\rangle \tag{2.20}$$

where a_i and b_i represent sums of the contributions of the energy and the entropy terms; they depend linearly on the temperature, and $\langle P_2\rangle$ ($=S$) and $\langle P_4\rangle$ are order parameters of the second and the fourth rank, respectively. In general, the second term may not be negligible in comparison with the S^2 term, depending on the value of $\langle P_4\rangle$. As the temperature increases, both S and $\langle P_4\rangle$ decrease. The $\langle P_4\rangle$ of 5CB changes sign at $T_c-T\simeq5°C$. The ratio of $\langle P_4\rangle/S$ is about 15% at $T=20°C$. If the $\langle P_4\rangle$ of a LC is much smaller than S in its nematic range, Equation 2.20 is reduced to the mean-field theory, i.e. $K_{ii}\simeq S^2$. The second term in Equation 2.20 is responsible for the difference observed between K_{11} and K_{33}.

For many LC compounds and mixtures, the magnitudes of the elastic constants follow the order $K_{33}>K_{11}>K_{22}$. Therefore, LC alignment also plays an essential role in achieving fast response time. For example, a homogeneous cell should have a faster response time than its corresponding twisted-nematic cell owing to the elastic constant effect, provided that all the other parameters such as cell gap and viscosity remain the same. The detailed performance of a homogeneous cell for reflective displays is discussed in Chapter 3.

6 LC Absorption

To the first order, LC absorption in the visible region is negligible for display applications [45]. However, in some large screen projection displays and polariser-free handheld displays for outdoor applications, light absorption could be an important issue depending on the power of the lamp and wavelength region of interest. In projection display using a high-power lamp, the UV and IR portions of the spectra need to be filtered out [46]. Otherwise, the absorbed UV light may decompose the LC molecules and degrade the lifetime of the LC device. On the other hand, the absorbed light is converted into thermal energy and consequently heats up the LC cell. Since the physical properties of a thermotropic LC (e.g. birefringence, viscosity and elastic constant) are sensitive to temperature, the display performance will be affected through the absorption of light.

6.1 UV/visible absorption

The major absorption of a LC compound occurs in two spectral regions: ultraviolet (UV) and infrared (IR). The $\sigma \to \sigma^*$ electronic transitions take place in the vacuum UV (100–180 nm) region, whereas the $\pi \rightarrow \pi^*$ electronic transitions occur in the UV (180–400 nm) region [47]. As the molecular conjugation gets longer, the absorption wavelengths extend towards the visible. UV absorption affects the photo-stability and lifetime of a LC device. A high-energy UV photon may break the chemical bonds of a long chain LC molecule and cause deterioration in the molecular alignment which, in turn, degrades the device performance [48]. In the visible region, most LC compounds are transparent. The major optical loss originates from light scattering due to the LC director fluctuations, rather than absorption.

6.2   IR absorption

In the near IR region, some harmonics of molecular vibration bands begin to appear. Figures 2.9 and 2.10 show the absorption spectra of 5CB in the visible, near IR and far IR regions, respectively. The saddle absorption minimum occurs at around $\lambda \simeq 0.8\,\mu m$; that is far from the electronic resonance and is right before the harmonics of molecular vibration begins.

In the mid (3–5 μm) and far IR (7–14 μm) regions, there exist several fundamental molecular vibration bands, such as CH, C=C, C≡C and CN [49]. As a result, the baseline absorption can be relatively large depending on the detailed molecular composition [50]. For laser application at a narrow frequency, it is possible to fine-tune the molecular structure and minimise the absorption [51]. However, for broadband application, especially in the 7–14 μm range, the baseline absorption is as large as $10\,cm^{-1}$. Thus, in order to keep transmittance greater than 90%, the LC layer thickness should be kept below $100\,\mu m$.

The absorption frequency of a vibration band depends on the reduced mass m and spring constant k. Generally speaking, the CH, CH_2 and CH_3 vibration bands are superimposed in the 3.2–3.8 μm region. Their transition intensities are rather strong. Such hydrocarbon associated vibration absorption is intrinsic to nearly all the thermotropic LC molecules discovered so far. An LC with a longer alkyl chain would enhance the absorption intensity of the CH_2 band because there are more CH_2 vibrations available.

Also present in the 3–5 μm mid-IR band are the well-known CN terminal group and C≡C linking group. Their reduced mass is larger than the CH group, thus, their vibrations

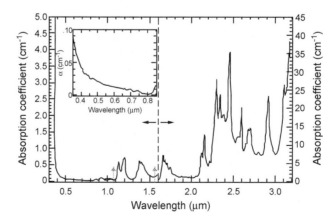

Figure 2.9 The absorption coefficient of 5CB in the visible and near IR regions. Measurement temperature $T = 50°C$

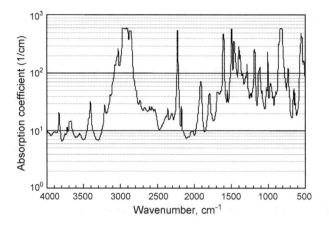

Figure 2.10 The absorption coefficient of 5CB in the IR region. Measurement temperature $T = 50°C$

occur at longer wavelengths. However, the reduced masses of C≡N and C≡C only differ slightly, so that their vibration wavelengths both occur in the 4.5 μm region. Unlike the cyano group whose stretching vibration is sharp and its intensity strong, the acetylene exhibits much weaker transition intensity. On the other hand, the diacetylene (C≡C)$_2$ linking group exhibits two vibration transitions (one at 4.54 and another at 4.66 μm); both have a modest intensity. Another frequently employed isothiocyanate (NCS) polar group is known to yield a higher Δn and a lower viscosity than the corresponding cyano compound. However, its absorption is not only strong but also has a wide range (4.3–5.3 and 10.3–11.2 μm). This makes the NCS group undesirable for CO_2 laser application.

Although the C=C group has the same reduced mass as an acetylene, the double bond has a weaker spring constant than the triple bond, resulting in a longer vibration wavelength (6.25 μm). From 7 to 10 μm, the C—O stretching vibration, in-plane C—H, and in-plane

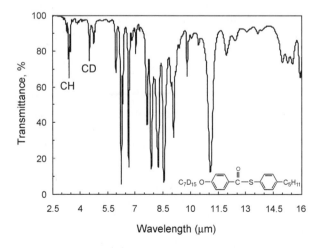

Figure 2.11 The IR absorption spectra of a post-deuterated compound (courtesy of L. Su of KSU)

C—C deformations are responsible for the observed absorption. At 10.6 μm, the absorption is affected by the shoulder of in-plane C—C deformations (centered at 9.95 μm) of the aromatic rings. Therefore, as the wavelength moves away from this peak, transmission increases gradually before the next strong band (centered at 12.3 μm; the out-of-plane C—H deformation) occurs.

The CH, CH_2 and CH_3 vibration bands have strong absorption centered at 3.4 μm. Per-deuteration moves this absorption peak to above 4.5 μm owing to increased reduced mass. Figure 2.11 shows a partially deuterated LC compound. The CH and CD absorption bands are clearly indicated in Figure 2.11. This post-deuteration process is useful for shifting IR absorption to a longer wavelength regime.

7 Some Commonly Used LC Compounds

Several reflective display modes, e.g. TN, STN, guest-host and homeotropic cells, have been developed using two, one or zero polarisers. The required LC properties could vary depending on the intended application. Some commonly desirable LC properties are good material stability, a wide temperature range for storage and a low viscosity for fast response. With regard to other properties such as dielectric anisotropy and birefringence, each display may have its own preference. For example, for a TFT-LCD employing twisted or homeotropic LC alignment, high resistivity is more important than a large dielectric anisotropy. However, for super-twisted nematic and cholesteric displays, high dielectric anisotropy is more attractive to reduce operating voltage than high resistivity.

7.1 LCs for TN

Table 2.2 lists some key LC structures used for the TN display applications. In general, for low information content displays such as wristwatches and calculators, manufacturing yield and cost are the major concerns. The second minimum TN display (see Chapter 4) provides a large cell gap tolerance. Under this circumstance, high birefringence and low operation voltage LC mixtures are especially desired. Cyano-biphenyl LCs [52] possess a large $\Delta\varepsilon$ and relatively high Δn, thus, they are commonly used for these applications.

7.2 LCs for STN

For a reflective STN display, the required $d\Delta n$ of the LC layer is around 0.7 μm. In order to maintain a cell gap in the 5 μm range, the required LC birefringence is about 0.14. For a given twist angle, the sharpness of the electro-optic transition curve is determined by the K_{33}/K_{11} ratio. The alkenyl LCs

exhibit a larger K_{33}/K_{11} ratio and lower viscosity than the corresponding alkyl compound [53]. Therefore, they are attractive for reflective STN applications.

Table 2.2   Physical properties of some phenyl-cyclohexane LCs

LCs	Phase transitions (°C)	$\Delta\varepsilon$	Δn	η (mPas)
R₅⟨⟩⟨⟩-OCH₃	K 41 N (31) I	−0.5	0.08	8
R₅⟨⟩⟨⟩-F	K 34 I	3.2	0.050	3
R₅⟨⟩⟨⟩-OCF₃	K 14 I	7.1	0.046	4
R₅⟨⟩⟨⟩-OCHF₂	K 1 N (−17) I	7.6	0.058	7
R₅⟨⟩⟨⟩-CF₃	K 10 I	11	0.040	9
R₅⟨⟩⟨⟩-CN	K 31 N 55 I	13	0.112	22

7.3   LCs for AMLCD

For reflective active-matrix addressed LCDs using TN cells (see Chapter 4), the required $d\Delta n$ is in the 200–350 nm range. In order to maintain a cell gap above 3 μm the LC birefringence needs to be kept below 0.12. Low Δn LCs are particularly attractive for reflective displays. For such applications, high resistivity and low viscosity are the primary concerns. The impurity ions are often trapped at the interfaces of polyimide alignment layers and the contamination depends heavily on the LC structures. From a model calculation, cyano compounds have the largest heat of interaction for trapping impurity ions [54]. By contrast, the alkoxy-trifluoromethyl compound

R⟨⟩⟨⟩⟨⟩-OCF₃

has the lowest heat of interaction. Meanwhile, it exhibits quite a low viscosity. Although the compounds containing the trifluoro-methyl group normally do not possess an LC phase, they are still useful for forming eutectic mixtures. Their mixing rates are 10–20%, depending on the melting point and enthalpy of fusion.

In some displays, low operating voltage is preferred and a higher dielectric anisotropy is needed. From the principles illustrated in Figure 2.5, more dipoles could be added to enhance the dielectric anisotropy. Tables 2.3 and 2.4 depict some difluoro [55, 56] and trifluoro [24] compounds. Their $\Delta\varepsilon$ is found in the 6–10 range.

Table 2.3 Phase transition temperature and enthalpy of fusion of some difluoro LCs

LCs	Phase transitions (°C)	ΔH (kcal/mol)
R_3—⬡—⬡(F)—F (with lateral F)	K −22.3 I	4.02
R_2—⬡—CH$_2$CH$_2$—⬡(F)—F	K −35.0 I	3.77
R_2—⬡—⬡—⬡(F)—F	K 51.8 N 85.4 I	3.64
R_2—⬡—⬡—CH$_2$CH$_2$—⬡(F)—F	K 39.8 N 87.9 I	6.78
R_2—⬡—CH$_2$CH$_2$—⬡—⬡(F)—F	K 28.3 N 74.2 I	3.36
R_2—⬡—CH=CH—⬡—⬡(F)—F	K 36.9 N 95.8 I	3.96
R_2—⬡—CH$_2$CH$_2$—⬡—⬡(F)—F	K 35.2 N 61.6 I	3.37
R_2—⬡—⬡—⬡(F)—F	K 69.0 N (60.9) I	5.89

7.4 Negative $\Delta\varepsilon$ LCs

To realise the useful electro-optic effect of a homeotropic cell, the LC employed normally possesses a negative dielectric anisotropy. In some special cases, such as in-plane switching using the fringing field effect, positive $\Delta\varepsilon$ LCs can also be used [57, 58]. From Equation 2.7, in order to obtain negative dielectric anisotropy, the dipoles should be in the lateral positions. Table 2.5 lists a few compounds with negative dielectric anisotropy. Again, when cyano is placed in a lateral position, the compound has a large but negative $\Delta\varepsilon$. In the interest of obtaining high resistivity, the lateral difluoro group is a favorable choice.

For an LCD projector using homeotropic alignment [59], the cell gap is often controlled at around 4 μm. Owing to the thermal effect of the lamp, the LC operation temperature often reaches 50–60°C so that response time is within 20 ms. To achieve a faster response time,

Table 2.4 Physical properties of some trifluoro LCs

LCs	Phase transitions (°C)	$\Delta\varepsilon$	Δn	η (mPas)
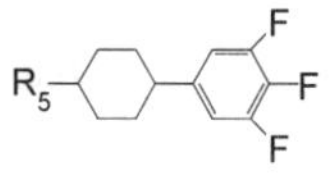	K 16.7 I	—	—	—
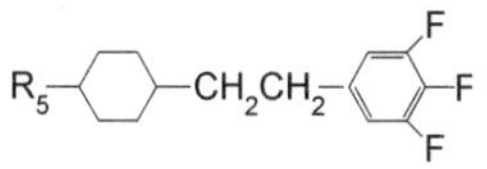	K −0.6 I	6.8	0.029	—
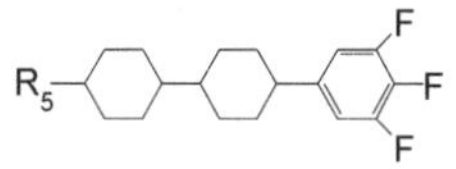	K 87.3 N 101.2 I	7.8	0.078	26.6
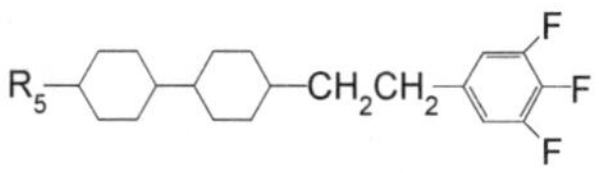	K 54.6 S$_B$ 57.0 N 103.9 I	7.8	0.074	26.6
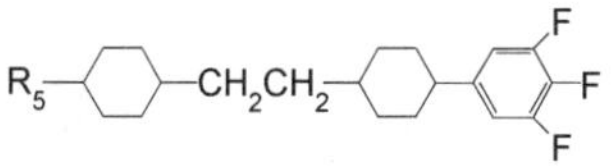	K 46.5 N 91.1 I	6.3	0.070	30.1
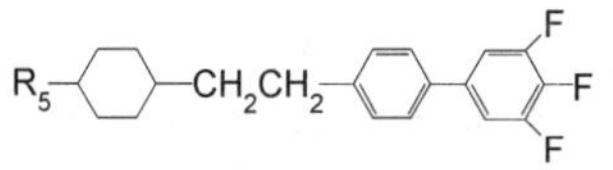	K 40.2 N 65.2 I	10.8	0.124	18.6
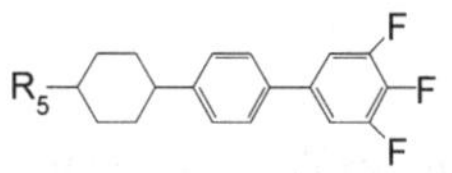	K 30.4 N 58.0 I	11.3	0.129	32.1

the cell gap should be reduced and a higher Δn LC mixture is used. The compounds listed in Table 2.5 have low birefringence. If a higher birefringence is required, molecular conjugation should be increased. Table 2.6 lists some (2,3) difluoro tolanes with the structure shown below [60]. The $\Delta\varepsilon$ of these tolanes is about −6 and $\Delta n\approx0.17$.

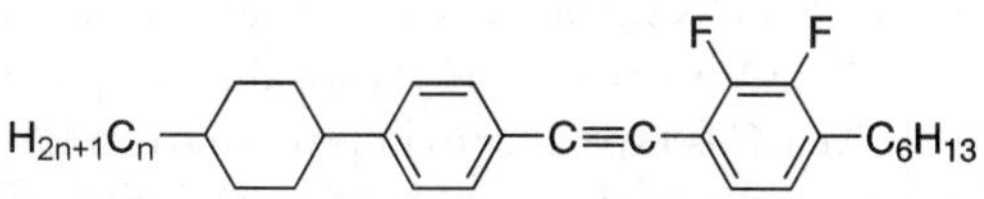

From Table 2.6, these laterally difluoro tolanes exhibit relatively low melting temperatures and high clearing temperatures. The observed low melting temperature is mainly attributed

Table 2.5 Physical properties of some negative $\Delta\varepsilon$ LCs

LCs	T_c	$\Delta\varepsilon$	Δn
R_5—⬡—⬡(CN)—R_3	18.2	−8.2	0.03
R_3—⬡—CO_2—(F,F)—OR_2	52.7	−7.0	0.09
R_3—⬡—⬡—(F,F)—R_1	138.3	−2.7	0.09
R_3—⬡—⬡—(F,F)—OR_2	172.4	−5.9	0.09
R_3—⬡—⬡(N,F)—OR_2	170.6	−8.4	0.08
R_5—⬡—⬡—(F,F)—R_2	110.8	−2.2	0.14
R_5—⬡—⬡—(F,F)—OR_2	165.1	−5.3	0.15

Note: these $\Delta\varepsilon$ and Δn are extrapolated results.

to the molecular breadth. The lateral difluoro group widens the molecular separation and weakens the inter-molecular association, resulting in low melting temperatures. On the other hand, the high clearing temperature originates from the cyclohexane ring. The small ΔH of these homologues is quite desirable to form eutectic mixtures.

7.5 High birefringence LCs

High birefringence liquid crystals are particularly attractive for reflective displays employing cholesteric and holographic PDLC structures (see Chapter 7). In a cholesteric display, the reflectance spectral bandwidth $\Delta\lambda$ is linearly proportional to the pitch length p and Δn of the LC employed. Thus, a high Δn LC is favorable to improve the display brightness. In addition, the use of a high Δn LC could reduce the number of helical pitches required for the Bragg reflection to occur. That means that the LC cell gap can be reduced. For a cholesteric display with uniform pitch length, the device operation voltage is proportional to the cell gap [61]. As

Table 2.6 Phase transition temperatures (in °C) and enthalpy of fusion (ΔH in kcal/mol) of the laterally difluoro-tolane LC homologues. Here K, N and I stand for the crystalline, nematic and isotropic phases, respectively

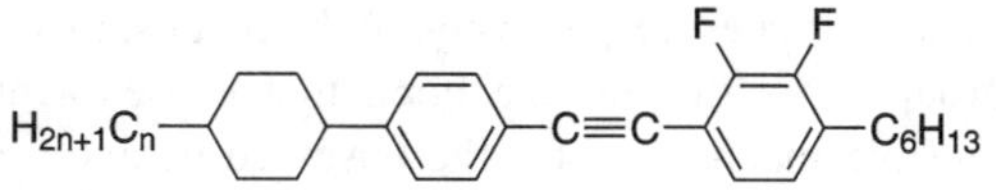

n	Phase transitions	ΔH
2	K 34.6 N 75.9 I	3.52
3	K 38.5 N 94.2 I	2.28
4	K 22.5 N 105.8 I	3.05
5	K 38.3 N 104.6 I	5.22
6	K 36.8 N 68.9 I	2.60
7	K 37.2 N 86.2 I	2.85

a result, a high Δn LC not only enhances display brightness but also reduces operation voltage. For a H-PDLC, the LC droplets have diameters in the 100 nm region and are randomly distributed at a voltage-off state. As the external voltage is turned on, the LC droplets are aligned along the electric field direction. In a randomly distributed state, the average LC refractive index is $(n_e + 2n_o)/3$. For an aligned state, its index n_o is close to that of the host polymer matrix, resulting in a transparent state. Thus, the effective birefringence change is $\Delta n/3$. This means that an LC with $\Delta n = 0.3$ is equivalent to 0.1 in the H-PDLC system. Moreover, high Δn also helps to reduce operating voltage as the LC droplets do not need to be driven so hard in order to get the required phase change.

Based on the birefringence dispersion theory described in Equation 2.1, a linearly conjugated molecule should exhibit high birefringence and low viscosity. To obtain $\Delta n > 0.35$ in the visible region, non-polar and polar diphenyl-diacetylenes [62], diphenyl-hexendiynes [63] and tolanes [64, 65] have been developed. In the following, we briefly describe their phase transition behavior and birefringence properties.

Diphenyl-diacetylenes

Diphenyldiacetylene liquid crystals (abbreviated as PTTP, where P stands for a phenyl ring and T for a carbon-carbon triple bond) exhibit a linearly conjugated structure. Thus, the birefringence is high and viscosity is relatively low. However, the nematic range of fluorinated PTTP compounds is quite narrow (less than 10°C). Replacing the alkyl by an alkenyl side chain group greatly enhances the clearing point and widens the nematic range. The molecular structures of the alkenyl diphenyldiacetylenes are shown below [66].

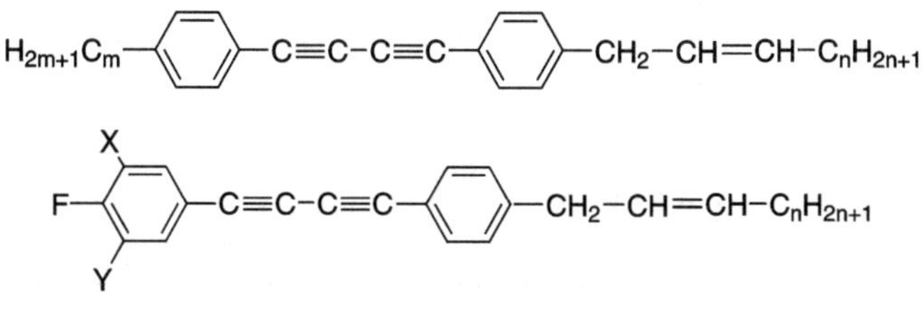

As compared to their alkyl parent compounds, the alkenyl PTTPs shown above have a double bond located between the second and third carbons from the right phenyl ring. For simplicity, we denote these alkenyl compounds as PTTP-m,$(n+3)d2$ and PTTP-$(n+3)d2$,F for the alkyl and fluoro homologues, respectively. In the polar series, X and Y can be either a hydrogen or fluoro group. It is this double bond that causes a dramatic nematic range broadening. The responsible mechanism for the observed nematic range broadening is not yet completely understood. It is speculated that the double bond alters the side-chain flexibility, leads to denser molecular packing, and consequently enhances the melting and clearing temperatures.

The phase transition temperatures and measured figure-of-merit (FoM) of some alkenyl fluoro PTTP compounds are listed in Table 2.7 for comparison with their corresponding alkyl compounds. From Table 2.7, the alkenyl PTTPs have much a wider nematic range than their alkyl counterparts. To overcome the high melting temperature problem of alkenyl PTTPs, one could form eutectic mixtures using multiple components from among these homologues.

Although PTTP liquid crystals have several attractive physical properties, their photo and thermal stability is insufficient for long-term device operation [67]. This is because the diacetylene group is relatively easily polymerised by UV photons or thermal agitation. One way to prevent such polymerisation is to add a scavenger such as a nitro-amino tolane, shown below [68]:

$$C_5H_{11}-\overset{\overset{\displaystyle H}{|}}{N}-\!\!\left\langle\!\!\bigcirc\!\!\right\rangle\!\!=\!\!\left\langle\!\!\bigcirc\!\!\right\rangle\!\!-NO_2$$

Figure 2.12 depicts the scavenger effect on the UV stability of PTTP-24/36. Without a scavenger, a PTTP-24/36 mixture (open circles) can only survive 20 seconds of UV exposure ($\lambda = 350$ nm, $I \simeq 100$ mw/cm^2). The UV-induced polymerisation is believed to degrade the molecular alignment. The process is believed to start from the entrance interface and gradually migrating to the bulk. As a result, the threshold voltage is smeared and the

Table 2.7 Phase transition temperatures (in °C), nematic range (ΔT), and measured figure-of-merit ($K_{11}\Delta n^2/\gamma_1$) of the fluoro and olefin fluoro PTTP liquid crystals. Also listed for comparison are two cyano PTTPs. () denotes monotropic transition

PTTP-nX	T_{mp}	T_c	ΔT	FoM
4F	85.4	87.7	2.3	
4d2,F	93.3	164.8	71.5	95 at 95°C
5F	88.4	95.6	7.2	
5d2,F	91.6	163.0	71.4	90 at 95°C
5d2, FF	86.3	126.5	40.2	
5d2, FFF	69.0	91.2	22.2	
6F	76.0	80.1	3.9	
6d2,F	82.0	149.2	67.2	50 at 95°C
6d2,FF	78.9	117.2	38.3	
6d2,FFF	79.1	80.3	1.2	
6CN	(144.0)	145.3	1.3	
5d2,CN	169.3	225.1	55.8	

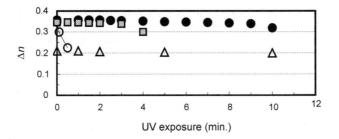

Figure 2.12 The UV stability of E63 and PTTP-24/36 with and without the nitro-amino-tolane scavenger. Open circles PTTP-24/36, squares PTTP-24/36 + 5% nitro-amino-tolane, solid circles PTTP-24/36 + 10% nitro-amino-tolane, triangles E63

birefringence is reduced. With 5% and 10% of nitro-amino-tolane, the lifetime of PTTP-24/36 is improved to 3 and 10 minutes, as shown in Figure 2.12. Also included in Figure 2.12 is the UV stability result (triangles) of E63. The 10% scavenger added PTTP-24/36 (closed circles) has about the same stability as E63 except that PTTP has a higher birefringence and a lower visco-elastic coefficient than E63.

The nitro-amino-tolane scavenger is also found to be equally effective for improving the thermal stability of the PTTP liquid crystals. Without a scavenger, PTTP-24/36 would gradually turn to a yellowish color and degrade performance (e.g. decreased birefringence and increased viscosity) at $T = 80°C$ storage for about two days. With 10% nitro-amino-tolane, the thermal stability is extended to beyond 2000 hours under the same operation conditions. Another advantage of using a nitro-amino-tolane scavenger is the reduced operation voltage. Owing to its huge dielectric anisotropy ($\Delta\varepsilon \approx 60$), the nitro-amino-tolane also helps to reduce the threshold voltage of the host mixture significantly.

Diphenyl-hexendiynes

To further increase molecular conjugation, diphenyl-hexendiynes (abbreviated as PTDTP) with structures shown below have been studied [61]:

$$H_{2n+1}C_n - \text{benzene} - C \equiv C - \cdots - C \equiv C - \text{benzene} - X$$

Here, X can be an alkyl (C_mH_{2m+1}), alkoxy (OC_mH_{2m+1}) or a fluoro (F) group. The phase transition temperatures and extrapolated birefringence of some PTDTP compounds are listed in Table 2.8. The double bond in PTDTP liquid crystals serves at least two purposes: it increases the conjugation length, and it bends the core molecule which leads to a lower melting point. Although the diphenyl-hexendiyne liquid crystals have a larger birefringence than diphenyl-diacetylenes, their photo-stability is worsened because of the linking double bond conjugation.

Table 2.8 Phase transition temperatures (in °C) and extrapolated birefringence (Δn) of the alkyl-alkyl and alkyl-fluoro diphenyl- hexendiyne (PTDTP) liquid crystals. Here the symbols O1 and O2 represent the OCH_3 and OC_2H_5 alkoxy groups

$$H_{2n+1}C_n \—\!\!\bigcirc\!\!\—C{\equiv}C\—\—C{\equiv}C\—\!\!\bigcirc\!\!\—X$$

PTDTP-nX	T_{mp}	T_c	Δn
24	49.9	131.8	
25	49.3	138.6	
32	81.2	150.0	
33	95.3	162.3	
34	56.3	143.8	0.46
35	64.1	146.0	
3,O1	87.5	188.0	
3,O2	111.0	200.0	
3F	66.3	141.2	
4F	72.2	123.9	0.40
5F	77.8	129.1	

Tolanes

Among linearly conjugated compounds, tolanes have reasonably good UV and thermal stability. Many types of tolane with different substituents have been studied. Table 2.9 lists a few examples. From Table 2.9, cyano-tolane exhibits a monotropic phase transition behavior with a high clearing point, which is unfavorable for forming eutectic mixtures. Replacing the alkyl side chain with an alkenyl group [69], the monotropic phase is converted to the enantiotropic nematic phase, which possesses a wide nematic range and a lower melting point, as compared to counterparts containing a triple bond. The extrapolated birefringence of these tolanes at $T = 20°C$ and $\lambda = 589$ nm exceeds 0.4.

The isothiocyanato (NCS) tolanes tend to exhibit smectic phase. This is not desirable for forming nematic mixtures, Substituting fluoro groups in the (3, 5) positions not only lowers the melting point but also suppresses the smectic phase, as shown in Table 2.9. Surprisingly, the naphthalene NCS tolanes developed by Hull University [70] exhibit a pure nematic phase. Its birefringence is as high as 0.54 at $\lambda = 589$ nm. The NCS group contributes to the electron conjugation resulting in improved birefringence. It also possesses a large dipole moment. Thus, the dielectric anisotropy of the NCS compounds is about 10. Moreover, the NCS compounds usually have lower viscosity than their cyano counterparts. A shortcoming of the NCS compounds is that they are more difficult to synthesise.

Bistolanes

Bistolanes have a high Δn due to their long molecular conjugation. However, as the conjugation length increases, the melting point also increases. An effective way to lower the melting point is to substitute a lateral alkyl group in the middle phenyl ring to widen the

Table 2.9 Physical properties of some substituted tolane derivatives

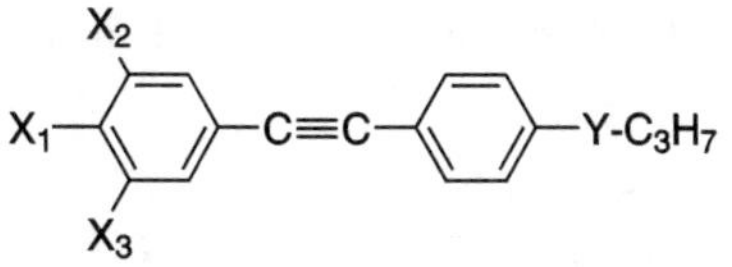

X_1	X_2	X_3	Y	Phase transitions	Δn
CN	H	H	—	K 87 N (78) I	
CN	H	H	O	K 98 N 103 I	
CN	H	H	≡	K 119 N (87) I	0.46
CN	H	H	=	K 100 N 150 I	0.47
CN	F	H	≡	K 109 I	0.414
CN	F	H	=	K 74 N 102 I	0.426
NCS	H	H	C_2H_4	K 92 Sm(82) I	
NCS	F	F	OCH_2	K 71 N(62) I	0.41
F	H	H	≡	K 84 I	
F	H	H	=	K 88 I	
C_3H_7	H	H	≡	K 99 I	0.351
C_3H_7	H	H	=	K 73 N 113 I	0.359
CF_3O	H	H	≡	K 121 I	
CF_3O	H	H	=	K 122 N 127 I	
F	F	H	=	K 61 I	0.305
F	F	F	=	K 34 I	0.27

molecular separation, as shown below [71, 72]:

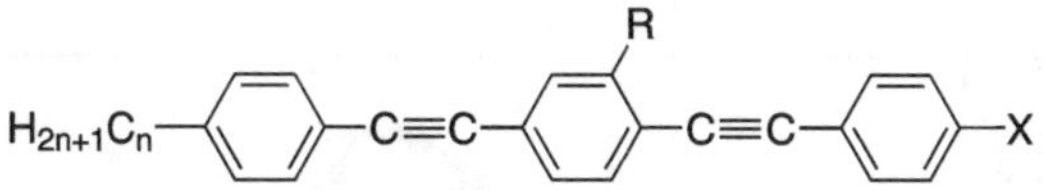

Here, X can be an alkyl, F or CN group, and R can be a hydrogen, methyl, ethyl or fluoro group. Table 2.10 shows the phase transition temperatures, heat fusion enthalpy, dielectric anisotropy and birefringence of some fluoro and cyano bistolanes [73].

From Table 2.10, the alkyl-fluoro bistolane has a relatively high melting point, as expected from the linear molecular structure. With the lateral fluoro substitution, both melting and clearing points drop by about 30 degrees. With a methyl or ethyl substitution, the melting point is decreased more substantially. The lateral alkyl group increases the molecular breadth and reduces the intermolecular associations. The thermal energy required to separate the molecules is reduced, resulting in lower melting and clearing temperatures.

8 Cholesteric Displays

For cholesteric displays, high birefringence liquid crystals would broaden the reflection spectra and increase the display brightness. Figure 2.13 shows such an example. A high birefringence ($\Delta n \approx 0.35$) mixture PTTP-24/36 + 10% nitro-amino-tolane was used for a

Table 2.10 Phase transition temperatures (in °C) and heat fusion enthalpy (ΔH, in kcal/mol) of the methyl and ethyl-substituted fluoro and cyano bistolane liquid crystals. $\Delta \varepsilon$ and Δn are extrapolated from the E63 host at 1 kHz and $\lambda = 515$ nm at $T = 22$°C

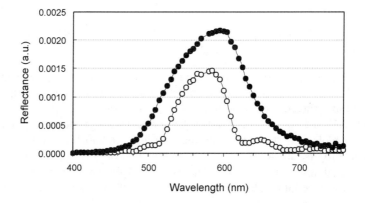

nX	R	T_m	T_c	ΔH	$\Delta \varepsilon$	Δn
42	H	140.7	211.9	2.97	0.8	0.4
4F	H	172.5	207.1	5.73	8	
4F	F	145.7	170.3	6.49	10	
2F	CH_3	103.7	183.2	5.73		
3F	''	101.7	191.2	7.37		
5F	''	69.4	162.8	4.97		
6F	''	73.5	150.5	8.40		
5CN	''	114.0	203.0	5.46		
6CN	''	100.7	215.2	7.10	15	0.53
3F	C_2H_5	65.0	112.9	5.10		
4F	''	69.1	98.3	5.25		
5F	''	60.9	103.2	5.20		
6F	''	46.5	89.1	4.77	7	0.37
3CN	''	102.9	161.9	4.56		
4CN	''	118.5	157.4	7.32		
5CN	''	85.2	159.3	3.51		
6CN	''	77.2	131.7	4.64	13	0.51

Figure 2.13 Brightness enhancement of a cholesteric display. Circles host LC mixture, ZLI-5400-100. Dots PTTP-24/36 + 10% nitro-amino-tolane

demonstration. A 2 μm cholesteric cell was prepared with a pitch length of about 0.34 μm so that its peak reflection occurred at $\lambda \simeq 600$ nm [74].

The open circles in Figure 2.13 represent the measured reflectance of the ZLI-5400-100 cell and the dots are the data of the high Δn LC mixture. From Figure 2.13, the overall display brightness is nearly doubled for the high birefringence mixture. The PTTP mixture also exhibits a relatively low viscosity. The response time of the 2 μm cell is less than 1 ms. The saturation voltage of the PTTP cell is about 30 V.

High birefringence is not the only requirement for high-brightness cholesteric displays. Other desirable properties of LC mixtures include a wide nematic range, high $\Delta\varepsilon$ and low γ_1/K_{11}. More high-birefringence, low-viscosity and high $\Delta\varepsilon$ compounds need to be developed; then eutectic mixtures with a wide nematic range, $\Delta\varepsilon>10$, and modest viscosity could be formulated for high-brightness cholesteric display applications. Owing to its bistable nature (i.e. low power consumption), the cholesteric display is a potential candidate for electronic books. However, how to achieve a full color display remains a technical challenge.

To achieve multi-color display, Kent Display Inc. [2] has developed a stacked approach. Two or three cholesteric panels are stacked together with each layer designed to reflect a primary color. An eight-color cholesteric display has been demonstrated using three stacked LC layers. A shortcoming of this approach is that stacking-induced parallax and pixel registration problems will limit the device resolution and viewing angle.

Another approach for producing multicolor display is to use tunable chiral material (TCM) developed by Kent State University [75]. The chirality of TCM can be photochemically tuned by controlling the UV exposure time. In the fabrication process, TCMs are included in the cholesteric formulation. The RGB pixel arrays are then subject to UV exposure subsequently. As a result, the chirality of the tunable chiral dopant are altered which, in turn, adjusts the chirality of the cholesteric mixture. To prevent material diffusion during UV exposure, pixel walls should be partitioned. For a cholesteric display, the operating voltage is inversely proportional to the pitch length. Therefore, different voltages are needed to activate the RGB pixels.

Advanced Display System developed an in-situ controlled pitch length method [76] for generating color cholesteric display. The methods include four steps: 1. Depositing a twist agent on a first substrate, 2. Bringing a second substrate into proximity with the first substrate to form at least one interstitial region, 3. Introducing LC having an initial pitch into the interstitial region, and 4. Stimulating the mixing of the LC and the in-situ twist agent. The in-situ agent will change the initial pitch length of the LC to form RGB pixels. A decent multicolor cholesteric display has been demonstrated.

References

1 M. Schadt and W. Helfrich, Voltage-dependent optical activity of a twisted nematic liquid crystal, *Appl. Phys. Lett.* **18**, 127 (1971).

2 D. Davis, A. Khan, C. Jones, X. Y. Huang, and J. W. Doane, Multiple color high resolution reflective cholesteric liquid crystal displays, *J. SID* **7**, 43 (1999).

3 R. Q. Ma and D. K. Yang, Optimization of polymer-stabilized bistable black-white cholesteric reflective display, *J. SID* **7**, 61 (1999).

4 M. Wand, W. N. Thurmes, R. T. Vohra, and K. M. More, Advances in ferroelectric liquid crystals for microdisplay applications, *SID Tech. Digest* **27**, 157 (1996).

5 V. Freedericksz and V. Zolina, Forces causing the orientation of an anisotropic liquid, *Trans. Faraday Soc.* **29**, 919 (1933).

6 M. F. Schiekel and K. Fahrenschon, Deformation of nematic liquid crystals with vertical orientation in electric fields, *Appl. Phys. Lett.* **19**, 391 (1971).

7 C. H. Gooch and H. A. Tarry, The optical properties of twisted nematic liquid crystal structures with twisted angles $\leq 90°$, *J. Phys.* **D8**, 1575 (1975).

8   E. Jakeman and E. P. Raynes, Electro-optic response times of liquid crystals, *Phys. Lett.* **A39**, 69 (1972).

9   T. J. Scheffer and J. Nehring, A new highly multiplexable liquid crystal display, *Appl. Phys. Lett.* **45**, 1021 (1984).

10   D. K. Yang, L. C. Chien, and J. W. Doane, Cholesteric liquid crystal polymer gel dispersion bistable at zero field, *Conference record of the 11th Int'l Display Research Conf.*, p. 49 (1991).

11   J. L. Fergason, Polymer encapsulated nematic liquid crystals for display and light control applications, *SID Tech. Digest* **16**, 68 (1985).

12   R. L. Sutherland, V. P. Tondiglia, L. V. Natarajan, T. J. Bunning, and W. W. Adams, Electrically switchable volume gratings in polymer-dispersed liquid crystals, *Appl. Phys. Lett.* **64**, 1074 (1994).

13   S. T. Wu, U. Efron, and L. D. Hess, Infrared birefringence of liquid crystals, *Appl. Phys. Lett.* **44**, 1033 (1984).

14   D. Dolfi, M. Labeyrie, P. Joffre, and J. P. Huignard, Liquid crystal microwave phase shifter, *Electron. Lett.* **29**, 926 (1993).

15   K. C. Lim, J. D. Margerum, and A. M. Lackner, Liquid crystal millimeter wave electronic phase shifter, *Appl. Phys. Lett.* **62**, 1065 (1993).

16   F. Guerin, J. M. Chappe, P. Joffre, and D. Dolfi, Modeling, synthesis and characterization of a millimeter-wave multilayer microstrip liquid crystal phase shifter, *Jpn. J. Appl. Phys.* **36**, 4409 (1997).

17   S. T. Wu, Birefringence dispersion of liquid crystals, *Phys. Rev.* **A30**, 1270 (1986).

18   S. T. Wu, C. S. Wu, M. Warengham, and M. Ismaili, Refractive index dispersions of liquid crystals, *Opt. Eng.* **32**, 1775 (1993).

19   L. Schroder, *Z. Phys. Chem.* **11**, 449 (1893).

20   J. J. Van Laar, *Z. Phys. Chem.* **63**, 216 (1908).

21   I. Haller, Thermodynamic and static properties of liquid crystals, *Prog. Solid State Chem.* **10**, 103 (1975).

22   I. C. Khoo and S. T. Wu, "Optics and Nonlinear Optics of Liquid Crystals" (World Scientific, Singapore, 1993).

23   H. J. Deuling, *Solid State Phys.* Suppl. 14, "Liquid Crystals", ed. L. Liebert (Academic, New York, 1978).

24   S. Naemura, Recent progress in LC materials for AMLCDs, *Proc. Asia Display'98*, 3 (1998).

25   R. Tarao, H. Saito, S. Sawada, and Y. Goto, Advances in liquid crystals for TFT displays, *SID Tech. Digest* **25**, 233 (1994).

26   Y. Nakazono, H. Ichinose, A. Sawada, S. Naemura, and K. Tarumi, *Int'l Display Research Conference*, p. 65 (1997).

27   W. Maier and G. Meier, A simple theory of the dielectric characteristics of homogeneous oriented crystalline-liquid phases of the nematic type, *Z. Naturforsch. Teil A* **16**, 262 (1961).

28   M. Schadt, Low frequency dielectric relaxation in nematic and dual frequency addressing of field effect, *Mol. Cryst. Liq. Cryst.* **89**, 77 (1982).

29   S. Naemura, Liquid crystals for multimedia display use: materials and their physical properties, *Mat. Res. Soc. Symposium Proceedings* **424**, 295 (1997).

30   W. H. De Jeu, "Physical properties of liquid crystalline materials" (Gordon and Breach, New York, 1980).

31   S. Hess, Irreversible thermodynamics of nonequilibrium alignment phenomena in molecular liquids and in liquid crystals, *Z. Naturforsh. Teil A* **30**, 1224 (1975).

32   H. Kneppe, F. Schneider, and N. K. Sharma, Rotational viscosity of nematic liquid crystals, *J. Chem. Phys.* **77**, 3203 (1982).

33   A. F. Martins and A. C. Diogo, Simple molecular statistical interpretation of the nematic viscosity, *Portgal. Phys.* **9**, 1 (1975).

34 A. C. Diogo and A. F. Martins, Thermal behavior of the twist viscosity in a series of homologous nematic liquid crystals, *Mol. Cryst. Liq. Cryst.* **66**, 133 (1981).

35 S. T. Wu and C. S. Wu, Rotational viscosity of nematic liquid crystals, *Liq. Cryst.* **8**, 171 (1990).

36 H. Imura and K. Okano, Temperature dependence of the viscosity coefficients of liquid crystals, *Jpn. J. Appl. Phys.* **11**, 1440 (1972).

37 V. V. Belyaev, S. Ivanov, and M. F. Grebenkin, Temperature dependence of rotational viscosity of nematic liquid crystals, *Sov. Phys. Crystallogr.* **30**, 674 (1985).

38 M. A. Osipov and E. M. Terentjev, Rotational diffusion and rheological properties of liquid crystals, *Z. Naturforsch. Teil A* **44**, 785 (1989).

39 S. T. Wu and C. S. Wu, Experimental confirmation of Osipov-Terentjev theory on the viscosity of liquid crystals, *Phys. Rev. A* **42**, 2219 (1990).

40 P. G. de Gennes, "The Physics of Liquid Crystals" (Clarendon, Oxford, 1974).

41 W. Maier and A. Saupe, A simple molecular statistical theory for nematic liquid crystal phase, Part II, *Z. Naturforsh. Teil A* **15**, 287 (1960).

42 H. Gruler, The elastic constants of a nematic liquid crystal, *Z. Naturforsh. Teil A* **30**, 230   (1975).

43 D. Demus, Y. Goto, S. Sawada, E. Nakagawa, H. Saito, and R. Tarao, Trifluorinated liquid crystals for displays, *Mol. Cryst. Liq. Cryst.* **260**, 1 (1995).

44 W. H. Gelbert and A. Ben-Shaul, Molecular theory of curvature elasticity in nematic liquid crystals, *J. Chem. Phys.* **77**, 916 (1981).

45 S. T. Wu and K. C. Lim, Absorption and scattering measurements of nematic liquid crystals, *Appl. Opt.* **26**, 1722 (1987).

46 F. J. Kahn, "Projection Displays", *SID Seminar Lecture Notes*, Vol. 2 (1994).

47 H. H. Jaffe and M. Orchin, "Theory and Applications of Ultraviolet Spectroscopy" (Wiley, New York, 1962).

48 A. M. Lackner, J. D. Margerum, and C. I. Van Ast, *Mol. Cryst. Liq. Cryst.* **141**, 289 (1986).

49 L. J. Bellamy, "The Infrared Spectra of Complex Molecules" (Wiley, New York, 1958).

50 S. T. Wu, Absorption measurements of liquid crystals in the UV, visible and infrared, *J. Appl. Phys.* **84**, 4462 (1998).

51 S. T. Wu, J. D. Margerum, H. B. Meng, C. S. Hsu, and L. R. Dalton, Potential liquid crystal mixtures for CO_2 laser application, *Appl. Phys. Lett.* **64**, 1204 (1994).

52 G. Gray, K. J. Harrison, and J. A. Nash, New family of nematic liquid crystals for displays, *Electron. Lett.* **9**, 130 (1973).

53 T. Geelhaar, K. Tarumi, and H. Hirschmann, Trends in LC materials, *SID Tech. Digest* **27**, 167 (1996).

54 H. Saito, E. Nakagawa, T. Matsushita, F. Takeshita, Y. Kubo, S. Matsui, K. Miyazawa, and Y. Goto, *IEICE Trans. Electron.* **E79-C**, 1027 (1996).

55 Y. Goto, T. Ogawa, S. Sawada, and S. Sugimori, Fluorinated liquid crystals for active matrix displays, *Mol. Cryst. Liq. Cryst.* **209**, 1 (1991).

56 A. I. Pavluchenko, N. I. Smirnova, and V. F. Petrov, Synthesis and properties of liquid crystals with fluorinated terminal substituents, *Mol. Cryst. Liq. Cryst.* **209**, 225 (1991).

57 M. Oh-e, M. Yoneya, and K. Kondo, Switching of negative and positive dielectrio-anisotropic liquid crystals by in-plane electric fields, *J. Appl. Phys.* **82**, 528 (1997).

58 S. H. Lee, H. Y. Kim, I. C. Park, B. G. Rho, J. S. Park, H. S. Park, and C. H. Lee, Rubbing-free, vertically aligned nematic liquid crystal display controlled by in-plane field, *Appl. Phys. Lett.* **71**, 2851 (1997).

59 S. E. Shield and W. P. Bleha, Liquid Crystals: Applications and Uses, ed. B. Bahadur, (World Scientific, Singapore, 1990). Vol. 1, Ch.16.

60 S. T. Wu, C. S. Hsu, and J. M. Chen, Room temperature lateral difluoro-tolanes for high contrast displays, *Mol. Cryst. Liq. Cryst.* **304**, 441 (1997).

61 P. G. de Gennes and J. Prost, "The physics of liquid crystals" (Oxford University Press, New York, 1995).

62 S. T. Wu, J. D. Margerum, H. B. Meng, L. R. Dalton, C. S. Hsu, and S. H. Lung, Asymmetric diphenyl-diacetylene liquid crystals, *Appl. Phys. Lett.* **61**, 630 (1992).

63 Y. Goto, T. Inukai, A. Fijita, and D. Demus, New nematics with high birefringence, *Mol. Cryst. Liq. Cryst.* **260**, 23 (1995).

64 G. W. Gray and A. Mosley, Mesomorphic transition temperatures for the homologous series of 4-n-alkyl-4′-cyanotolanes and other related compounds, *Mol. Cryst. Liq. Cryst.* **37**, 213 (1976).

65 H. Takatsu, K. Makoto, S. Urawa, Y. Tanaka, and H. Sato, Tolan-type nematic liquid crystalline compounds, U. S. Patent 4,726,910 (Feb.23, 1988).

66 S. T. Wu, M. E. Neubert, S. S. Keast, D. G. Abdallah, S. N. Lee, M. E. Walsh, and T. A. Dorschner, Wide nematic range alkenyl diphenyldiacetylene liquid crystals, *Appl. Phys. Lett.* **77**, 957 (2000).

67 M. D. Wand, R. T. Vohra, W. N. Thurmes, and K. M. More, *Ferroelectrics* **180**, 333 (1996).

68 S. T. Wu, E. Sherman, J. D. Margerum, K. Funkhouser, and B. M. Fung, High solubility and low viscosity dyes for guest-host displays, *Asia Display'95*, p. 567 (1995).

69 C. Sekine, K. Fujisawa, Y. Fujimoto, and M. Minai, Optical anisotropy of phenylacetylene homologues, *Mol. Cryst. Liq. Cryst.* **332**, 235 (1999).

70 A. J. Seed, K. J. Toyne, J. W. Goodby, and M. Hird, Synthesis, transition temperatures, and optical properties of various 2,6-disubstituted naphthalenes and related 1-benzothiophenes with butylsulfanyl and cyano or isothiocyanoto terminal groups, *J. Mater. Chem.* **10**, 2069 (2000).

71 S. T. Wu, C. S. Hsu, and K. F. Shyu, High birefringence and wide nematic range bistolane liquid crystals, *Appl. Phys. Lett.* **74**, 344 (1999).

72 S. T. Wu, C. S. Hsu, and Y. Y. Chuang, Room temperature bistolane liquid crystals, *Jpn. J. Appl. Phys.* **38**, L286 (1999).

73 S. T. Wu, C. S. Hsu, Y. Y. Chuang, and H. B. Cheng, Physical properties of polar bistolane liquid crystals, *Jpn. J. Appl. Phys.* **39**, L38 (2000).

74 S. T. Wu, J. A. Hou, and B. G. Wu, High brightness and low voltage cholesteric liquid crystal displays, *The 6th Asia SID Proceedings*, pp. 154–158 (2000).

75 L. C. Chien, U. Muller, M. F. Nabor, and J. W. Doane, Multicolor reflective cholesteric displays, *SID Tech. Digest* **26**, 169 (1995)

76 Y. D. Ma and B. G. Wu, Methods of manufacturing multicolor liquid crystal displays using in situ mixing technique, US patent 5,949,513 (1999).

3

Phase Retardation Effect

1 Introduction

For a reflective liquid crystal display, high brightness, high contrast ratio, low power consumption, good color purity and a fast response time are the key technical challenges. Among them, high brightness and high contrast are the most critical for direct-view and projection displays, while a fast response time is especially important for virtual projection display employing a color sequential approach.

The brightness of a display is affected by the transmittance of the polariser, color filters, the LC operation mode and the aperture ratio of the active driving element, e.g. thin-film transistors. In a reflective LCD, the TFT arrays are often hidden beneath the reflector. Thus, the aperture ratio can be made greater than 90%. This is a significant advantage over a transmissive display, particularly when a high-resolution device, such as a microdisplay, is needed. In a transmission-type TFT-LCD, its aperture ratio could decrease from 80% to 40% as the resolution increases from VGA to SXGA. Also, the LC cell normally provides 90–100% light modulation efficiency for the three primary (red, green and blue), colors. Therefore, using a polariser is a decisive factor for display brightness.

A display without a polariser can produce greater than 50% reflectance; however, the contrast ratio is normally limited to 5–10 : 1. A low contrast implies that the dark state is not sufficiently black. Thus, the displayed color is unsaturated and it is difficult to achieve full color. Even for a reflective display using one polariser, the device contrast ratio still struggles to exceed 50 : 1 due to surface reflections or glare induced by the ambient light.

Several physical mechanisms have been developed for modulating light using liquid crystals. These include phase retardation [1] (the electrically controlled birefringence effect), polarisation rotation [2], mixed mode between polarisation rotation and phase retardation effects [3], absorption [4], light scattering [5], and Bragg reflection [6]. Reflective display devices employing absorption, light scattering or Bragg reflection do not require a polariser. As a result, the display reflectance could exceed 50%. This is particularly essential for mobile communication devices designed for indoor and outdoor applications. On the other hand, display devices employing polarisation rotation, phase retardation or mixed mode effects need a polariser in order to obtain high contrast. When the incident light passes through the polariser twice, the reflectance is reduced to $\leq 40\%$, not to mention the losses from aperture ratio and the reflector. Therefore, the brightness of such displays is always a concern. Extensive efforts have been put into reflector technologies such that the reflected light is directed to a finite viewing cone ($30 \pm 20°$), instead of the usual $180°$ Lambertian distribution. In such circumstances, the display within this viewing cone can be brighter than the MgO standard white.

In this chapter we introduce the homogeneous and homeotropic cells that use the phase retardation effect, sometimes called the electrically controlled birefringence (ECB) effect, for reflective displays. The reflective twisted nematic effect will be described in Chapter 4, the super-twisted nematic effect in Chapter 5, the guest-host effect in Chapter 6, the light scattering effect in Chapter 7, Bragg reflection in Chapter 8, and the bistable TN in Chapter 9.

2   Homogeneous Cell

A homogeneous cell is not suitable for a transmissive display because of the narrow viewing angle and the lack of a common dark state for RGB colors [7]. However, in a reflective display, the viewing angle is equivalent to a two-domain cell due to the mirror image effect discussed in Chapter 1. For many handheld displays, the viewing angle requirement is not so stringent. To obtain a common dark state for a full-color display, a phase compensation film can be used to cancel the residual phase retardation of the cell resulting from boundary layers. Thus, the film-compensated homogeneous cell has potential for reflective direct-view and projection displays, as it exhibits a fast response time, wide viewing angle, high contrast, weak color dispersion, and relatively low operating voltage [8].

2.1   LC alignment

In a homogeneous cell, the bulk LC directors are parallel to each other, except at the boundary layers. The two boundary layers are in parallel directions, as depicted in Figure 3.1(a). If the pretilt angles are antiparallel as shown in Figure 3.1(b), a π-cell is formed [9]. From the line drawing shown in Figure 3.1(b), the π-cell has wide and symmetric viewing characteristics. A more detailed device structure using the π-cell will be discussed in Chapter 12.

2.2   Phase retardation effect

When a plane wave is incident normally to a uniaxial liquid crystal layer sandwiched between two polarisers, the outgoing beam will experience a phase retardation δ due to the different propagation velocities of the extraordinary and ordinary rays inside the LC [1]:

$$\delta = \frac{2\pi d}{\lambda}(n_e - n_o) = 2\pi d \Delta n / \lambda \tag{3.1}$$

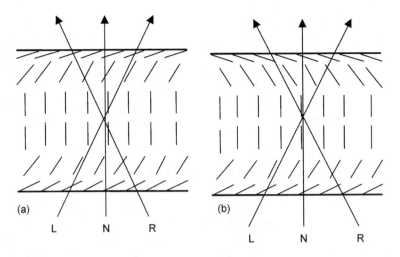

Figure 3.1 Schematic drawing of (a) a homogeneous and (b) a π-cell at $V > V_{th}$. The pretilt angle directions of a homogeneous cell are parallel, and antiparallel for a π-cell

In Equation 3.1 d is the cell gap, Δn is the birefringence and λ is the wavelength.

When a homogeneous cell is sandwiched between two polarisers, the normalised light transmittance is governed by the following equation:

$$T = \cos^2 \chi - \sin 2\beta \sin 2(\beta - \chi) \sin^2 (\delta/2) \tag{3.2}$$

Here χ is the angle between the polariser and the analyser, β is the angle between the polariser and the LC directors, and $\delta = 2\pi d \Delta n / \lambda$ is the phase retardation. For the simplest case that $\beta = 45°$ and the two polarisers are either parallel ($\chi = 0$) or crossed ($\chi = 90°$), the normalised light transmittances are simplified to:

$$T_{//} = \cos^2 (\delta/2) \tag{3.3}$$

$$T_{\perp} = \sin^2 (\delta/2) \tag{3.4}$$

For a homogeneous cell, the effective phase retardation depends on the wavelength and the applied voltage. In the visible region, the birefringence decreases as the wavelength increases. Birefringence dispersion was discussed in detail in Chapter 2, as it is closely related to the LC structures and temperature. Here, we first demonstrate the voltage effect.

For a homogeneous aligned LC cell, the Freedericksz threshold voltage V_{th} exists [10]. This threshold voltage is related to the splay elastic constant K_{11} and dielectric anisotropy $\Delta \varepsilon$ as [11]:

$$V_{th} = \pi (K_{11}/\varepsilon_0 \Delta \varepsilon)^{1/2} \tag{3.5}$$

When the voltage exceeds the Freedericksz threshold voltage, the LC directors are reoriented along the electric field direction. As a result, the effective birefringence and, in turn, the phase retardation are decreased. In a high voltage regime where $V \gg V_{th}$, virtually all the LC directors in the bulk are aligned normal to the substrates except for the boundary layers. The remaining phase retardation is small owing to the vanishing birefringence. However, these boundary layers cannot easily be completely reoriented by the external field. This implies that a good dark state is difficult to achieve in a homogeneous cell without a compensation film.

3 Film-compensated Homogeneous Cells

For a reflective display using a homogeneous cell, the total phase retardation is doubled ($\delta = 4\pi d\Delta n/\lambda$) because the incident light traverses the LC layer twice. To simulate the crossed-polariser condition, a $\lambda/4$ film sitting between the polariser and the front LC substrate is needed. Without the $\lambda/4$ film, the polariser functions as two parallel polarisers for a reflective display. Under crossed polarisers, the dark state normally occurs in the high-voltage regime. Conversely, the dark state of a full color display is difficult to achieve in the parallel-polariser condition. Thus, unless otherwise mentioned, we will assume the crossed-polariser condition.

3.1 Computer simulations

Owing to the extensive development of computer simulation programs, the simulation results are generally reliable. Several commercial simulation packages, such as DiMos and LCD Master, are available. In the computer simulation results presented here, the LC director distributions and the corresponding electro-optic properties of the homogeneous cells were calculated by using Oseen–Frank elastic continuum theory [12, 13] and extended Jones matrix method [14], respectively.

To design a homogeneous cell for a reflective display, we first calculate the required $d\Delta n$ of the cell. For intensity modulation, we need a π phase change and a good dark state for a high contrast ratio. This leads to $d\Delta n = \lambda/4$, that is to say the LC cell functions like a quarter-wave plate. Using $\lambda = 550$ nm, we find $d\Delta n = 137.5$ nm. The cell gap and birefringence can be chosen independently, depending on the need. One could choose a thinner cell gap to obtain a faster response time or choose a lower birefringence LC mixture to maintain a reasonable cell gap for high-yield manufacturing. Another design with $d\Delta n = \lambda/2$ and a $\lambda/4$ compensation film has been found to have weak color dispersion [15]. The tradeoff is in the narrow viewing angle.

For the purpose of illustrating the design procedures, a Merck LC mixture MLC-6297-000 has been chosen for the computer simulations. The LC parameters are listed as follows: $\beta = 45°$, pretilt angle $\alpha = 2°$, elastic constants $K_{11} = 13.4$, $K_{22} = 6.0$ and $K_{33} = 19.0$ pN, dielectric constants $\varepsilon_{//} = 10.5$ and $\Delta\varepsilon = 6.9$, $\Delta n = 0.125$, 0.127 and 0.129 for R = 650, G = 550, and B = 450 nm.

Figure 3.2 depicts the voltage-dependent light reflectance of a homogeneous cell with $d\Delta n = 137.5$ nm under the crossed-polariser condition. For the purpose of comparing intrinsic LC performance, we only consider the normalised reflectance; the optical losses from the polariser, substrate surfaces, indium tin oxide and any other compensation film are

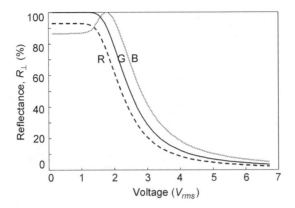

Figure 3.2 Computer simulation results of the voltage-dependent normalised reflectance of a homogeneous cell. $d\Delta n = 137.5\,\text{nm}$, $\beta = 45°$ and pretilt angle $\alpha = 2°$. R = 650, G = 550 and B = 450 nm.

neglected. From Figure 3.2, the bright state intensity variation among RGB colors is around 10%. This is very similar to that of the 90°-twisted nematic cell. In the high-voltage regime, the reflectance is monotonically decreasing. However, it is difficult to obtain a common dark state for the RGB colors within the voltage supply limits.

3.2 Compensation principles

In order to obtain a common dark state, a uniaxial compensation film with birefringence in the x–y plane (assuming z is the cell thickness direction) should be added [16]. The principal axis of the film is oriented to be orthogonal to the LC axis such that their phase retardations are subtractive:

$$\delta = 4\pi(d_1\Delta n_1 - d_2\Delta n_2)/\lambda \tag{3.6}$$

In Equation 3.6 $d_{1,2}$ and $\Delta n_{1,2}$ represent the thickness and birefringence of the LC layer and compensation film, respectively. The $d_2\Delta n_2$ value has an important effect on the dark state voltage. A small $d_2\Delta n_2$ means that the phase cancellation (i.e. the dark state) would take place at a higher voltage. As $d_2\Delta n_2$ increases, the dark state gradually shifts toward a lower voltage. When $d_1\Delta n_1 = d_2\Delta n_2$, a normally black mode is obtained. However, this dark state is too sensitive to variations of cell gap, film thickness and temperature, so that it is not suitable for practical applications.

3.3 Electro-optic effect

Figure 3.3 plots the voltage-dependent reflectance of a homogeneous cell with $d_1\Delta n_1 = 184\,\text{nm}$ and $d_2\Delta n_2 = -48\,\text{nm}$. This normally white mode has a relatively weak color dispersion (around 10%) and a low dark state voltage. In the high-voltage regime,

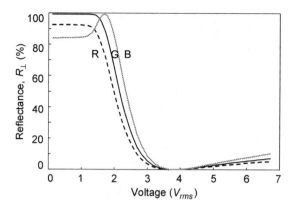

Figure 3.3   Calculated results of the voltage-dependent normalised reflectance of a film-compensated homogeneous cell. $d_1\Delta n_1 = 184\,\mathrm{nm}$ and $d_2\Delta n_2 = 48\,\mathrm{nm}$

the residual LC phase is diminishing but the phase of the compensation film remains. As a result, some light leakage is observed. For display applications, the dark state voltage can be controlled by the driving circuit. An important consideration concerns the width of the dark state minimum so that when the temperature fluctuates the display contrast is not significantly affected. From simulations, a smaller $d_2\Delta n_2$ value would lead to a broader dark state at a higher voltage. This means that the required voltage swing is larger.

3.4   Viewing angle

For a reflective display using a homogeneous cell, the incident and reflected light beams see opposite tilts of the LC directors [17]. Thus, its viewing angle is equivalent to what is observed in a two-domain transmissive cell [18, 19]. If the $d\Delta n$ is close to a first-order half-wave plate, i.e. $d\Delta n \simeq \lambda/4$, then its viewing angle is quite wide and the color dispersion is as good as in a TN cell [20, 21]. In addition, a low operation voltage can be achieved by adding a phase compensation film. Thus, a film-compensated homogeneous reflective cell overcomes the problems of viewing angle, operating voltage, and color dispersion that are encountered in a transmissive displays while preserving the advantage of a fast response time. This operation mode is particularly attractive for those reflective displays with a strong requirement for a fast response time, such as color-sequential projection displays [22].

The horizontal and vertical viewing angles of the film-compensated homogeneous cell have been calculated and the results are shown in Figures 3.4(a) and (b), respectively. From Figure 3.4(a), such a reflective display exhibits a good dark state within a $\pm 20°$ viewing cone. Thus, it is useful for a high-contrast projection display. From Figure 3.4(a), the horizontal viewing angle of such a reflective cell is wider than $\pm 60°$. However, reverse contrast still exists beyond $-35/+45°$ in the vertical viewing direction (Figure 3.4(b)). For a palm-sized reflective display (e.g. personal digital assistant or hand held personal computer), such a viewing range is good enough. Thus, this normally white reflective mode has potential for both projection and direct-view displays.

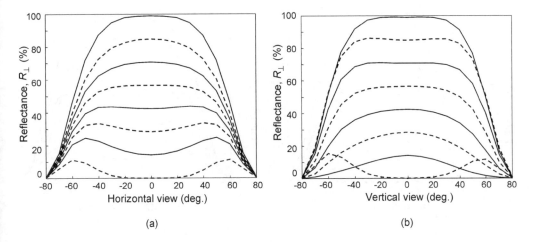

Figure 3.4 Calculated (a) horizontal and (b) vertical viewing angles of a uniaxial film-compensated reflective homogeneous LC cell. The LC has $d_1\Delta n_1 = 184$ nm and the uniaxial film has $d_2\Delta n_2 = 48$ nm

The dark state shown in Figure 3.4(a) corresponds to $V = 4.0V_{rms}$. At this voltage, the central part of the LC directors is aligned nearly perpendicular to the substrates, as in a homeotropic cell. In the small angle approximation (Chapter 12), the effective phase retardation of a homeotropic cell is proportional to the square of the viewing angle. Adding a negative birefringence film could cancel the phase retardation at oblique angles and improve the viewing cone to 60°. An even wider viewing angle has been demonstrated by using two domains and two compensation films [23].

To reduce the dark state reflectance as observed in Figure 3.4(b), a negative birefringence $(n_x = n_y > n_z)$ film with $d(n_z - n_x) = -100$ nm was added. Due to the isotropic refractive indices on the film surface (x–y plane), the voltage-dependent reflectance at normal incidence is unchanged. The calculated horizontal and vertical viewing angle of such a negative and positive film compensated homogeneous cell are shown in Figures 3.5(a) and (b), respectively. From Figure 3.5(a), the dark state leakage is reduced significantly and the horizontal viewing angle exceeds $\pm 70°$. From Figure 3.5(b), a good dark state is also obtained over a very wide vertical viewing range. The only drawback is that the eighth gray level (at $V = 0$) shows reverse contrast at greater than $\pm 30°$ viewing angle. Since both the eighth and seventh gray levels are bright, this reverse contrast will not be as noticeable as the dark state inversion depicted in Figure 3.5(b). For practical applications, the positive and negative films can be replaced by a biaxial film.

3.5 Cell gap tolerance

In Figure 3.3, the LC used has $d_1\Delta n_1 = 184$ nm and the compensation film (polycarbonate) has $d_2\Delta n_2 = 48$ nm. For high-yield manufacturing, it is desirable to have a large cell gap tolerance. Let us keep $d_2\Delta n_2$ unchanged and investigate how the variation of $d_1\Delta n_1$ affects the display performance. Figure 3.6 shows the voltage-dependent reflectance at

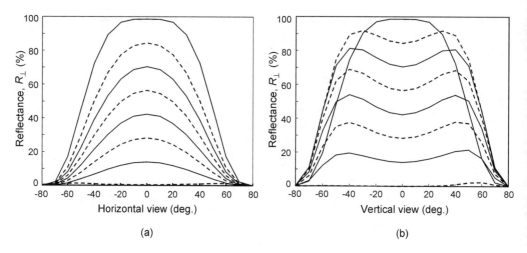

Figure 3.5 Calculated (a) horizontal and (b) vertical viewing angles of a film-compensated reflective homogeneous LC cell. The LC has $d_1\Delta n_1 = 184$ nm, the positive uniaxial film has $d_2\Delta n_2 = +48$ nm and the negative film has $d_3\Delta n_3 = -50$ nm. A biaxial film could be used to replace these two films

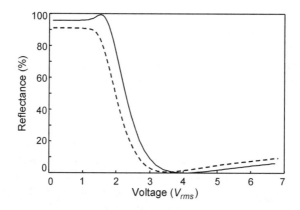

Figure 3.6 Cell gap tolerance of a film-compensated homogeneous cell. Solid line: $d\Delta n = 200$ nm; broken lines: $d\Delta n = 160$ nm. Compensation film $d_2\Delta n_2 = 48$ nm

$\lambda = 550$ nm of a cell with the $d\Delta n$ value deviated by $\pm 10\%$ (solid and dashed lines are for $d\Delta n = 200$ and 160 nm, respectively) from its optimal value. From Figure 3.6, a large $d\Delta n$ variation only causes a relatively small reflectance change and about a 0.5 V decrease in the dark state voltage. Thus, the film-compensated homogeneous cell has a large cell gap tolerance.

3.6 Temperature effect

Next, let us estimate how many degrees of temperature change would lead to a 10% change in Δn, assuming the cell gap is unchanged. To do so, let us use Haller's empirical formula for temperature-dependent birefringence [24]:

$$\Delta n = \Delta n_o (1 - T/T_c)^{\beta} \tag{3.7}$$

where Δn_o stands for the birefringence at $T = 0$, T_c is the clearing point of the LC mixture and β is a material parameter. For the MLC 6297-000 mixture, $T_c \simeq 371$ K and $\beta \simeq 0.25$. As the temperature increases from 22 to 48°C, the LC Δn drops by 10%. Therefore, in a projection display, the intended operation temperature has to be taken into consideration beforehand in order to optimise the display performance.

3.7 Response time

The fast response time is a direct consequence of the thin LC layer employed. From the dynamics of the LC directors, the response time is proportional to the square of the LC layer thickness. In a reflective cell with $d\Delta n = 184$ nm, if $\Delta n = 0.1$, then $d = 1.84$ μm. The response time (rise and decay) of such a thin cell is about 3 ms. Color sequential projection display using a film-compensated homogeneous cell would be feasible.

Two common difficulties for the thin cell approach are in the cell gap uniformity and lower manufacturing yield. If a display device is not so demanding on response time, a lower Δn LC can be used to maintain a high manufacturing yield. For example, if a LC mixture has $\Delta n = 0.06$, then the cell gap can be maintained at $d \simeq 3$ μm. Such a cell gap is relatively easy to manufacture.

To design a low Δn LC mixture for thin cell application, saturated LC materials should be considered. From the birefringence dispersion model (Equation 2.1) the $\sigma \rightarrow \sigma*$ electronic transitions (the λ_o-band) make around a 20% contribution to the overall birefringence of the cyano biphenyl (5CB) LC compound. Based on this extrapolation, a fully saturated LC compound should possess Δn as low as 0.03. Indeed, some dioxane LC compounds with $\Delta n \simeq 0.032$ have been synthesised [25]. However, to formulate such a low Δn mixture while exhibiting a wide nematic range (−40 to 80°C) and preserving a reasonable dielectric anisotropy, a high voltage-holding ratio and a low viscosity, a dozen of such low Δn LC compounds need to be prepared.

3.8 Operating voltage

Low operating voltage is desirable because it not only reduces the power consumption but also improves the TFT stability. As technology advances, the operating voltage will gradually decrease from 5 to 3.3 and then 2.5 V_{rms}. For a homogeneous cell, two approaches can be taken to reduce the operation voltage: to increase the dielectric anisotropy $\Delta \varepsilon$ of an LC, and to increase the compensation film thickness. Extensive efforts in molecular engineering have been made to optimise the dielectric anisotropy without losing other desirable properties.

Phase compensation turns out to be a simple approach for lowering the operation voltage. From Equation 3.6, if the $d_2 \Delta n_2$ of the compensation film is increased, the dark state shifts

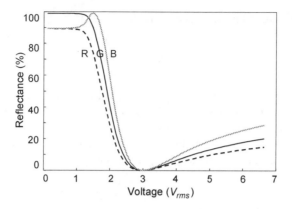

Figure 3.7   The voltage-dependent light reflectance of a film-compensated homogeneous cell. The LC employed is MLC-11800-100 with $\Delta\varepsilon = 8.2$, $d_1\Delta n_1 = 194$ nm and film $d_2\Delta n_2 = 65$ nm. The operation voltage is as low as 1.8 V

to a lower voltage. This provides a simple way to control the operating voltage of a display device.

Figure 3.7 demonstrates a 1.8 V voltage swing using the combined high $\Delta\varepsilon$ LC and a thicker phase compensation film. $\Delta\varepsilon = 8.2$ (e.g. MLC-11800-100), $d\Delta n = 194$ nm, and $d_2\Delta n_2 = 65$ nm were used for simulations. The threshold voltage occurs at around 1.2 V and the dark state voltage at around 3 V. Above 3 V, light leakage gradually increases due to the non-zero net phase retardation. In principle, one can continue to lower the operation voltage by increasing the $d_2\Delta n_2$ of the compensation film. The problem is that the minimum for the dark state might be too narrow to tolerate temperature fluctuation and cell gap non-uniformity.

3.9 Gray scale

For an eight-bit per color display, the voltage-dependent reflectance curve is divided into eight equal intensity gray levels. A shallower voltage-dependent reflectance curve gives easier gray scale controllability. To compare the thin homogeneous cell investigated here with a commonly employed transmissive TN cell, we have used the same LC parameters and the same dark state voltage (say, 6.0 V). For the first-minimum TN cell, its $d\Delta n \simeq 480$ nm at $\lambda = 550$ nm. For a reflective film-compensated homogeneous cell, we have used $d_1\Delta n_1 = 162$ nm, and $d_2\Delta n_2 = 25$ nm. Results are compared in Figure 3.8 at $\lambda = 550$ nm.

From Figure 3.8, the gray scale controllability of the film-compensated thin homogeneous cell is somewhat better than the 90°-TN cell. Due to its twisted nature, the TN cell has a slightly higher threshold voltage. Also, due to the self-phase compensation of the orthogonal boundary layers of a TN cell, its transmittance starts to diminish at a lower voltage, as compared with the homogeneous cell. As a result, the film-compensated homogeneous cell shows a wider voltage interval per gray scale.

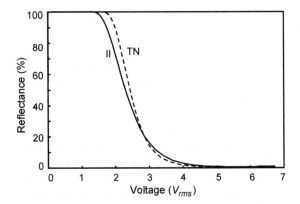

Figure 3.8 Comparison of gray scale capability between a film-compensated reflective homogeneous cell and a transmissive TN cell. The TN cell has $d\Delta n = 480$ nm at $\lambda = 550$ nm

3.10 Phase-only modulation

In addition to amplitude modulation, the homogeneous cell is also an efficient pure phase modulator. If the polarisation axis of the incoming beam is parallel to the LC directors ($\beta = 0$), the outgoing light will experience phase-only modulation; no amplitude modulation occurs. Under this configuration, no phase compensation film or polarisers are needed, if the incoming laser beam is already linearly polarised. One can construct two-dimensional phased arrays and control the spatial voltage of each pixel to form a phase grating [26]. The incoming light will then be diffracted a different direction. This optical phase in array has a potential application for agile laser beam steering and network switching [27].

4 Homeotropic Cell

Homeotropic alignment [28] is used in another LC cell utilising the ECB effect for transmissive and reflective displays. The homeotropic cell, also known as a perpendicular or vertically aligned cell, exhibits the highest contrast ratio of all the LC modes developed. Moreover, its contrast ratio is insensitive to the incident light wavelength, LC layer thickness and operating temperature. Both projection [29, 30] and direct-view displays using homeotropic LC cells [31, 32] have been demonstrated.

The homeotropic cell also exhibits a faster response time than its corresponding homogeneous or twisted nematic cell. Two factors that contribute to the faster response time are the elastic constant and the cell gap. To achieve one π phase retardation for a transmissive display, the required $d\Delta n$ for homogeneous and homeotropic cells is the same, that is $d\Delta n = \lambda/2$. However, for a 90° TN cell, the required $d\Delta n = 0.866\,\lambda$. On the other hand, the governing elastic constants for homogeneous, twisted and homeotropic cells are splay (K_{11}), twist (K_{22}) and bend (K_{33}), respectively. From Chapter 2, the order $K_{33} > K_{11} > K_{22}$ holds for most LC mixtures. The response time of an LC layer is proportional to $\gamma_1 d^2/K\pi^2$, where γ_1 is the rotational viscosity and K is the corresponding elastic constant. Therefore, the homeotropic cell has the best response time and contrast ratio of the three compared. This has been proven by some wide angle direct-view displays employing a homeotropic cell [33, 34].

One difficulty of the homeotropic cell is the need for high resistivity LC mixtures with a negative dielectric anisotropy. High resistivity is required for an active-matrix LCD in order to avoid image flickering. Negative $\Delta\varepsilon$ is required to obtain a useful electro-optic effect. From Chapter 2, to obtain negative $\Delta\varepsilon$ the dipoles, in particular the fluoro groups, need to be in the lateral positions. Significant progress in material development has been achieved in the past decade. Nevertheless, the use of negative $\Delta\varepsilon$ LC compounds is still much less common than positive ones. In addition, the lateral dipole groups often exhibit higher viscosity than the axial compounds owing to the larger moment of inertia.

A special device design that utilises a positive $\Delta\varepsilon$ LC mixture in homeotropic alignment has been proposed [35]. The LC directors are driven by a lateral fringing field for achieving a wide viewing angle. Unless specially mentioned, we assume that a negative $\Delta\varepsilon$ LC mixture is used for the homeotropic alignment discussed in this chapter.

4.1 Cell preparation

Several methods have been developed to align homeotropic cells including dipping, alcohol-treated SiO$_2$ film, polyimide rubbing and photoalignment.

Dipping method

For the dipping method, the glass substrates are dipped in a surfactant that, in turn, induces LC molecular alignment. Several surfactants have been studied [36, 37]. For example, trialkoxy-silanes can self-assemble to glass, ITO, CdS, CdZnS and Al surfaces and produce LC alignment. Two other well known surfactants are DMOAP (N,N-Dimethy-N-Octadecyl-3-Aminopropyl trimethoxy silyl chloride) for homeotropic, and MAP (N-Methyl-3-Aminopropyl-trimethoxysilane) for homogeneous alignment. A major problem of such dipping-induced LC alignment is thermal stability. As the temperature increases, the pretilt angle also increases significantly. Since the electro-optic behavior of an LCD is strongly dependent on the pretilt angle, the practical application of such dipping-induced alignment is therefore limited.

SiO$_x$ evaporation

Medium-angle and shallow-angle deposition (MAD/SAD) of SiO$_x$ layers has been used for aligning homogeneous LC cells [38]. To extend this MAD/SAD method to produce a tilted-homeotropic cell, an extra alcohol layer is needed. In the alcohol treatment method, the ITO glass substrates are first deposited with 150 nm thick SiO$_x$ layers at oblique angles (30° MAD and 5° SAD) and then thermally evaporated with a long chain aliphatic alcohol (C$_{18}$H$_{37}$OH) to bond alkoxy groups to the surface [39]. The LC molecules are aligned nearly perpendicularly to the substrate surfaces with a small pretilt angle varying from 2 to 5°. Such alcohol-treated homeotropic cells are found to have good temperature and photo stability. Moreover, a contrast ratio higher than 1000 : 1 has been routinely obtained. An imperfection of the MAD/SAD deposition method is that a small splay is observed. This means that the LC directors are not exactly parallel to each other. This does not affect the dark state since all the LC directors are perpendicular to the substrate surfaces. However, the on-state transmittance might not reach 100%.

To eliminate the small splay in the LC directors, a simplified process called moving deposition (MD) has been developed [40]. The MD alignment provides a two-step treatment technique to prepare a substrate that induces a tilt to the LC directors. First, a 10 nm thick SiO_2 layer is deposited over the ITO-coated glass by moving the substrate past a magnetron in-line sputtering source. After the deposition of the silica layer, the substrate is exposed to an alcohol at a sufficiently high temperature to react with the hydroxyl groups on the surface of the silica layer. The long chain alcohol is typically evaporated at 140°C. The pretilt angle is from 0.5 to 4 degrees, and is of uniform azimuth such that the projections of the directors onto the substrate all lie approximately parallel. Such an MD alignment is stable to temperature variation.

A simplified technique for producing tilted-perpendicular LC alignment by a single oblique evaporation of SiO_2 without surfactant has been developed [41]. This single oblique evaporation of SiO_2 alone can produce either homogeneous or homeotropic alignment depending on the dielectric anisotropy of the LC materials employed. A pretilt angle of 2–3° is obtained at a 30–60° evaporation angle. Since this method does not require alcohol treatment, the process time is shortened and the cost reduced. Without alcohol, the voltage holding ratio is also improved.

In general, the SiO_x evaporation technique produces excellent alignment with good uniformity, high contrast ratio and sharp threshold. It has been used in some small LCD panels. However, due to the need for a vacuum, it is inconvenient for processing large substrates. For large panels, polyimide rubbing is more convenient.

Mechanical buffing

Unidirectional buffing of a spin-coated polyimide (PI) film produces sufficient anchoring energy to align nematic liquid crystals [42]. For homeotropic alignment, polyimide SE-7511L and SE-1211 (Nissan Chemicals) and JALS-203 (Japan Synthetic Rubber Co.) are available commercially. The rubbing process is so simple that it has been commonly utilised to fabricate large LCD panels. However, three problems are encountered, as follows. The mechanical rubbing process may damage, contaminate or cause static charges on the thin film transistors (TFTs); as a result, the production yield is reduced. Secondly, the post-curing temperature of a commercial-grade PI is around 200–300°C; this curing temperature may be too high for some displays, such as low-temperature amorphous [43] and poly-silicon [44] TFT-LCDs. Thirdly, the thickness of a spin-coated PI layer over a large substrate may not be sufficiently uniform, the different PI thickness could result in a non-uniform voltage drop across the film.

Photo-induced alignment

Several non-rubbing methods such as Langmuir-Blodgett films [45], stretched polymers [46], photolithography [47], stamped polymers [48] and linearly polarised ultraviolet light

(LPUV) exposure of polymers [49, 50] have been developed. Among them, photo-alignment is particularly attractive because it not only eliminates surface scratches induced by mechanical rubbing but also simplifies the multi-domain fabrication processes for widening the viewing angle.

Many polymeric materials, such as photo-isomerisation of azo-compounds doped in polymers [51, 52], poly(vinyl-cinnamate) derivatives [53, 54], and polyimides [55, 56], have been used for photo-alignment studies. Among these alignment agents, poly(vinyl-cinnamate) film was found to have insufficient thermal stability. On the other hand, polyimides possess excellent thermal stability, but deep UV light ($\lambda = 254$ nm) is needed to induce liquid crystal alignment. The high-energy UV photons could decompose the polymer backbones and form charge-trapping centers on the LC and polyimide interfaces. As a result, undesirable image sticking and flickering effects could occur during active-matrix addressing [57].

A polyimide that incorporates the photo-reactive cinnamate group into its side chain (abbreviated as PICA) has been developed [58]. PICA is photo-crosslinkable and exhibits good solubility in polar solvents such as N-methylpyrrolidone, N,N-dimethyl acetamide, γ-butyrolactone, 2-pentanone and tetrahydrofuran. It has two outstanding features: it uses long-wave UV ($\lambda \simeq 350$ nm) for exposure and as a result the PI backbone is not decomposed and no charge trapping centers are formed; also the curing temperature is relatively low. The latter is particularly attractive for low-temperature poly-silicon TFT-LCD application.

4.2 Transmissive displays

For computer calculations, a Merck high resistivity mixture MLC-6608 was used. The physical properties of MLC-6608 are as follows: $n_e = 1.558$, $n_o = 1.476$ (at $\lambda = 589$ nm and $T = 20°C$); clearing point $T_c = 90°C$; dielectric anisotropy $\Delta\varepsilon = -4.2$, and rotational viscosity $\gamma_1 = 186$ mPas at 20°C.

Voltage-dependent light transmission

In principle, to obtain 100% transmittance for a transmissive homeotropic cell only that requires $d\Delta n \sim \lambda/2$. If $\lambda = 550$ nm is chosen, then $d\Delta n \sim 275$ nm. However, this is the minimum $d\Delta n$ value required because under such a condition, 100% transmittance would occur at $V \gg V_{th}$. Due to the finite voltage swing from TFT, the $d\Delta n$ value needs to be increased.

Voltage-dependent optical transmittance of a homeotropic cell with $d\Delta n = 350$ nm between crossed polarisers is shown in Figure 3.9. From Figure 3.9, an excellent dark state is observed at normal incidence. As the applied voltage exceeds the threshold voltage ($V_{th} \simeq 2.1$ V_{rms}), the LC directors are reoriented by the applied electric field, resulting in light transmission. At 6 V the normalised transmission reaches 100% for green light. For blue, the light is around 18% shorter in wavelength and the LC is around 8% higher in birefringence, so that the overall phase retardation is about 26% larger than that for green. It takes less director rotation or requires lower voltage to reach 100% transmittance. On the other hand, the red light is 18% longer in wavelength and the LC is 4% lower in birefringence than for green. Thus, it may not reach 100% transmittance, depending on the $d\Delta n$

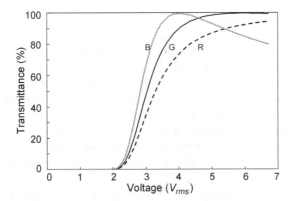

Figure 3.9 Voltage-dependent light transmittance of a homeotropic LC cell. $d\Delta n = 350\,nm$ and $\beta = 45°$

value. From Figure 3.9, the color dispersion at 5.5 V is about 12%, which is comparable to the 90°-TN cell.

Viewing angle

The computer simulated horizontal and vertical viewing angles of a 4.2 μm thick homeotropic cell are depicted in Figure 3.10. During calculations, the electro-optic curve shown in Figure 3.9 is divided into eight gray levels. For the first gray level corresponding to $V = 0$, light leakage begins to increase beyond $\pm 20°$. As the light leakage increases, the display contrast ratio degrades accordingly. Therefore, for a homeotropic cell, its viewing zone is limited to about $\pm 20°$. This is adequate for projection displays, but insufficient for direct-view displays. For direct-view displays, the viewing angle needs to be greater than $\pm 80°$ in order to compete with emissive displays, such as CRT and plasma.

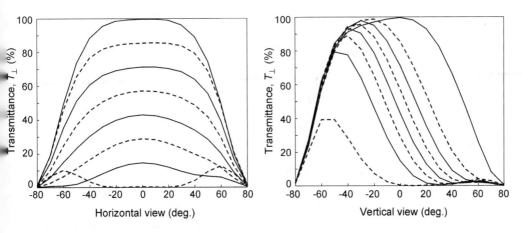

Figure 3.10 Horizontal and vertical viewing angles of a homeotropic LC cell. $d\Delta n = 350\,nm$

Film-compensated homeotropic cell

To improve the viewing angle, a negative phase retardation film can be added to cancel the light at oblique angles. In a homeotropic LC cell, the LC mixture employed has a positive birefringence, i.e. $n_e > n_o$. Therefore, in order to compensate for the phase retardation (in the voltage-off state) at oblique incidence, an ideal compensation film should possess the following properties: negative birefringence and the same $d\Delta n$ as the LC cell at any incident angle; and the same birefringence dispersion as the LC material used. The former condition would cancel the phase retardation of the outgoing light after traversing through the LC cell over a wide range of incident angles. The latter would result in perfect phase cancellation for all three primary colors of a full-color display.

Figure 3.11 shows the horizontal and vertical viewing angles of the same $d = 4.2\,\mu$m homeotropic cell, after adding a negative birefringence film with $d\Delta n = -100$ nm. In the horizontal direction, the viewing angle is extended to $\pm 60°$ without gray scale inversion. However, in the vertical direction, the contrast beyond $\pm 20°$ is still poor. Moreover, reverse contrast is observed when the viewing angle exceeds $+20°$. Thus, a simple negative birefringence film improves the viewing angle in three directions but worsens it in one of the vertical ones, say the bottom-up direction.

Several promising approaches to improve the viewing angle of a transmissive homeotropic cell have been developed. Details are given in Chapter 12.

4.3 Reflective displays

Both crossed and parallel polariser configurations can be considered for a reflective display. A $\lambda/4$ film is used in conjunction with a linear polariser in the crossed-polariser, but not in the parallel-polariser condition. The crossed-polariser configuration has the advantage of a high contrast ratio, while a parallel-polariser configuration offers lower costs because of the elimination of the $\lambda/4$ film.

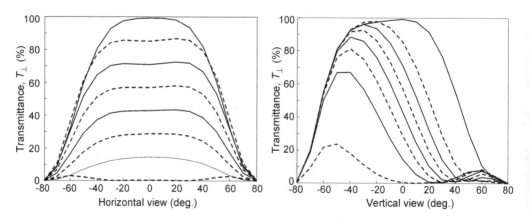

Figure 3.11 Horizontal and vertical viewing angles of a film-compensated transmission-type homeotropic LC cell. LC: $d\Delta n = 350$ nm; film: $d\Delta n = -100$ nm

Crossed-polariser configuration

For reflective displays, the incident light traverses the LC medium twice. Thus, the cell gap is reduced to half that of the corresponding transmissive display. To reproduce the same EO curve as shown in Figure 3.9, the cell gap is $d = 2.1\,\mu m$, instead of 4.2 µm. The response time is, therefore, four times faster.

Due to the mirror image effect discussed in Chapter 1, the viewing angle is widened, as shown in Figure 3.12. A very high contrast ratio is preserved within the $\pm 20°$ cone and no gray scale inversion is observed within the $\pm 30°$ cone. Thus, the reflective homeotropic cell can be used for projection displays without using any compensation film.

For a direct-view display using a reflective homeotropic cell, its viewing angle is inadequate. To widen the viewing angle, a negative birefringence film with $d\Delta n = -75\,nm$ is added. Because the negative birefringence film is isotropic in the x–y plane, the EO curve at a normal incidence angle is not affected. However, such a compensation film cancels the phase arising from oblique angles. Figure 3.13 shows the horizontal and vertical viewing angles of such a film-compensated homeotropic reflective cell. In the horizontal direction, the viewing angle is as wide as $\pm 80°$. In the vertical direction, the first seven gray levels also have a $\pm 80°$ viewing angle. The eighth level has reverse contrast beyond $\pm 30°$. However, because the eighth level is the brightest, the reverse contrast will not look so bad as if it were dark.

Parallel-polariser configuration

In this configuration, a linear polariser, but no $\lambda/4$ film, is required. Figure 3.14 shows the voltage-dependent reflectance of a 2.5 µm reflective homeotropic cell employing a MLC-6608 LC mixture. In the voltage-off state, all colors have 100% normalised reflectance. As the voltage increases, the blue light reaches the $R_{\|} = 0$ first, followed by green and red. From Figure 3.14, there is no common dark state for the RGB colors. At around 4.1 volts the green light is quite dark, but blue has reverse contrast. Sunlight is rich in green and weak in blue. Therefore, the voltage-dependent reflectance curve for white light resembles the green color.

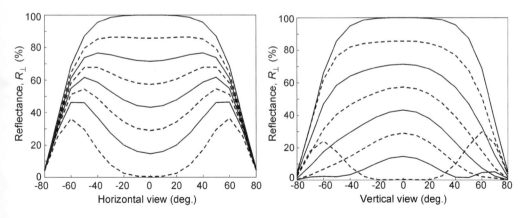

Figure 3.12 Horizontal and vertical viewing angles of a reflective homeotropic LC cell. $d\Delta n = 174\,nm$

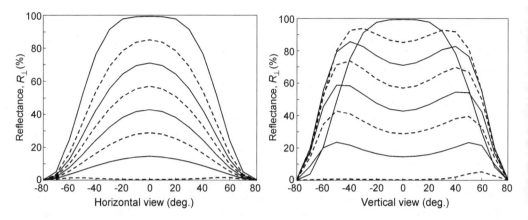

Figure 3.13 Horizontal and vertical viewing angles of a film-compensated reflective homeotropic LC cell. LC: $d\Delta n = 174$ nm; film: $d\Delta n = -75$ nm

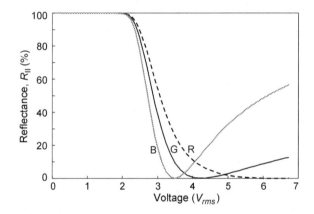

Figure 3.14 Voltage-dependent reflectance of a homeotropic LC cell under the parallel-polariser condition. $d\Delta n = 208$ nm, $\beta = 45°$

A rough estimate indicates that the average contrast ratio at 4 volts is about 20 : 1. When such an LC cell is used in a reflective display, the contrast ratio will drop to below 10 : 1 due to the surface reflection of the substrates.

Another advantage of this normally white mode is a low operating voltage. The threshold is around 2.1 V and the dark state voltage is around 4.1 V. Thus, the required voltage swing is only 2 V. To further reduce the operation voltage, one can simply increase the $d\Delta n$ of the cell.

5 Chiral-homeotropic Cell

The operation mechanism of a homeotropic cell is based on the phase retardation effect. Thus, its color dispersion is large. On the other hand, a 90°-twisted nematic cell utilises the polarisation rotation effect and, thus, its transmittance is less sensitive to the wavelength. However, for the polarisation rotation effect to take place, the required $d\Delta n$ value $(0.866\,\lambda)$ is higher than that of the phase retardation effect $(d\Delta n = 0.5\,\lambda)$. If the same LC is used, the TN cell needs a thicker LC layer than the corresponding homogeneous or homeotropic cell.

The chiral-homeotropic cell [59, 60] is also known as the homeotropic to twisted-planar (HTP) mode [61]. The structure of a chiral-homeotropic cell is basically a homeotropic cell doped with chiral molecules, such as ZLI-811 or CB-15 (from Merck) [62]. For a pure homeotropic cell without chiral molecules, LC directors tilt without twist throughout the cell as the applied voltage exceeds the threshold. In a chiral-homeotropic cell, the LC directors possess both tilt and twist from one surface to the other based on the balanced torque exerted by the applied electric field, chiral molecules and aligning surfactants [63]. The ratio of the pitch length p to the LC layer thickness d determines the twist angle of the LC directors. For example, to make a 90° twisted chiral-homeotropic cell, the d/p ratio should be equal to 0.25.

5.1 Transmissive displays

The voltage-dependent transmittance of a 90° chiral-homeotropic LC cell is depicted in Figure 3.15(a). The LC employed for simulation is MLC-6608, cell gap $d = 6.5\,\mu m$ (or $d\Delta n = 540\,nm$), $d/p = 0.25$ and $\beta = 0°$. In the voltage-off state, the transmittance is as dark as a homeotropic cell between crossed polarisers. When the applied voltage exceeds $2.1\,V_{rms}$, light transmittance starts to increase, and reaches 100% at about $3.5\,V_{rms}$ for the green wavelength. For each color, the voltage required to reach maximum transmittance is different, owing to the phase retardation effect. However, in the high-voltage regime $(V > 5\,V_{rms})$, the transmittance of each color does not vary with voltage. This is the unique feature of the chiral-homeotropic cell.

Without the chiral dopant, the homeotropic cell operates under the pure phase retardation effect. Based on Equation 3.2, the beta angle needs to be set at 45° in order to see the maximum phase retardation effect. Figure 3.15(b) shows the calculated transmission of a $6.5\,\mu m$ thick homeotropic cell with no chiral dopant. Comparing Figures 3.15(a) and (b), we find that the wavelength dispersion effect of the chiral-homeotropic cell is greatly suppressed in the high-voltage regime. This is because in a chiral-homeotropic cell the LC directors are tilted by the electric field, while the twist is preserved. Thus, both phase retardation and polarisation rotation effects coexist. The phase retardation effect boosts the transmission whereas the polarisation rotation effect reduces the wavelength dependency. The observed achromatic behavior depends on the twist angle, the beta angle and $d\Delta n$ value of the LC layer, similar to Mauguin's principle for a conventional TN cell [64].

In a chiral-homeotropic cell, the dark state preserves the high contrast ratio feature of a homeotropic cell, and the bright state has similar achromatic features to a TN cell. Thus, the chiral-homeotropic cell is sometimes called the inverse TN cell [65]. A significant disadvantage of the inverse TN cell is its large $d\Delta n$ requirement. In order to obtain the achromatic feature in the voltage-on state, the required $d\Delta n$ value is nearly two times greater than for a corresponding homeotropic cell. A large $d\Delta n$ not only leads to a slow response time but also to a narrow viewing angle. In addition, from the dynamic response analysis, the chiral-homeotropic cell geometry tends to produce backflow in the high voltage regime [66]. The backflow-associated optical bounce slows down the response time even more.

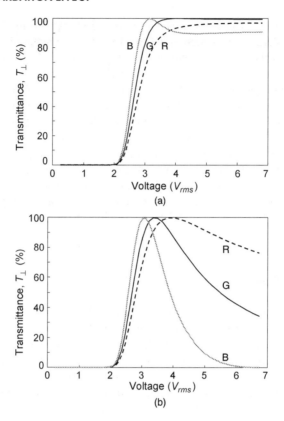

Figure 3.15 Voltage-dependent reflectance of (a) a chiral-homeotropic cell with $d\Delta n = 540$ nm at $\beta = 0°$, and (b) a homeotropic LC cell with same thickness but at $\beta = 45°$

5.2 Reflective displays

In a reflective display, the cell gap is reduced by approximately a factor of two so that the response time is shortened by four times. The mirror image effect helps to widen the viewing angle. The remaining challenges are high brightness, high contrast ratio and low cost. In this section, we demonstrate two reflective display modes using chiral-homeotropic cells.

90° Chiral-homeotropic cell

In reflective displays, the major difference between crossed and parallel polarisers is in the use of a $\lambda/4$ film. Under the crossed-polariser condition, the chiral-homeotropic cell has an excellent dark state. This means that under the parallel-polariser condition, the cell will have 100% reflectance in the voltage-off state. This normally white mode will also exhibit excellent achromatic behavior, i.e. its reflectance is independent of wavelength.

Figure 3.16 plots the voltage-dependent reflectance of a 90° chiral-homeotropic cell. The cell gap is $d = 3.5\,\mu$m and $\beta = 15°$. At $V \simeq 4\,V_{rms}$, the average contrast ratio is about 20:1.

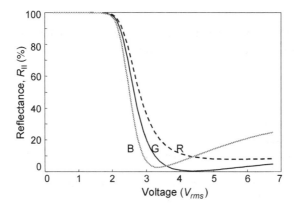

Figure 3.16 Voltage-dependent reflectance of a 90° chiral-homeotropic cell under the parallel-polariser condition. $d\Delta n = 290$ nm, $\beta = 15°$ and $d/p = 1/4$

63.6° Chiral-homeotropic cell

Since the voltage-on state of a chiral-homeotropic cell is like a TN cell in its voltage-off state, the optimisation method developed for the TN cell can be applied to the chiral-homeotropic cell. Such optimisation methods will be described in Chapter 4. Here we just mention the results. When the twist angle is at 63.6°, reflectance under the crossed-polariser condition can reach 100% in the voltage-off state [67]. However, the voltage-on state has a relatively low (about 20 : 1) contrast ratio because the two boundary layers are not orthogonal and do not compensate each other.

In the 63.6° chiral-homeotropic cell, under the crossed-polariser condition, the voltage-off state is as dark as for a homeotropic cell. The voltage-on state should, in principle, have the same achromatic behavior as the 63.6° TN cell. Thus, high contrast, high reflectance and weak color dispersion can be achieved.

Figure 3.17 shows the voltage-dependent reflectance of a 63.6° chiral-homeotropic cell, under the parallel-polariser condition [68]. The cell gap is $d = 3$ μm, the chiral pitch length

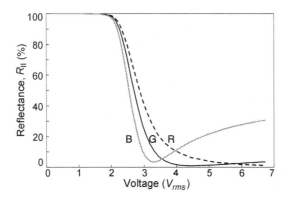

Figure 3.17 Voltage-dependent reflectance of a 63.6° chiral-homeotropic cell under the parallel-polariser condition. $d\Delta n = 249$ nm, $\beta = -2°$ and $p = 17$ μm

$p = 17 \,\mu\text{m}$, and $\beta = -2°$. As compared to the results of the $90°$ chiral-homeotropic cell shown in Figure 3.17, the dark state for the red color is improved, but is somewhat worsened for the blue. The overall contrast ratio is also in the vicinity of $20:1$.

References

1 M. Born and E. Wolf, Principles of Optics (Pergamon Press, New York, 1980).
2 M. Schadt and W. Helfrich, Voltage-dependent optical activity of a twisted nematic liquid crystal, *Appl. Phys. Lett.* **18**, 127 (1971).
3 S. T. Wu and C. S. Wu, Mixed-mode twisted nematic cells for reflective liquid crystal displays, *Appl. Phys. Lett.* **68**, 1455 (1996).
4 G. H. Heilmeier and L. A. Zanoni, Guest-host interactions in nematic liquid crystals, *Appl. Phys. Lett.* **13**, 91 (1968).
5 J. L. Fergason, Polymer encapsulated nematic liquid crystals for display and light control applications, *SID Tech. Digest* **16**, 68 (1985).
6 D. K. Yang, L. C. Chen and J. W. Doane, Cholesteric liquid crystal/polymer dispersion for haze-free light shutters, *Appl. Phys. Lett.* **60**, 3102 (1992).
7 S. T. Wu and C. S. Wu, Optimization of film-compensated homogeneous cells for liquid crystal displays, *Liq. Cryst.* **24**, 811 (1998).
8 S. T. Wu and C. S. Wu, Comparative studies of single-polarizer reflective liquid crystal displays, *J. SID* **7**, 119 (1999).
9 P. J. Bos and K. R. Koechler/Beran, The pi-cell: a fast liquid crystal optical switching device, *Mol. Cryst. Liq. Cryst.* **113**, 329 (1984).
10 V. Freedericksz and V. Zolina, Forces causing the orientation of an anisotropic liquid, *Trans. Faraday Soc.* **29**, 919 (1933).
11 H. J. Deuling, Elasticity of nematic liquid crystals, Solid State Phys. Suppl. **14**: Liquid Crystals, ed. L. Liebert (Academic, New York, 1978).
12 C. W. Oseen, The theory of liquid crystals, *Trans. Faraday Soc.* **29**, 883 (1933).
13 F. C. Frank, Liquid crystals: theory of liquid crystals, *Discuss. Faraday Soc.* **25**, 19 (1958).
14 R. C. Jones, New calculus for the treatment of optical systems, *J. Opt. Soc. Am.* **32**, 486 (1942).
15 T. H. Yoon, G. D. Lee, G. H. Kim, S. C. Kim, W. S. Park, J. C. Kim, and S. H. Lee, Reflective liquid crystal display using 0o-twisted half-wave cell, *SID Tech. Digest* **31**, 750 (2000).
16 Y. Fujimura, T. Nagatsuka, H. Yoshimi, S. Umemoto, and T. Shimomura, Optical properties of retardation film, *SID Tech. Digest* **23**, 397 (1992).
17 C. L. Kuo, C. K. Wei, S. T. Wu, and C. S. Wu, Reflective display using mixed-mode twisted nematic liquid crystal cell, *Jpn. J. Appl. Phys.*, Part I, **36**, 1077 (1997).
18 S. Tanuma et al. Japanese patent JP 63-106624 (1988).
19 K. H. Yang, Two-domain twisted nematic and tilted homeotropic liquid crystal displays for active matrix applications, *Conference Record of the 1991 International Display Research Conference*, San Diego, CA. Oct. 15–17, 1991, p. 68.
20 K. Lu and B. E. A. Saleh, A single-polarizer reflective LCD with wide-viewing-angle range for gray scales, *SID Tech. Digest of Application Papers* **27**, 63 (1996).
21 S. T. Wu, A biaxial film compensated thin homogeneous cell for reflective liquid crystal displays, *J. Appl. Phys.* **83**, 4096 (1998).
22 T. Uchida, Field sequential full color LCD without color filter by using fast response LC cell, *Proc. Int'l Display Workshop*, pp. 151–4 (1998).
23 K. Ohmuro, S. Kataoka, T. Sasaki, and Y. Koike, Development of super-high-image-quality vertical-alignment-mode LCD, *SID Tech. Digest* **28**, 845 (1997).

24   I. Haller, Thermodynamic and static properties of liquid crystals, *Prog. Solid State Chem.* **10**, 103 (1975).

25   P. Kirsch and E. Poetsch, Novel polar liquid crystals with very low birefringence based on trans-1,3-dioxane building blocks, *Adv. Mater.* **10**, 602 (1998).

26   D. P. Resler, D. S. Hobbs, R. C. Sharp, L. J. Friedman, and T. A. Dorschner, High efficiency liquid crystal optical phased array beam steering, *Opt. Lett.* **21**, 689 (1996).

27   P. F. McManamon, T. A. Dorschner, D. L. Corkum, L. J. Friedman, D. S. Hobbs, M. Holz, S. Liberman, H. Q. Nguyen, D. P. Resler, R. C. Sharp, and E. A. Watson, Optical phased array technology, *Proc. IEEE* **84**, 268 (1996).

28   M. F. Schiekel and K. Fahrenschon, Deformation of nematic liquid crystals with vertical orientation in electric fields, *Appl. Phys. Lett.* **19**, 391 (1971).

29   J. Grinberg, W. P. Bleha, A. D. Jacobson, A. M. Lackner, G. D. Myer, L. J. Miller, J. D. Margerum, L. M. Fraas, and D. D. Boswell, Photoactivated birefringence liquid crystal light valve for color symbology display, *IEEE Trans. Electron Devices ED -22*, 775 (1975).

30   R. D. Sterling and W. P. Bleha, D-ILA technology for electronic cinema, *Soc. Infor. Display, Tech. Digest* **31**, 310 (2000).

31   A. Takeda, S. Kataoka, T. Sasaki, H. Chida, H. Tsuda, K. Ohmuro, Y. Koike, T. Sasabayashi, and K. Okamoto, A super-high-image-quality multi-domain vertical alignment LCD by new rubbing-less technology, *SID Tech. Digest* **29**, 1077 (1997).

32   M. Oh-e, M. Yoneya, and K. Kondo, Switching of negative and positive dielectric anisotropic liquid crystals by in-plane electric fields, *J. Appl. Phys.* **82**, 528 (1997).

33   C. K. Wei, Y. H. Lu, C. L. Kuo, C. Y. Liu, H. D. Liu, W. C. Chang, H. P. Huang, C. M. Cheng, and D. C. Yan, A wide-viewing angle polymer-stabilized homeotropically aligned LCD, *SID Tech. Digest* **29**, 1081 (1998).

34   Y. Kume, N. Yamada, S. Kozaki, H. Kisishita, F. Funada, and M. Hijiligawa, Advanced ASM mode: Improvement of display performance by using a negative-dielectric liquid crystal, *SID Tech. Digest* **29**, 1089 (1998).

35   K. H. Kim, S. B. Park, J. U. Shim, J. H. Souk, and J. Chen, New LCD modes for wide-viewing-angle applications, *SID Tech. Digest* **29**, 1085 (1998).

36   F. J. Kahn, Orientation of liquid crystals by surface coupling agents, *Appl. Phys. Lett.* **22**, 386 (1973).

37   S. Matsumoto, K. Kawamoto, and N. Kaneko, Surface-induced molecular orientation of liquid crystals by carboxylatochromium complexes, *Appl. Phys. Lett.* **27**, 268 (1975).

38   J. L. Janning, Thin film surface orientation for liquid crystals, *Appl. Phys. Lett.* **21**, 173 (1972).

39   A. M. Lackner, J. D. Margerum, L. J. Miller, and W. H. Smith, Jr., Photostable tilted-perpendicular alignment of liquid crystals for light valves, *Proc. SID* **31**, 321 (1990).

40   W. H. Smith, Jr., H. L. Garvin, K. Robinson, and L. J. Miller, Inducing tilted perpendicular alignment in liquid crystals, U.S. Patent **5**, 350, 498 (1994).

41   M. Lu, K. H. Yang, T. Nakasogi, and S. J. Chey, Homeotropic alignment by single oblique evaporation of SiO2 and its application to high resolution microdisplays, *SID Tech. Digest* **31**, 446 (2000).

42   J. Cognard, Alignment of nematic liquid crystals and their mixtures, *Mol. Cryst. Liq. Cryst.* Suppl. **1**, 1 (1982).

43   S. He, H. Nishiki, J. Hartzell, and Y. Nakata, Low temperature PECVD a-Si:H TFT for plastic substrates, *SID Tech. Digest* **31**, 278 (2000).

44   S. Utsunomiya, S. Inoue, and T. Shimoda, Low temperature poly-si TFT on plastic substrate using surface free technology by laser ablation/annealing, *SID Tech. Digest* **31**, 916 (2000).

45   T. Sasaki, H. Fuji, and M. Nishikawa, Polyatomic acid Langmuir-Blodgett film for liquid crystal alignment, *Jpn. J. Appl. Phys. Part 2*, **31**, L632 (1992).

46   H. Aoyama, Y. Yamazaki, N. Matsuura, H. Mada, and S. Kobayashi, Alignment of liquid crystals on the stretched polymer films, *Mol. Cryst. Liq. Cryst.* **72**, 127 (1981).

47   Y. Kawata, K. Takatoh, M. Hasegawa, and M. Sakamoto, The alignment of nematic liquid crystals on photolithographic micro-groove patterns, *Liq. Cryst.* **16**, 1027 (1994).

48   E. S. Lee, P. Vetter, T. Miyashita, T. Uchida, M. Kano, M. Abe, and K. Sugawara, Control of liquid crystal alignment using stamped-morphology method, *Jpn. J. Appl. Phys. Part 2*, **32**, L1436 (1993).

49   K. Ichimura, Y. Suzuki, T. Seki, A. Hosoki, and K. Aoki, Reversible change in alignment mode of nematic liquid crystals regulated photochemically by "command surfaces" modified with an azobenzene monolayer, *Langmuir* **4**, 1214 (1988).

50   W. M. Gibbons, P. J. Shannon, S. T. Sun, and B. J. Swetlin, Surface-mediated alignment of nematic liquid crystal with polarized laser light, *Nature* **351**, 49 (1991).

51   P. J. Shannon, W. M. Gibbons and S. T. Sun, Patterned optical-properties in photopolymerized surface-aligned liquid-crystal films, *Nature* **368**, 532 (1994).

52   Y. Iimura, J. Kusano, S. Kobayashi, Y. Aoyagi, and T. Sugano, Alignment control of a liquid crystal on a photosensitive polyvinylalcohol film, *Jpn. J. Appl. Phys. Part 2*, **32**, L93 (1993).

53   M. Schadt, M. Schmitt, V. Kizinkov, and V. Chigrinov, Surface-induced parallel alignment of liquid crystals by linearly polymerized photopolymers, *Jpn. J. Appl. Phys. Part 1*, **31**, 2155 (1992).

54   Y. Iimura, T. Saitoh, S. Kobayashi, and T. Hashimoto, *J. Photopolym. Sci. Technology* **8**, 258 (1995).

55   J. L. West, X. Wang, Y. Ji and J. R. Kelly, Polarized UV exposed polyimide films for liquid crystal alignment, *SID Tech. Digest* **26**, 703 (1995).

56   M. Nishikawa, B. Taheri, and J. L. West, Mechanism of unidirectional liquid-crystal alignment on polyimides with linearly polarized ultraviolet light exposure, *Appl. Phys. Lett.* **72**, 2403 (1998).

57   K. H. Yang, K. Tajima, A. Takenaka, and H. Takano, Charge trapping properties of uv-exposed polyimide films for the alignment of liquid crystals, *Jpn. J. Appl. Phys. Part 2*, **35**, L561 (1996).

58   W. C. Lee, C. S. Hsu, and S. T. Wu, Liquid crystal alignment with a photo-crosslinkable and solvent-soluble polyimide film, *Jpn. J. Appl. Phys. Part 2*, **39**, L471 (2000).

59   R. A. Kashnow, U.S. Patent **3**, 914, 022 (1975).

60   K. A. Crandall, M. R. Fisch, R. G. Petschek, and C. Rosenblatt, Homeotropic rub-free liquid crystal light shutter, *Appl. Phys. Lett.* **65**, 118 (1994).

61   S. W. Suh, S. T. Shin, and S. D. Lee, Novel electro-optic effect associated with a homeotropic to twisted–planar transition in nematic liquid crystals, *Appl. Phys. Lett.* **68**, 2819 (1996).

62   S. T. Wu and C. S. Wu, High brightness projection displays using mixed-mode twisted-nematic liquid crystal cell, *SID Tech. Digest* **27**, 763 (1996).

63   S. T. Wu, C. S. Wu, and K. W. Lin, Chiral-homeotropic liquid crystal cells for high contrast and low voltage displays, *J. Appl. Phys.* **82**, 4795 (1997).

64   C. H. Gooch and H. A. Tarry, The optical properties of twisted nematic liquid crystal structures with twisted angles $\leq 90°$, *J. Phys. D* **8**, 1575 (1975).

65   J. Patel and G. B. Cohen, Inverse twisted nematic liquid-crystal device, *Appl. Phys. Lett.* **68**, 3564 (1996).

66   L. Y. Chen and S. H. Chen, Dynamics of chiral-homeotropic LCDs, *SID Tech. Digest* **30**, 636 (1999).

67   T. Sonehara, Photo-addressed liquid crystal SLM with twisted nematic ECB (ECB-TN) mode, *Jpn. J. Appl. Phys.* **29**, L1232 (1990).

68   S. W. Suh, J. S. Patel, and S. D. Lee, Reflective homeotropic mode in a twisted nematic liquid crystal, *Appl. Phys. Lett.* **73**, 1062 (1998).

4

Twisted Nematic Cells

1 Introduction

Twisted nematic (TN) cells have been widely used for both transmissive and reflective displays employing either one or two polarisers. To satisfy the wide-ranging application requirements, various TN configurations have been investigated. For instance, the 90° TN cell [1] sandwiched between two crossed polarisers has been used in many transmissive direct-view, projection and reflective direct-view displays. These displays have a white appearance at the voltage-off state and gradually turn dark as the voltage exceeds the threshold. The key features of these normally white displays are a high contrast ratio (greater than 300 : 1), low operating voltage (lesser than 5 V_{rms}), weak color dispersion (lesser than 8%), and sufficiently fast response time (around 30 ms). However, when the 90° TN cell is used for a reflective display, such as a wrist watch, the second polariser situated between the rear substrate and the reflector not only reduces display reflectivity but also causes parallax. Thus, this device configuration is not suitable for high-resolution display devices. To overcome these problems, the TN cell with one polariser has been tried [2]. However, due to the large cell gap employed, the device contrast ratio was not sufficient for practical applications.

As the cell gap (or $d\Delta n$) is reduced to below the Gooch–Tarry first minimum [3], the mixed-mode twisted nematic (MTN) cell has been shown to exhibit a high contrast ratio, high reflectivity, low voltage, and weak color dispersion under the single polariser configuration [4]. In short, the performance of a MTN cell for reflective display is equivalent to that of a TN cell for a transmissive display. Both reflective direct-view and projection displays using MTN cells have been demonstrated. In direct-view displays, the MTN cell requires only one front polariser and a $\lambda/4$ film to achieve a normally white mode [5]. Thus, the parallax is eliminated and reflectivity is enhanced. In the projection displays, a polarising beam splitter (PBS) functions as two crossed polarisers. The MTN cell basically preserves all the good features of a transmissive TN cell. Owing to the smaller $d\Delta n$ requirement of a MTN cell, its response time could be four times faster than a TN cell if the same Δn is considered. By using a thin cell gap, sequential color using a single MTN cell can be achieved. This method, in conjunction with CMOS transistors fabricated on a silicon-wafer, has found useful applications as projection monitors and virtual microdisplays for wireless communications [6].

For normally black reflective projection displays, the 45° TN [7,8] and homeotropic [9] cells are the two favorable choices. In principle, the normally black display exhibits a lower operation voltage, as described in Chapter 3. A major difference is that the 45° TN cell needs a positive $\Delta\varepsilon$ LC mixture, while the homeotropic cell uses a negative $\Delta\varepsilon$ LC mixture. The molecular designs for positive and negative LC mixtures are quite different, as described in Chapter 2.

In Section 2 we first describe the analytical solutions of TN cells for transmission displays. In Section 3, we extend the Jones matrix treatment to MTN cells for reflective displays.

2 Transmissive TN Cells

Several approaches, such as the Berreman 4×4 matrix [10–12], the General Geometrical Optics Approximation [13], and the extended Jones matrix [14, 15], have been developed to analyse the transmittance of a TN cell. In this section we use the Jones matrix method [16]. An advantage of the Jones matrix method is that an analytical solution can be obtained. The basic idea is to design a TN cell so that it behaves like an achromatic half-wave plate. The impinging light will follow the twist of the LC directors. After traversing the cell, the polarisation remains linear, but the axis is rotated 90°.

The normalised transmittance $T_\perp$ of a TN cell can be described by the following Jones matrices as $T_\perp = |M|^2$ [17].

$$M = |\cos\beta \quad \sin\beta| \begin{vmatrix} \cos\phi & -\sin\phi \\ \sin\phi & \cos\phi \end{vmatrix}$$
$$\begin{vmatrix} \cos X - i(\Gamma/2)\,(\sin X/X) & \phi(\sin X/X) \\ -\phi(\sin X/X) & \cos X + i(\Gamma/2)\,(\sin X/X) \end{vmatrix} \begin{vmatrix} -\sin\beta \\ \cos\beta \end{vmatrix} \qquad (4.1)$$

Here β is the angle between the polarisation axis and the front LC director, ϕ is the twist angle, $X = [\phi^2 + (\Gamma/2)^2]^{1/2}$ and $\Gamma = 2\pi d\Delta n/\lambda$, where d is the cell gap. By simple algebraic calculations, the following analytical expressions for M and $|M|^2$ are derived:

$$M = \left(\frac{\phi}{X}\cos\phi\sin X - \sin\phi\cos X\right) - i\frac{\Gamma}{2}\frac{\sin X}{X}\sin(\phi - 2\beta) \qquad (4.2)$$

$$|M|^2 = T_\perp = \left(\frac{\phi}{X}\cos\phi\sin X - \sin\phi\cos X\right)^2 + \left(\frac{\Gamma}{2}\frac{\sin X}{X}\right)^2 \sin^2(\phi - 2\beta) \qquad (4.3)$$

Equation 4.3 is a general formula that describes the light transmittance of a TN cell as a function of the twist angle, β angle and $d\Delta n/\lambda$.

2.1 Bisector effect

To search for the maximum transmittance in a general twisted cell, we set $\partial T_\perp/\partial\beta = 0$, keeping the second derivative negative, and obtain the following equation:

$$4\beta_{max} = 2\phi - (2n + 1)\pi, \quad n = 0, 1, 2, \ldots \tag{4.4}$$

This result is consistent with that derived by Ong from a general geometrical-optics approach [18]. For the lowest order $n = 0$, β_{max} decreases from 15, 0, −15 to −30° as the twist angle decreases from 120, 90, 60 to 30°, respectively. (See Figure 4.1). The maximum $T_\perp$ of each TN cell reaches 100% at these beta angles. The 90° TN cell at $\beta = 0°$ shows the weakest color dispersion.

From the same derivation procedures, the minimum transmittance of a general TN cell occurs when

$$4\beta_{min} = 2\phi - (2n)\pi, \quad n = 0, 1, 2, \ldots \tag{4.5}$$

This minimum is described by the first term shown in Equation 4.3. Thus, at the bisector $(\beta = \phi/2)$ and the proper phase matching condition, i.e. $\tan(X)/X = \tan(\phi)/\phi$, the $T_\perp = 0$ state can be achieved.

Next, let us consider an intermediate voltage state such that the LC directors are tilted, but the twist components are not yet completely unwound. Based on such director profiles, the cell can be treated as a bulk TN sandwiched between two boundary layers. Due to the high tilt, the effective Δn in the bulk area is sufficiently small that $\Gamma \to 0$ and $X \to \phi$. As the residual phase retardation originating from boundary layers is so small, the majority of the light transmittance is still governed by the bulk. Under this

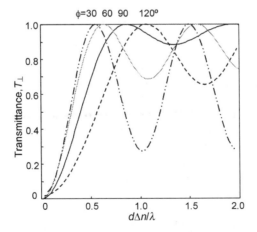

Figure 4.1 Computer simulation results of the normalised transmittance $T_\perp$ of the 120, 90, 60 and 30° twisted cells as a function of $d\Delta n/\lambda$. Their corresponding β_{max} are 15, 0, −15 and −30°, respectively

approximation, the first term in Equation 4.3 can be neglected and the on-state transmittance is reduced to

$$(T_\perp)_{on} \sim \left(\frac{\Gamma}{2} \frac{\sin \phi}{\phi} \right)^2 \sin^2 (\phi - 2\beta) \tag{4.6}$$

Equation 4.6 shows that a minimum state is reached when $\beta = (\phi - n\pi)/2$, as the phase condition $(\tan X)/X = (\tan \phi)/\phi$ is always approximately matched. Thus, the bisector effect leads to the darkest black state under the same director distribution; this is equivalent to having the lowest dark state voltage.

2.2   90° TN cell

The 90° TN cell has been widely used for transmissive direct-view and projection displays and reflective direct-view displays using two polarisers. In this section we derive the analytical expressions for $T_\perp$ as a function of beta angle and $d\Delta n/\lambda$. By substituting $\phi = \pi/2$ into Equation 4.3, we obtain the following expression for the normalised transmittance:

$$T_\perp = \cos^2 X + \left(\frac{\Gamma}{2X} \cos 2\beta \right)^2 \sin^2 X \tag{4.7}$$

When $\cos X = \pm 1$ (i.e. $X = m\pi$; $m =$ integer), then $\sin X = 0$ and the second term in Equation 4.7 vanishes. Therefore, $T_\perp = 1$, independently of β. Setting $X = m\pi$ and knowing that $\Gamma = 2\pi d\Delta n/\lambda$, we deduce the Gooch–Tarry condition [3]:

$$\frac{d\Delta n}{\lambda} = \left(m^2 - \frac{1}{4} \right)^{1/2} \tag{4.8}$$

When $m = 1$, $d\Delta n/\lambda = \sqrt{3}/2$. This is the Gooch–Tarry first minimum condition for 90° TN cells.

Figure 4.2 depicts the normalised transmittance $(T_\perp)$ of the 90° TN cell as a function of $d\Delta n/\lambda$ at $\beta = 0$, 15, 30 and 45°. From Equations 4.7 and 4.8, the first $T_\perp = 1$ occurs at $d\Delta n/\lambda = \sqrt{3}/2$, independently of β. However, the color dispersion (i.e. the wavelength dependency of the transmittance) strongly depends on β. At $\beta = 0$, $T_\perp$ is insensitive to $d\Delta n/\lambda$ beyond the first minimum. As β increases, the color dispersion is also increased.

The observed color dispersion can be explained by Equation 4.7. When $(\Gamma \cos 2\beta/ 2X)^2$ is close to unity, then $T_\perp$ approaches unity and is insensitive to $d\Delta n/\lambda$. In this case, the 90° TN cell behaves like a broadband achromatic half-wave plate. This condition happens when $\beta \simeq 0°$ (or 90°) and $\Gamma \simeq 2X$. The latter condition implies a large Γ such that $\Gamma/2 \gg \phi$. A large Γ means a large cell gap which leads to a slow response time if this half-wave plate needs to be switched. Therefore, a more effective way to restrain color dispersion is to keep $\beta \simeq 0°$.

From Equation 4.7, the $(\Gamma \cos 2\beta/2X)^2$ term oscillates wildly as β increases from 0 to 45°. At $\beta = 45°$, $\cos 2\beta = 0$; the right-hand term in Equation 4.7 vanishes and the transmittance is solely determined by the $(\cos^2 X)$ term. The first $T_\perp = 1$ occurs at $d\Delta n/\lambda = \sqrt{3}/2$ and the first $T_\perp = 0$ occurs when $d\Delta n/\lambda = \sqrt{2}$, as shown in Figure 4.2. Please note that $\beta = 45°$ is

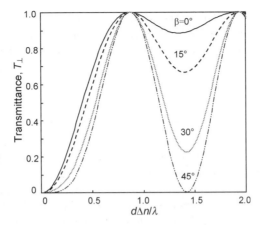

Figure 4.2 Computer simulation results of the normalised transmittance $T_\perp$ of a 90° twisted cell as a function of $d\Delta n/\lambda$ and β

the bisector of the 90° TN cell. As expected, at the bisector the color dispersion becomes more pronounced than that at $\beta = 0°$.

Figure 4.3 plots the voltage-dependent light transmittance (V–T) of the 90° TN cell at $\beta = 0$, 22.5 and 45°. The cell gap is chosen to be $d = 4.8\,\mu\text{m}$ and $\Delta n = 0.1$ to satisfy the first minimum condition at $\lambda = 550\,\text{nm}$. The maximum transmittance always reaches unity, independently of the beta angle. As β increases from 0 to 45°, the dark state voltage is gradually decreased. At the bisector ($\beta = \phi/2$), the dark state occurs at the lowest voltage.

For projection displays using three 90° TN LC panels, each cell can be optimised with a different LC while keeping the same cell gap. Under this circumstance, the advantage of the bisector approach is clear. In the two-bottle LC system that Merck developed, the LC mixtures have very similar physical properties in terms of phase transition temperatures, dielectric anisotropy and viscosity; the only difference is their birefringence. For example, the red panel can be filled with a higher Δn mixture (MLC-9000-100; $\Delta n = 0.114$), the blue panel with a lower Δn mixture (MLC-9000-000; $\Delta n = 0.0874$), and the green panel with 1 : 1 ratio of these

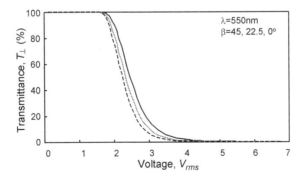

Figure 4.3 Computer simulation results of the voltage-dependent normalised transmittance $T_\perp$ of the 90° TN cell with $d\Delta n = 480\,\text{nm}$ and $\lambda = 550\,\text{nm}$. From right to left, $\beta = 0$, 22.5 and 45°

two mixtures so that its $\Delta n = 0.1$. The results are obvious; the V–T curves of the RGB wavelengths overlap quite well and the color dispersion is no longer a problem. In such circumstances, the bisector approach exhibits a clear advantage in lowering the operation voltage.

2.3   75° TN cell

As the twist angle is reduced from 90°, the first maximum $T_\perp$ occurs at a smaller $d\Delta n/\lambda$ and at a negative beta angle, as described by Equation 4.4. A thinner cell gap is attractive to improve the response time. For instance, the peak transmittance of the 75° TN cell occurs at $\beta_{max} = -7.5°$ and $d\Delta n/\lambda \simeq 0.73$. The required $d\Delta n$ is about 20% smaller than that of a 90° TN cell. Thus, its response time would be 40% faster if the same LC mixture were used. A general drawback of a non-90° TN cell is a lower contrast ratio within the available voltage limit. This is attributed to the absence of self-phase compensation from the boundary layers. From Equation 4.6, we could adjust the beta angle to achieve a dark voltage-on state, as long as the light transmittance is not sacrificed too much. The beta angle effect provides another degree of freedom for optimising non-90° TN cells.

Figure 4.4 depicts the $T_\perp$ of a 75° TN cell as a function of $d\Delta n/\lambda$ at $\beta = -7.5, 7.5, 22.5$ and 37.5°. For the case of $\beta = -7.5°$, $T_\perp = 1$ occurs at $d\Delta n \simeq 400$ nm. The smaller $d\Delta n$ is attractive for a faster response time and wider viewing angle. From Figure 4.4, the maximum $T_\perp$ gradually declines as the β angle increases. Although at the bisector ($\beta = 37.5°$) the dark state voltage is the lowest, its maximum transmittance is reduced to $T_\perp \simeq 93\%$ and the required $d\Delta n$ is about 5% larger. Thus, it may not be a preferred choice.

Figure 4.5 plots the V–T curves of the 75° TN cell at $\beta = -7.5°$. The light transmittance and color dispersion at the voltage-off state are as good as for the 90° TN cell, except that its dark state voltage is higher. To reduce this dark state voltage, one could employ a LC mixture with a higher dielectric anisotropy or use a slightly larger beta angle, e.g. 0 or 7.5° where the transmittance is still high and the color dispersion is insignificant.

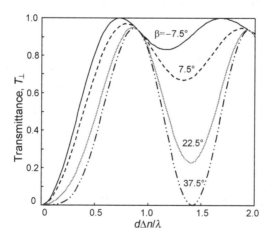

Figure 4.4 Computer simulation results of the normalised transmittance $T_\perp$ of the 75° TN cell as a function of $d\Delta n/\lambda$ and β

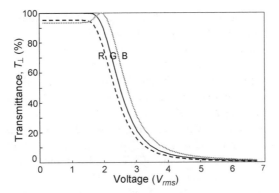

Figure 4.5 Computer simulation results of the voltage-dependent normalised transmittance $T_\perp$ of the 75° TN cell at $\beta = -7.5°$. $d\Delta n = 400\,\text{nm}$

Is the viewing angle of the 75° TN cell better than that of the 90° TN cell owing to its smaller $d\Delta n$ value? As discussed in Chapter 3, the light leakage increases for the cell, involving more birefringence effect and possessing a larger $d\Delta n$. Although the 75° TN cell uses more birefringence effect than the 90° TN cell, its required $d\Delta n$ value is smaller. As a result, these positive and negative factors balance out, resulting in a similar viewing angle.

2.4   60° TN cell

From Equation 4.4, the maximum $T_\perp$ of the 60° TN cell takes place at $\beta = -15°$ and $d\Delta n/\lambda \simeq 0.618$. This device configuration was first discussed by Leehouts et al. [19] Figure 4.6 plots $T_\perp$ as a function of $d\Delta n/\lambda$ for the 60° TN cell at $\beta = -15, 0, 15$ and 30°. The maximum $T_\perp$ gradually declines as the beta angle deviates from $-15°$. At the bisector ($\beta = 30°$), the

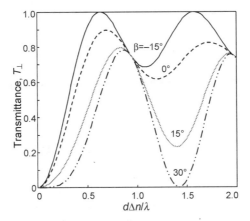

Figure 4.6 Computer simulation results of the normalised transmittance $T_\perp$ of the 60° TN cell as a function of $d\Delta n/\lambda$ and β

maximum $T_\perp$ drops to about 78% which is unacceptable for major display applications. Thus, we need to evaluate the V–T curves at $\beta = -15°$.

The V–T curve of the 60° TN cell at $\lambda = 550$ nm is plotted in Figure 4.7 for comparison with those of $\phi = 110$, 90 and 30° at their optimal beta angle and $d\Delta n$ value. The corresponding $(\beta, d\Delta n)$ values for the 110, 90, 60 and 30° TN cells are (5°, 520 nm), (0°, 480 nm), ($-15°$, 340 nm), and ($-30°$, 285 nm), respectively. As the twist angle decreases, the required $d\Delta n$ to achieve $T_\perp \simeq 1$ also decreases. As a result, the low twist cells have an advantage in fast response time while retaining weak color dispersion. However, their contrast ratio is reduced owing to the lack of self-phase compensation effect.

From Figure 4.7, the dark state of the 30° TN cell is rather poor. The contrast ratio at 5 V_{rms} is only 20:1. As the twist angle increases, the black state becomes darker and the required voltage is lower. At $\phi = 110°$ a good dark state occurs at about 3.5 V_{rms} and the transmission gradually increases as the voltage continues to increase. Thus, we have to control the dark state voltage in order to obtain a high contrast ratio.

2.5 100° TN cell

A high twist ($90 < \phi < 180°$) cell possesses a sharper transmission curve and, therefore, is useful for enhancing the multiplexing duty ratio. The voltage-dependent reflectance curve of a normal 90° TN cell is too shallow for multiplexing purposes. On the other hand, the 210 or 240° super twisted nematic (STN) cell can have a duty ratio of 1/240 or 1/480 [20]. Some general difficulties of an STN cell are a small cell gap tolerance, stringent ITO resistivity and slow response time. To make a TN display with a duty ratio between 1/16 and 1/64, a twist angle $90 < \phi < 180°$ can be considered.

Figure 4.8 shows the calculated $T_\perp$ as a function of $d\Delta n/\lambda$ for the 100° twisted cell at $\beta = 5$, 20, 35 and 50°. The maximum $T_\perp$ occurs at $\beta = 5°$ and $d\Delta n/\lambda \simeq 1$. Unlike the low twist cell, the maximum $T_\perp$ of the 100° TN cell at the bisector ($\beta = 50°$) remains reasonably

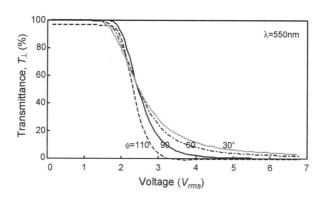

Figure 4.7   The V–T curve of the 60° TN cell at $\lambda = 550$ nm for comparison with those of $\phi = 110$, 90 and 30° at their optimal beta angle and $d\Delta n$ value. The corresponding $(\beta, d\Delta n)$ values for the 110, 90, 60 and 30° TN cells are (5°, 520 nm), (0°, 480 nm), ($-15°$, 340 nm) and ($-30°$, 285 nm), respectively

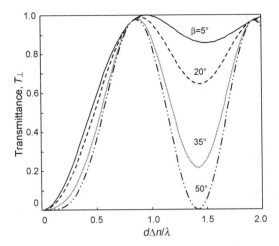

Figure 4.8 Computer simulation results of the normalised transmittance $T_\perp$ of the $100°$ TN cell as a function of $d\Delta n/\lambda$ and β

high (about 96%) and the corresponding $d\Delta n$ value is lower than that for $\beta_{max} = 5°$ (where $T_\perp \simeq 100\%$). Thus, the bisector approach is more feasible.

Figure 4.9 compares the V–T curves of the $100°$ TN cell at $\beta = 5°$ (with $d\Delta n = 520$ nm) and $\beta = 50°$ ($d\Delta n = 480$ nm) at $\lambda = 550$ nm. At $\beta = 5°$, the transmittance reaches 100%, color dispersion remains very weak and the dark state voltage is about 4 V. At $\beta = 50°$, the maximum transmittance is limited to about 98% and color dispersion is increased. A major advantage of the bisector effect is that the dark state voltage is reduced.

As demonstrated in Equation 4.5, setting the beta angle close to the bisector is an efficient way to reduce the dark state voltage and improve the contrast ratio. However, the maximum transmission efficiency is another important factor to be kept in mind. For a low twist cell ($\phi < 90°$), its maximum transmittance is greatly reduced as β approaches $\phi/2$. In contrast, a high twist cell ($\phi > 90°$) preserves a reasonably steep V–T curve and low operation

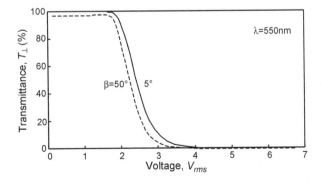

Figure 4.9 The voltage-dependent light transmittance $T_\perp$ of a $100°$ TN cell. Solid line: $d\Delta n = 520$ nm, $\beta = 5°$; and dashed lines: $d\Delta n = 480$ nm, $\beta = 50°$. $\lambda = 550$ nm

at the bisector except that its $d\Delta n$ is larger. A larger $d\Delta n$ has disadvantages with a slower response time and narrower viewing angle.

Continuing to increase the twist angle will further sharpen the V–T curves. However, when the twist angle exceeds 120°, the dark state under the simple crossed-polariser condition is gradually degraded so that the display contrast ratio is reduced. Thus, the twist angle of these high twist cells should be limited to 120°. The electro-optic effect of super-twisted nematic (STN) cells ($\phi > 180°$) will be discussed in Chapter 5.

3 Reflective MTN Cells

In a reflective cell, the incident light traverses the linear polariser, $\lambda/4$ film and LC layer, and is reflected back by the embedded mirror in the inner side of the rear substrate. Under the crossed-polariser configuration, to obtain high brightness in the voltage-off state the LC cell needs to function as an achromatic $\lambda/4$ plate. Therefore, the incoming linearly polarised light passes the two cascaded $\lambda/4$ films twice upon reflection and is transmitted by the polariser.

Under the above-mentioned reflective display configuration, the Jones matrix N is expressed as follows [21]:

$$N = |\cos\beta \quad \sin\beta| \begin{vmatrix} \cos X - i\dfrac{\Gamma}{2}\dfrac{\sin X}{X} & -\phi\dfrac{\sin X}{X} \\[2ex] \phi\dfrac{\sin X}{X} & \cos X + i\dfrac{\Gamma}{2}\dfrac{\sin X}{X} \end{vmatrix} \begin{vmatrix} \cos X - i\dfrac{\Gamma}{2}\dfrac{\sin X}{X} & \phi\dfrac{\sin X}{X} \\[2ex] -\phi\dfrac{\sin X}{X} & \cos X + i\dfrac{\Gamma}{2}\dfrac{\sin X}{X} \end{vmatrix} \begin{vmatrix} -\sin\beta \\[2ex] \cos\beta \end{vmatrix} \tag{4.9}$$

By expanding Equation 4.9 we derive the following expressions for N and the normalised reflectance $R_\perp = |N|^2$:

$$N = i\left(\Gamma\frac{\sin X}{X}\right)\left(\sin 2\beta\cos X - \frac{\phi}{X}\cos 2\beta\sin X\right) \tag{4.10}$$

$$|N|^2 = \left(\Gamma\frac{\sin X}{X}\right)^2\left(\sin 2\beta\cos X - \frac{\phi}{X}\cos 2\beta\sin X\right)^2 \tag{4.11}$$

To find the maximum and minimum of $R_\perp$, in principle we could take the first and second derivatives of Equation 4.11 with respect to β and Γ and solve the miscellaneous equations. However, the equation resulting from $\partial R_\perp/\partial\Gamma = 0$ is too complicated to solve analytically. It is better to plot Equation 4.11 using the numerical method.

Several approaches, such as parameter space [22], dynamic parameter space [23], the Berreman 4 × 4 matrix [24], and the equivalent retarder method [25], have been developed to analyse reflective LC cells. The results are consistent between the various methods.

When $\beta = 0$, Equation 4.11 has special solutions for $R_\perp = 1$. In this circumstance, the normalised reflectance $R_\perp$ is reduced to

$$R_\perp = \left[\frac{2\phi(\Gamma/2)}{\phi^2 + (\Gamma/2)^2} \right]^2 \sin^4 X \qquad (4.12)$$

From Equation 4.12, if $\sin X = 1$ (i.e. $X = \pi/2$, $3\pi/2$, etc.) and $\phi = \Gamma/2$, then $R_\perp = 1$. For the case that $X = \pi/2$, we find $\phi = \sqrt{2}\pi/4$ (or 63.6°) and $d\Delta n/\lambda = \sqrt{2}/4$. When $X = 3\pi/2$, we obtain $\phi = 3\pi/(2\sqrt{2})$ (or 190.9°) and $d\Delta n/\lambda = 3/(2\sqrt{2})$. These cell configurations were first reported by Sonehara [26]. These cells have a high reflectance; however, their contrast ratio is limited to about $20:1$ due to the absence of intrinsic phase compensation. The bisector effect should be helpful in lowering the operation voltage and boosting the contrast ratio.

From Equation 4.11, the bisector effect is not immediately obvious owing to the non-unity factor of (ϕ/X). However, if we make the same approximation that Δn is small in the intermediate voltage regime, then X is reduced to ϕ and Equation 4.11 is simplified to

$$(R_\perp)_{on} \sim \left(\frac{\Gamma \sin \phi}{\phi} \right)^2 \sin^2(\phi - 2\beta) \qquad (4.13)$$

From Equation 4.13, the bisector effect also exists in the reflective twisted cells. The darkest black state appears at $\beta = \phi/2$. Comparing Equation 4.13 with Equation 4.6, we find that the reflective cell experiences residual phase retardation twice as large as that of the transmissive one. This is because the incident light traverses the boundary layers twice in a reflective cell.

Low voltage is desirable to reduce the power consumption and the costs of the electronics. However, it is not the only important parameter for display applications. In reflective displays, brightness is more important than low operation voltage. Thus, the priority between light efficiency and operation voltage needs to be taken into account. In general, the maximum reflectance does not occur at $\beta = \phi/2$. It may be necessary to optimise reflectance and compromise for the operation voltage. This has been shown in several mixed-mode twisted nematic (MTN) cells for reflective displays to be described in this section. For example, the optimal beta angle to achieve maximum reflectance of the 90° MTN cell occurs at $\beta = 20°$. At the bisector, the maximum reflectance drops to 57%, which is too low to be practically useful, although it does lower the operation–voltage. For some MTN cells with a 60–70° twist angle, their reflectance remains high at $\beta \simeq \phi/2$. Thus, both high reflectance and low operation voltage can be achieved simultaneously. The trade-off is in more pronounced color dispersion.

In the following sections we will focus on the LC operation modes developed for reflective displays using only one polariser. Let us start the analysis from the nominal transmissive 90° TN cell which is being used in wrist watches. However, this time we will replace the second polariser with a quarter-wave film to simulate the crossed polariser condition. Afterwards, we will introduce some high performance MTN cells for reflective direct-view and projection displays.

3.1   90° TN cell

The 90° TN cell has been widely used for transmissive displays due to its high efficiency for white light modulation, high contrast ratio and low operation voltage. However, when a 90° TN cell is used for a reflective display (such as in a wrist watch), two crossed polarisers are required in order to obtain a high contrast ratio. The second polariser situated between the back substrate and the reflector not only reduces brightness but also causes parallax which limits the resolution of the device. Removing the second polariser would eliminate the parallax. Nevertheless, its reflectance is greatly reduced, as shown in Figure 4.10. In Figure 4.10 $R_\perp$ stands for the normalised reflectance under the crossed-polariser condition, i.e. incorporating a quarter-wave film with a sheet polariser.

   In the voltage-off state, the incident linearly polarised light basically follows the twist of the LC directors and is then reflected back by the mirror. As the outgoing light has the same polarisation as the incident one, it is blocked by the crossed analyser, resulting in a dark state. In a high-voltage state, the LC directors are aligned along the electric field direction, except at the boundary layers. The incident light sees little phase retardation. The outgoing light preserves the same polarisation state as the incident one. Thus, a dark state is observed again. In an intermediate state, both phase retardation and polarisation effects coexist. From Figure 4.10 the light modulation efficiency is low and there is no common dark state for the RGB colors even in the crossed-polariser configuration. Thus, the conventional 90° TN cell is not suitable for the single-polariser reflective display.

3.2   90° MTN cell

To overcome simultaneously the brightness, contrast ratio and parallax problems of a conventional 90° TN cell, the mixed-mode TN (MTN) cell has been developed. Basically, the reflective MTN cell preserves most of the desirable features of a transmissive TN cell, such as high contrast, achromatic behavior and low operation voltage. Solely from the cell structure standpoint, a MTN cell is identical to a TN except for a smaller $d\Delta n$ and different

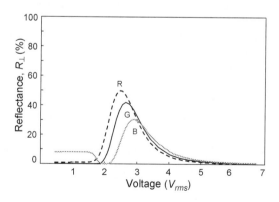

Figure 4.10   Normalised reflectance $R_\perp$ of a 90° TN cell used in a single-polariser reflective display. $d\Delta n = 480$ nm; $\beta = 0°$; $R = 650$, $G = 550$, and $B = 450$ nm

beta angle. Nevertheless, the underlying physical mechanism and performance are not quite the same. A MTN cell is different from a TN in the following aspects [27].

Smaller dΔn

In a reflective display, the incident light traverses the LC cell twice. Thus, the required $d\Delta n$ of a MTN cell is below the Gooch–Tarry first minimum $(d\Delta n/\lambda = \sqrt{3}/2)$ for a TN cell. As a result, the polarisation rotation effect in a MTN cell is incomplete. An additional phase retardation or birefringence effect is therefore needed to convert the elliptically polarised light back to linear more efficiently. Owing to its smaller cell gap requirement, the response time of a MTN cell can be four times faster than the first minimum TN cell if the same LC mixture is used.

Beta angle

To utilise effectively the polarisation rotation effect, the beta angle of a 90° transmissive TN cell is normally set at zero or 90°. On the other hand, the pure birefringence effect for a non-twisted homogeneous cell maximises at $\beta = 45°$, as discussed in Chapter 2. For an MTN cell, its optimal beta angle appears at about 20–30°, depending on the $d\Delta n$ value and twist angle of the cell. This beta angle determines the mixing ratio between the polarisation rotation and birefringence effects. A smaller beta angle would result in a weaker wavelength dispersion owing to the smaller birefringence effect. A larger beta angle will result in a lower operation voltage due to the influence of the bisector effect except that its color dispersion is more pronounced.

Wider viewing angle

The viewing angle of a reflective MTN cell is equivalent to a two-domain transmissive TN cell. This is because of the mirror image effect experienced by the incident and reflected beams, as described in Chapter 1. A typical 90° MTN viewing cone exceeds 40° without seeing a reversed gray scale.

To illustrate the design principles, let us begin with the 90° MTN cell. Firstly, we use Equation 4.11 to plot the normalised reflectance $(R_\perp)$ as a function of $d\Delta n/\lambda$ and β. Secondly, we calculate the voltage-dependent light reflectance as a function of twist angle, $d\Delta n/\lambda$ and beta angle.

Figure 4.11 depicts the computer simulation results of $R_\perp$ as a function of $d\Delta n/\lambda$ at $\beta = 0$, 20, 30 and 45°. At $\beta = 0°$ the peak reflectance only reaches about 70%. As β increases to 20° the maximum reflectance $(R_\perp = 88\%)$ occurs at $d\Delta n \simeq 0.45\lambda$. As β continues to increase the maximum reflectance gradually declines. At the bisector $(\beta = 45°)$, the maximum $R_\perp$ drops to 57%.

Figure 4.12 shows the voltage-dependent reflectance of a 90° MTN cell with $d\Delta n = 240$ nm and $\beta = 20°$. In the voltage-off state, the polarisation rotation and phase retardation effects coexist. A proper mixing between these two effects results in a high reflectance for the RGB colors studied, and thus the cell appears white. As the voltage

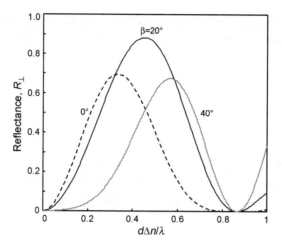

Figure 4.11 Computer simulation results of the normalised reflectance $R_\perp$ of a 90° MTN cell as a function of $d\Delta n/\lambda$ and β

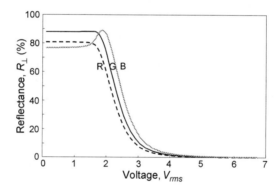

Figure 4.12 Normalised reflectance $R_\perp$ of a 90° MTN cell used in a single-polariser reflective display. $d\Delta n = 240$ nm and $\beta = 20°$

exceeds the optical threshold, molecular reorientation takes place and reflectance declines sharply. At 5 V_{rms}, a good common dark state is obtained for the RGB wavelengths studied. For a projection display employing a PBS, a contrast ratio of 300 : 1 can be obtained relatively easily. For a direct-view display, the contrast ratio is greatly influenced by the surface reflection and reflector design, as discussed in Chapter 1. A high contrast ratio and low operating voltage are two built-in features of a 90° MTN cell. The boundary layers of such a cell are orthogonal to each other so that their residual phase retardations compensate for each other. Consequently, its dark state takes place at a relatively low voltage.

The direct-view MTN result shown in Figure 4.12 is equivalent to a projection display employing a polarising beam splitter (PBS). The PBS acts as two crossed polarisers. Under this configuration, the normally white mode is obtained. Using a 3 μm MTN cell, its re-sponse time (rise + decay) is about 5 ms at $T = 50°$C. This is a typical operation temperature of a LCD projector due to the heating from a high-power metal-halide or xenon lamp. Therefore, a MTN cell is also attractive for color sequential projection displays using only one LCD panel.

The eight-gray level horizontal and vertical viewing characteristics of the 90° MTN cell have been calculated and results are shown in Figure 4.13. Within the 40° viewing cone, no gray scale inversion is observed.

3.3 80° and 70° MTN cells

From Figure 4.11, the peak reflectance ($R_\perp$) of the 90° MTN cell is only about 88%. To enhance reflectance, the 80° and 70° MTN cells have been studied. In the 80° MTN cell, $R_\perp$ is improved to about 97% at $d\Delta n = 272$ nm and $\beta \simeq 20°$. For a 70° MTN cell with $d\Delta n = 278$ nm and $\beta = 20°$, $R_\perp$ approaches 100%. The price paid for these non-90° MTN cells is in the reduced contrast ratio and higher operation voltage.

Figure 4.14 compares the contrast ratios of the 90, 80 and 70° MTN cells as functions of cell gap variation. At a given voltage, the 90° MTN offers the highest contrast ratio, followed by 80° and then 70°. This is because the natural phase compensation effect resulting from the boundary layers gradually disappears as the twist angle deviates from 90°. The contrast ratio of the MTN cells is quite insensitive to cell gap variation. This advantage is particularly important when rough reflectors are used for handheld displays [28]. The surfaces of these rough mirrors are not flat. The large MTN cell gap tolerance enables a reasonably high contrast ratio of such reflective displays to be obtained.

To enhance the contrast ratio, one could select a LC mixture with larger dielectric anisotropy, set the β angle toward the bisector, or add a phase compensation film. Adding a biaxial film or a hybrid LC polymer film [29] would simultaneously reduce the operation voltage and widen the viewing angle. A trade-off is in the increased cost. Tuning the beta angle toward the bisector is an effective approach except that the wavelength dispersion is increased. For instance, by setting $\beta = 35°$ for a 70° MTN or 40° for an 80° MTN cell, the dark state voltage is reduced to about 3 V except that the maximum reflectance is reduced to

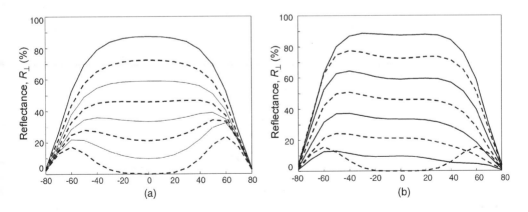

Figure 4.13 The eight-gray level horizontal (left) and vertical (right) viewing angle of a reflective 90° MTN cell. $d\Delta n = 240$ nm and $\beta = 20°$

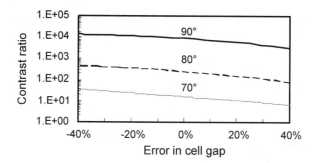

Figure 4.14 Simulated results of the contrast ratio as a function of cell gap variation for the 90, 80 and 70° MTN cells. The voltages for the bright and dark states are 0 and 5 V, respectively

95% and 80%, respectively. An 80% reflectance may not be acceptable for some reflective displays. Therefore, a proper compromise between the brightness and the contrast ratio is needed. For most direct-view reflective displays, a contrast ratio in the 30–50 range is considered sufficient due to glare from surface reflections. Thus, a twist angle ranging from 60 to 90° has been implemented in various commercial products. For example, ERSO reflective displays use $\phi = 80$–$90°$ to achieve high contrast and saturated colors [30], Sharp displays use $\phi = 75°$ to preserve reflectance of about 100% [31], and HKUST projection display uses $\phi = 70°$ and $\beta \simeq 25°$ to compromize reflectance and contrast ratio [32].

3.4   63.6° MTN cell

From Equation 4.12, a simple solution to obtain 100% reflectance at $\beta = 0$ is $\phi = 63.6°$. The 63.6° MTN cell (also called the TN-ECB mode) was first demonstrated by Sonehara and Okumura [26, 33]. Figure 4.15 depicts the voltage-dependent reflectance of the 63.6° MTN cell with $d\Delta n = 192$ nm and $\beta = 0°$. In a voltage-off state, the reflectance is high and the color

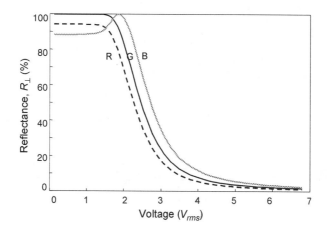

Figure 4.15 Normalised reflectance $R_{\perp}$ of a 63.6° MTN cell used in a single-polariser reflective display. $d\Delta n = 192$ nm and $\beta = 0°$

dispersion is small. The major problem of this operation mode is that its dark state voltage is quite high. This is because the inherent phase compensation effect of the 63.6° MTN cell no longer exists at $\beta = 0°$. The boundary layers cannot easily be oriented by the applied voltage. As a result, its operation voltage is high. At 5 V$_{rms}$, the contrast ratio is less than 20:1.

The eight-gray level viewing angles of the 63.6° MTN cell were calculated from 0 to 5 V$_{rms}$, and the results are shown in Figure 4.16. Despite its expected lower contrast, this thin LC cell exhibits a reasonably wide viewing angle; its horizontal view exceeds $\pm 60°$ and the vertical view is about $\pm 50°$ before the reversed contrast appears. It is known that a thinner LC layer would result in a wider viewing angle. In the small angle approximation, the effective phase retardation of an LC cell at an oblique angle is linearly proportional to the $d\Delta n$ of the LC layer and to the square of the incident angle.

A thin cell is also advantageous in response time. One problem related to a thin cell is that its cell gap is harder to control accurately. Therefore, the manufacturing yield may be reduced. One method to overcome the small $d\Delta n$ problem is to use a low Δn LC material. From the birefringence dispersion model, a fully saturated LC compound should possess a birefringence lower than 0.03. However, to formulate a practical LC mixture possessing a wide nematic range (-40 to 80°C) and other desirable physical properties (e.g. the dielectric anisotropy, voltage-holding ratio and viscosity), a dozen of such low Δn LC compounds need to be prepared. If a low birefringence LC mixture is used, say $\Delta n \simeq 0.05$, then the cell gap can be maintained at $d \simeq 3.8 \, \mu$m. This is not too difficult to manufacture.

3.5 60° MTN cell

As presented in Equation 4.13, the bisector effect is an effective way to lower the operation voltage for non-90° MTN cells. However, we need to bear in mind that brightness is always a priority parameter to be optimised. Thus, we need to find a MTN cell that preserves a high $R_\perp$ at $\beta = \phi/2$. From Equation 4.11, such a condition exists when the twist angle is in the vicinity of 60°. Figure 4.17 shows the $R_\perp$ versus $d\Delta n/\lambda$ plot of the 60° twisted cell at $\beta = 0$, 15, 30 and 45°. The reflectance has double peaks: one at $\beta = 0$ and $d\Delta n/\lambda \simeq 350$ nm, and

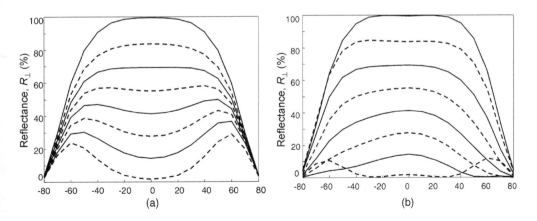

Figure 4.16 Calculated eight-gray level horizontal (left) and vertical (right) viewing angle of a reflective 63.6° MTN cell. $d\Delta n = 192$ nm and $\beta = 0°$

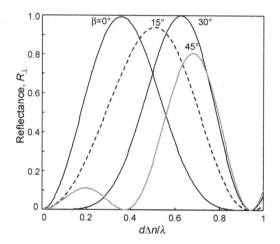

Figure 4.17 Computer simulation results of the normalised reflectance $R_\perp$ of the 60° twisted cell as a function of $d\Delta n/\lambda$ and β

another at $\beta = 30°$ and $d\Delta n/\lambda \simeq 625$ nm. The case of $\beta = 0$ is expected because the twist angle is so close to 63.6°. The case of $\beta = 30°$ is very interesting because it happens at the bisector [34, 35].

Figure 4.18 shows the voltage-dependent reflectance $R_\perp$ of a 60° MTN cell with $d\Delta n = 344$ nm and $\beta = 30°$. Indeed, a good dark state takes place at a relatively low voltage. Benefiting from the bisector effect, this dark state voltage is even lower than that of the 90° MTN cell.

The major problem of the 60° MTN cell is its relatively large color dispersion [36]. From Figure 4.18, the reflectance difference between the green and the blue light is as large as 45%. This means that the off-state is not a balanced white. As the voltage is increased, the blue light increases initially and then decreases, while the red and green decline monotonically. This implies that the display chromaticity changes very noticeably with the applied

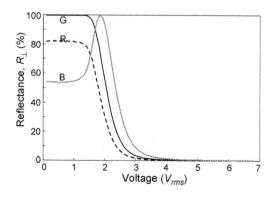

Figure 4.18 Normalised reflectance of a self phase compensation 60° MTN cell used in a single-polariser reflective display $d\Delta n = 344$ nm and $\beta = 30°$

voltage. Due to the severe dispersion effect, this cell may not be suitable for a direct-view reflective display using a single LCD panel.

For a projection display using three panels, each panel can be filled with different LC mixtures. One could use a higher Δn LC for the red channel and a lower Δn LC for the blue channel such that the $d\Delta n/\lambda$ for the RGB panels is nearly identical. This way, the red and blue curves shown in Figure 4.18 will overlap with the green. A high reflectance, low operation voltage and zero color dispersion can be achieved simultaneously.

The viewing characteristics of the 60° MTN cell are calculated and the results are shown in Figure 4.19. Within the $\pm 40°$ viewing cone no gray scale inversion is observed. However, the high contrast is confined in the $\pm 20°$ regime.

3.6   Low twist MTN cell

For a color sequential display, the LC response time (rise + decay) needs to be less than 5 ms. It is a challenging task for nematics, especially when low level gray scale switching is involved. In these circumstances, the thin cell becomes the most promising approach, despite the cell gap control accuracy and manufacturing yield problems. To search for the lower $d\Delta n$ requirement, twist angles below 60° are studied. We call these low twist cells.

Figure 4.20 plots $R_\perp$ versus $d\Delta n/\lambda$ for the $\phi = 60$, 45 and 30° cells at $\beta = 0$, −15 and −25°, respectively. At these beta angles, the cell has the maximum reflectance. As the twist angle decreases, the required $d\Delta n/\lambda$ value also decreases, but gradually saturates. To obtain a high contrast ratio for these low-twist cells, a uniaxial or biaxial film needs to be added. A uniaxial film could lower the operation voltage while a biaxial film could simultaneously reduce the operation voltage and increase the viewing angle.

Figure 4.21 shows the voltage-dependent light reflectance of a film-compensated 45° MTN cell. The LC has $d\Delta n = 170$ nm, film $d\Delta n = -15$ nm and $\beta = -15°$. This normally white mode exhibits a high reflectance, high contrast ratio and weak color dispersion. To further lower the operation voltage, a film with a larger $d\Delta n$ value should be considered.

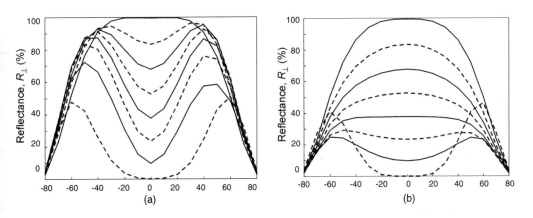

Figure 4.19   The eight-gray level horizontal (left) and vertical (right) viewing angle of a reflective 60° MTN cell. $d\Delta n = 344$ nm and $\beta = 30°$

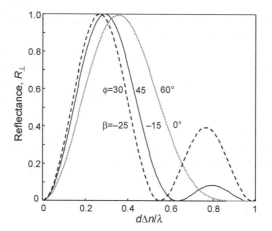

Figure 4.20 Computer simulation results of the normalised reflectance $R_\perp$ of some low twist MTN cells as a function of $d\Delta n/\lambda$ at the specified ϕ and β angles

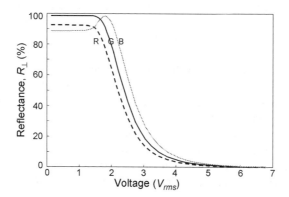

Figure 4.21 Film-compensated 45° MTN. LC $d\Delta n = 170$ nm, film $d\Delta n = -15$ nm and $\beta = -15°$

4 Reflective TN Cells

In this section we describe two TN cells (i.e. $d\Delta n/\lambda > \sqrt{3}/2$) for reflective displays. The first one is a 45° TN cell and the second one is a 53° TN cell. Recall that a TN cell differs from a MTN cell in two aspects: larger $d\Delta n$ and beta angle.

4.1 45° TN cell

The 45° TN cell (without specifying the beta angle) was developed in the 1970s for projection display using three light valves [37]. Computer simulations show that at $\beta = 0°$, the maximum reflectance of the 45° TN cell can only reach about 93%. However, by rotating the beta angle by around 10°, RGB can all reach 100% reflectance. Figure 4.22 plots the

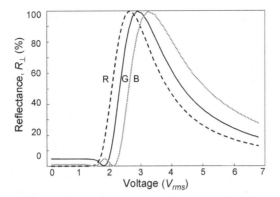

Figure 4.22 Voltage-dependent reflectance of a 45° TN cell. LC $d\Delta n = 443$ nm and $\beta = -12°$

voltage-dependent reflectance of a 45° TN cell with $d\Delta n = 443$ nm and $\beta = -12°$. The reason for requiring a relatively large $d\Delta n$ is to ensure a good dark state for the red wavelength. Because of this large $d\Delta n$, the voltage-dependent reflectance reaches 100% at a relatively low voltage.

From Figure 4.22, the dark and bright states of the RGB colors all occur at different voltages. For a projection display using three LCD panels, the dark state of each panel can be biased at a different voltage, or the $d\Delta n$ of each panel can be tuned to match with the wavelength by the two-bottle mixture. As a result, the overall contrast ratio could exceed 200 : 1 under the normally black operation condition. However, for direct-view display using a single LCD panel, the bias voltage is usually optimised for the green channel so that light leakage in the red and blue channels is inevitable. Thus, the contrast ratio (at the normal viewing direction) drops to about 30 : 1 under the crossed-polariser configuration. For the parallel-polariser condition, the contrast ratio is lowered to 10 : 1 due to the lack of a common dark state for the RGB wavelengths employed. Moreover, owing to the large $d\Delta n$ (443 nm) involved, the cell has a relatively narrow viewing angle.

4.2   53° TN cell

For a reflective direct-view display, if the achromatic quarter-wave film can be removed, then the panel cost will be reduced. In these circumstances, we are dealing with the parallel polariser condition [38] i.e. looking for the solution of $R_{\parallel} = 1$, or $R_{\perp} = 0$. From Equation 4.11, setting $\beta = 0°$ we find

$$R_{\perp} = \left(\Gamma \phi \frac{\sin^2 X}{X^2} \right)^2 \tag{4.14}$$

To search for $R_{\perp} = 0$, we find a trivial solution $X = m\pi$ for $\sin^2 X = 0$. Let us take the simplest case that $m = 1$, then $X = \pi$ and we derive the following relationship:

$$\frac{d\Delta n}{\lambda} = \left[1 - \left(\frac{\phi}{\pi} \right)^2 \right]^{1/2} \tag{4.15}$$

Caution should be exercised while using Equation 4.15 to design a reflective cell. Equation 4.15 has two unknowns: ϕ and $d\Delta n/\lambda$. For a given twist angle ϕ, we can always find a corresponding $d\Delta n$ value to satisfy Equation 4.15. This means that $R_\parallel = 1$ can be obtained. However, there is no guarantee that a good dark state in the voltage-on state can always be achieved. A special case at $\phi = 53°$ has been reported independently by two groups [39, 40]. If $\phi = 53°$, then $d\Delta n/\lambda = 0.956$, or $d\Delta n = 525$ nm for $\lambda = 550$ nm.

Figure 4.23 depicts the voltage-dependent reflectance of a 53° TN cell with $d\Delta n = 525$ nm and $\beta = 0°$. The reflectance $R_\parallel$ in the voltage-off state is high and is reasonably achromatic. Due to the wavelength-dependent phase retardation effect, the dark state occurs at different voltages for the RGB colors studied. For the red, its $d\Delta n/\lambda$ is the smallest so that its dark state occurs at the lowest voltage. For projection displays using three LCD panels, we could drive the RGB panels to their individual dark states. Thus, a high contrast ratio can be obtained for a full-color display. In the direct-view display using a single LCD panel, the voltage swing for the RGB pixels is normally adjusted to match the green color. The red and blue pixels will not share the same common dark state as the green. Thus, the overall contrast ratio is averaged to about $10:1$.

The performance of the 45° TN cell at $\beta = -12°$ is rather similar to that of the 53° TN cell at $\beta = 0°$. Their twist angles are not too far apart and the $d\Delta n$ values are similar. Each mode can be used for crossed- and parallel-polariser conditions. The crossed-polariser configuration is more suitable for a projection display using three LCD panels. The cell gap tolerance is large and the required operation voltage is relatively low. On the other hand, the parallel-polariser configuration shows high reflectance and weak color dispersion. It could be used for a direct-view display. However, both the 45 and 53° cells suffer from a narrow viewing angle due to their large $d\Delta n$ values.

4.3 Other TN cells

Besides 45 and 53° twisted cells, other twist angles such as 56.1 and 63° have also been analysed for parallel-polariser reflective displays [41]. The results are similar: no common dark state for the RGB colors is obtained. Thus, the contrast ratio is still limited to $10-15:1$.

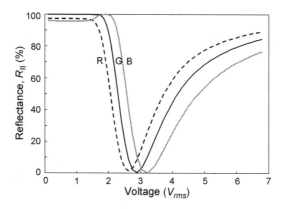

Figure 4.23 Normalised reflectance $R_\parallel$ of a 53° TN cell used in a single-polariser reflective display. LC $d\Delta n = 525$ nm and $\beta = 0°$. Parallel polarisers

An approach to achieve a common dark state for the RGB colors is to implement different cell gaps for different color pixels. For example, the red pixels will intentionally have a larger cell gap than the green and blue, such that their EO curves overlap. Such an uneven cell gap might be fabricated by controlling the thickness of the color filters. Undoubtedly, this would cause manufacturing difficulties and reduce the yield. How to achieve a high contrast ratio remains a technical challenge for any reflective display employing the parallel-polariser configuration.

References

1   M. Schadt and W. Helfrich, Voltage-dependent optical activity of a twisted nematic liquid crystal, *Appl. Phys. Lett.* **18**, 127 (1971).

2   A. Kmetz, A single-polarizer twisted nematic display, *Proc. SID* **21**, 63 (1980).

3   C. H. Gooch and H. A. Tarry, The optical properties of twisted nematic liquid crystal structures with twisted angles $\leq 90°$, *J. Phys. D* **8**, 1575 (1975).

4   S. T. Wu and C. S. Wu, Mixed twisted-nematic mode for reflective liquid crystal displays, *Appl. Phys. Lett.* **68**, 1455 (1996).

5   C. L. Kuo, C. K. Wei, S. T. Wu and C. S. Wu, Reflective display using mixed-mode twisted nematic liquid crystal cell, *Jpn. J. Appl. Phys.* **36**, 1077 (1997).

6   D. J. Schott, Reflective LCOS light valves and their applications to virtual displays, *Conference Records of the 1999 Int'l Display Research Conference* (Berlin, Germany, 1999), p. 485.

7   I. A. Shanks, Electro-optical color effects by twisted nematic liquid crystal, *Electron Lett.* **10**, 90 (1974).

8   J. Grinberg, A. Jacobson, W. Bleha, L. Miller, I. Fraas, D. Bosewell, and G. Myer, A new real-time noncoherent light image converter: the hybrid field effect liquid crystal light valve, *Opt. Eng.* **14**, 217 (1975).

9   M. F. Schiekel and K. Fahrenschon, Deformation of nematic liquid crystals with vertical orientation in electric fields, *Appl. Phys. Lett.* **19**, 391 (1971).

10   D. W. Berreman, Optics in stratified and anisotropic media: 4×4 matrix formulation, *J. Opt. Soc. Am.* **62**, 502 (1972).

11   R. J. Gagnon, Liquid crystal twist cell optics, *J. Opt. Soc. Am.* **71**, 348 (1981).

12   H. Wohler, G. Haas, M. Fritsch, and D. A. Mlynski, Faster 4×4 matrix method for uniaxial inhomogeneous media, *J. Opt. Soc. Am.* **A5**, 1554 (1988).

13   H. L. Ong, Origin and characteristics of the optical properties of general twisted nematic liquid crystal displays, *J. Appl. Phys.* **64**, 614 (1988).

14   P. Yeh, Extended Jones matrix method, *J. Opt. Soc. Am.* **72**, 507 (1982).

15   A. Lien, Extended Jones matrix representation for the twisted nematic liquid crystal display at oblique incidence, *Appl. Phys. Lett.* **57**, 2767 (1990).

16   R. C. Jones, New calculus for the treatment of optical systems, *J. Opt. Soc. Am.* **32**, 486 (1942).

17   S. T. Wu and C. S. Wu, Mixed-mode twisted-nematic cell for transmissive liquid crystal display, *Displays* **20**, 231 (1999).

18   H. L. Ong, Electro-optics of twisted nematic liquid crystal display by 2×2 propagation matrix at oblique angle, *Jpn. J. Appl. Phys.* **30**, L1028 (1991).

19   F. Leehouts, M. Schadt, and H. J. Fromm, Electro-optical characteristics of a new liquid-crystal display with an improved gray-scale capability, *Appl. Phys. Lett.* **50**, 1468 (1987).

20   T. Scheffer and J. Nehring, Twisted nematic and super-twisted nematic mode LCDs, *Liquid Crystals Applications and Uses*, ed. B. Bahadur (World Scientific, Singapore, 1990), Vol. 1, Ch. 10.

21 S. T. Wu and C. S. Wu, Mixed twisted-nematic mode for reflective liquid crystal displays, *Appl. Phys. Lett.* **68** 1455 (1996).

22 H. S. Kwok, Parameter space representation of liquid crystal display operating modes, *J. Appl. Phys.* **80**, 3687 (1996).

23 H. Cheng, H. Gao, and F. Zhou, Dynamic parameter space method to represent the operation modes of liquid crystal displays, *J. Appl. Phys.* **86**, 5935 (1999).

24 S. Stallinga, Berreman 4 × 4 method for reflective liquid crystal displays, *J. Appl. Phys.* **85**, 3023 (1999).

25 S. Stallinga, Equivalent retarder approach to reflective liquid crystal displays, *J. Appl. Phys.* **86**, 4756 (1999).

26 T. Sonehara and O. Okumura, A new twisted nematic ECB mode for a reflective light valve, *Proc. SID* **31**, 145 (1990).

27 S. T. Wu, C. S. Wu, and C. L. Kuo, Reflective direct-view and projection displays using twisted-nematic liquid crystal cells, *Jpn. J. Appl. Phys.* **36**, 2721 (1997).

28 Y. Itoh, S. Fujiwara, N. Kimura, S. Mizushima, F. Funada, and M. Hijikigawa, Influence of rough surface on the optical characteristics of reflective LCD with a polarizer, *SID Tech. Digest* **29**, 221 (1998).

29 T. Uesaka, T. Toyooka, and Y. Kobori, Wide-viewing-angle reflective TN-LCD with single polarizer and hybrid aligned nematic compensation films, *SID Tech. Digest* **30**, 94 (1999).

30 C. L. Kuo, C. L. Chen, D. L. Ting, C. K. Wei, C. K. Hsu, B. J. Liao, B. D. Liu, C. W. Hao, and S. T. Wu, A 10.4″ reflective MTN-mode TFT-LCD with video rate and full-color capabilities, *SID Tech. Digest* **28**, 79 (1997); also *J. SID* **7**, 109 (1998).

31 M. D. Tillin, M. J. Towler, K. A. Saynor, and E. J. Beynon, Reflective single-polarizer low and high twist LCDs, *SID Tech. Digest* **29**, 311 (1998).

32 H. S. Kwok, J. Chen, F. H. Yu, and H. C. Huang, Generalized mixed-mode reflective LCDs with large cell gaps and high contrast, *J. SID* **7**, 127 (1999).

33 T. Sonehara, Photo-addressed liquid crystal SLM with a twisted nematic ECB mode, *Jpn. J. Appl. Phys.* **29**, L1231 (1990).

34 H. A. Van Sprang, Reflective liquid crystal display device with twist angle between 50 and 68° and the polarizer at the bisectrix, US patent 5,490,003 (1996).

35 K. H. Yang, A self-compensated twisted-nematic liquid crystal mode for reflective light valves, *Euro Display'96*, p. 449 (1996).

36 K. H. Yang and M. Lu, Nematic liquid crystal modes for Si wafer-based reflective spatial light modulators, *Displays* **20**, 211 (1999).

37 J. Grinberg, A. Jacobson, W. Bleha, L. Miller, I. Fraas, D. Bosewell, and G. Myer, A new real-time noncoherent to coherent image converter: the hybrid field effect liquid crystal light valve, *Opt. Eng.* **14**, 217 (1975).

38 F. H. Yu, J. Chen, S. T. Tang, and H. S. Kwok, Reflective twisted nematic liquid crystal displays. II. Elimination of retardation film and rear polarizer, *J. Appl. Phys.* **82**, 5287 (1997).

39 J. Chen, F. H. Yu, S. T. Tang, H. C. Huang, and H. S. Kwok, New optimized reflective LCD modes for direct-view and projection displays, *SID Tech. Digest* **28**, 639 (1997).

40 Y. Saitoh, Y. Yoshida, and H. Kamiya, Reflective twisted-nematic mode color TFT-LCD panel, *SID Tech. Digest* **28**, 651 (1997).

41 E. Beynon, K. Saynor, M. Tillin, and M. Towler, Single-polarizer reflective twisted nematics, *J. SID* **7**, 71 (1999).

5

Super-twisted Nematic Displays

1 Introduction

A super-twisted nematic (STN) cell is defined as one in which the LC twist angle is greater than 90°. In order to make the twist angle larger than 90°, a small percentage of chiral agent needs to be doped into the LC mixture. The ratio of the cell gap d and pitch length p determines the twist angle $\phi : d/p = \phi/2\pi$. For example, to make a 240° STN cell, the d/p ratio should be 0.667. If the cell gap is intended to be $d = 6\,\mu m$, then the pitch length should be $9\,\mu m$. The pitch length of a chiral dopant is inversely proportional to its concentration (C in weight %) as $p = \zeta/C$. For a commonly used Merck chiral dopant ZLI-811, the proportionality constant $\zeta = 5\,\mu m$ in the ZLI-1132 host [1]. So, in order to obtain a $9\,\mu m$ pitch length using ZLI-811, its concentration needs to be 0.55%. Another common chiral agent CB-15 has $\zeta \simeq 12\,\mu m$.

A STN display is known to exhibit a much steeper electro-optic curve than TN so that the simple low-cost passive matrix addressing method can be applied [2]. A sharp EO curve enables more rows to be multiplexed. The duty ratio is often referred to as $1/N$, where N is the number of multiplexed rows. The sharpness of the EO curve is determined by the twist angle, pretilt angle, d/p ratio, bend/splay elastic constant ratio (K_{33}/K_{11}), twist/splay elastic constant ratio (K_{22}/K_{11}), and dielectric parameter $\gamma = (\varepsilon_\parallel - \varepsilon_\perp)/\varepsilon_\perp$ [3]. Although a higher twist leads to a steeper EO curve and, in turn, a larger number of multiplexing rows, it makes gray scale difficult to achieve. Conversely, a shallow EO curve, e.g. a TN, is ideal for gray scale applications except that its multiplexing number is too low. As a result, the most commonly used STN cells have twist angles ranging from 210 to 240°.

A typical transmission-type 210° STN display has $d\Delta n \simeq 0.9\,\mu m$ [4]. This is nearly twice the 90° TN cell ($d\Delta n \simeq 0.48\,\mu m$). If an LC mixture has $\Delta n = 0.15$, then the cell gap would be $d = 6\,\mu m$. A typical response time of a STN display is around $100{-}200\,ms$ which is too slow for video applications. The active addressing (AA) method has been developed to improve the response time and contrast ratio [5], however the cost of driving electronics is increased. The overall performance of the AA-STN is still lying between the passive-matrix STN and TFT-LCDs. As the manufacturing cost of TFT-LCDs continues to fall, the cost advantage of a STN display is gradually narrowed. Eventually, TFT-LCD is likely to dominate the high-end display markets and STN will find applications in portable displays, such as cellular phones, palm pilots, electronic books, and other personal digital assistant products. In these applications, low-power reflective and trans-reflective displays are highly desirable.

2 Reflective STN

Reflective STN displays with twist angles between 180 and 240° have found useful applications for modest and high-information content displays. A simple 180° STN exhibits an excellent black and white display with high contrast ratio. In addition, its cell gap tolerance is large. The major problem is that its multiplexing number is low. On the other hand, the 240° STN needs one or two phase compensation films, but its sharp EO curve is particularly suitable for high multiplexing displays.

2.1 Jones matrix

A general reflective STN display is depicted in Figure 5.1. It consists of a polariser, one or two phase retardation films, and a twisted cell with a reflector deposited in the inner side of the rear substrate. For simplicity, the front rubbing is taken to be in the horizontal direction. The axis of the polariser is at an angle β with respect to the front LC directors. In the crossed-polariser condition, the phase retardation film is a quarter-wave film, and its axis is

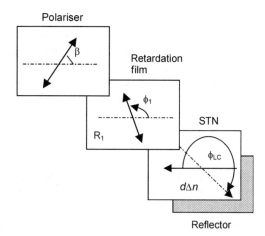

Figure 5.1 A general STN display employing a polariser and one phase compensation film

at 45° to the linear polariser. At null-voltage, the normalised reflectance of such a STN cell can be calculated by the Jones matrix method:

$$R_\perp = \left(\Gamma \frac{\sin X}{X}\right)^2 \left(\sin 2\beta \cos X - \frac{\phi}{X} \cos 2\beta \sin X\right)^2 \tag{5.1}$$

Here $\Gamma = 2\pi d\Delta n/\lambda$, d is the cell gap, $X = [\phi^2 + (\Gamma/2)^2]^{1/2}$ and ϕ is the twist angle. Using Equation 5.1, one should be able to find the optimal cell parameters (e.g. $d\Delta n$ and β) for a given twist cell and then calculate its voltage-dependent reflectance.

2.2 Special cases

When $\beta=0$, Equation 5.1 has the following special solutions [6, 7]:

$$\phi = \frac{(2m-1)\pi}{2\sqrt{2}} \tag{5.2}$$

$$d\Delta n = \frac{(2m-1)\lambda}{2\sqrt{2}} \tag{5.3}$$

Here, m is a positive integer. For the lowest three orders, $m=1$, 2 and 3, the $(\phi, d\Delta n)$ pair solutions are (63.6°, 194.5 nm), (190.8°, 583.4 nm) and (318.2°, 972.3 nm), respectively, assuming $\lambda=550$ nm. However, for computer simulations, the $\beta=0$ constraint can be removed and more options can be found.

2.3 Computer simulations

In the computer simulations presented in the following sections, unless specified, a Merck low viscosity mixture MLC-6297-000 is used. The parameters are as follows: $n_e = 1.613$, $n_o = 1.487$, $\varepsilon_\| = 10.5$, $\varepsilon_\perp = 3.6$, $K_{11} = 13.4$, $K_{22} = 5.4$, $K_{33} = 19.0$ pN, and clearing point 98°C. The pretilt angle of the cell is assumed to be $\alpha=6°$. We have chosen $\phi=180°$, 210° and 240° as examples. For the purpose of comparing cell performances, the optical losses from substrates, ITO and reflector (aluminum) are all ignored. Only the normalised reflectance is compared.

3 Reflective 180° STN

3.1 Normalised reflectance

The normalised reflectance of an 180° STN cell is plotted in Figure 5.2 for $\beta= -20$, -10, 0, 10 and 20°. It is found that when $-10 \leq \beta \leq 20°$, the null-voltage state has $R_\perp$ close to 100%. Thus, the determining factor is how to obtain the lowest dark state voltage. At $\beta=0°$,

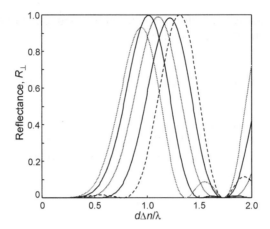

Figure 5.2 The normalised reflectance $R_\perp$ of an $180°$ STN cell. From left to right, $\beta = -20$, -10, 0, 10 and $20°$, respectively

the required $d\Delta n$ for achieving unit reflectivity is about $592\,\text{nm}$ which is lower than the corresponding transmissive STN cell. Since MLC-6297-000 has $\Delta n \simeq 0.126$ at $\lambda = 550\,\text{nm}$, this corresponds to a cell gap of $d \simeq 4.7\,\mu\text{m}$.

From Figure 5.2, there exists a common dark state ($R_\perp = 0$) at $d\Delta n/\lambda \simeq 1.7$, independent of the β angle. The exact $d\Delta n/\lambda$ value can be solved numerically by setting $R_\perp$ in Equation 5.1 equal to 0. An obvious solution is that $\sin X = 0$. This corresponds to $X = m\pi$, where $m = 1$, 2, etc. As the twist angle $\phi = \pi$, the lowest possible m is 2 and $d\Delta n/\lambda = \sqrt{3}$ is found. In principle, this cell design should lead to a normally black display. However, this normally black mode is too sensitive to the wavelength variation. The $d\Delta n/\lambda$ window to obtain the dark state is quite narrow, as shown in Figure 5.2. This means that a good dark state might be obtained for the green band, but the light leakage for the red and blue bands will be severe. In addition, the required $d\Delta n/\lambda$ is too large. A thick LC layer would result in a slow response time. Thus, this normally black mode is not favorable for full-color displays.

3.2 Electro-optic effects

The voltage-dependent reflectance of the $180°$ STN cell is shown in Figure 5.3 for the three primary colors, $R = 650$, $G = 550$ and $B = 450\,\text{nm}$. The $d\Delta n$ ($=592\,\text{nm}$) and β ($=0°$) values of the cell are designed so that in the off-state the peak reflectivity occurs at $\lambda = 550\,\text{nm}$. In this case, both blue and red have a lower reflectance in the voltage-off state. As shown in Figure 5.2, the blue wavelength is located at the right-hand side of the peak. This means that its effective phase retardation is too large. As the applied voltage exceeds the threshold, the LC directors are reoriented by the electric field, causing the effective Δn to reduce. As a result, the reflectivity increases and then decays owing to the vanishing Δn at high voltages. For the red band, its $d\Delta n/\lambda$ is insufficient to reach unit reflectivity even in the voltage-off state. As V increases, $R_\perp$ decreases monotonically. The dispersion of such a cell is relatively large. This is because the reflective STN also utilises the combination of polarisation rotation and birefringence effects.

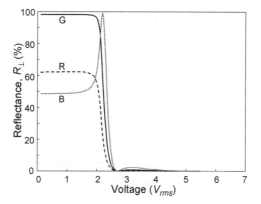

Figure 5.3 The voltage-dependent reflectance of the 180° STN. $d\Delta n = 592$ nm and $\beta = 0°$

The beta angle also has an important effect on the dark state voltage. Figure 5.4 shows the voltage-dependent reflectance of the 180° STN cell at $\beta = 3°$. The small bumps observed in Figure 5.3 in the high voltage regime are suppressed totally. The contrast ratio measured using an Argon laser line at $\lambda = 514.5$ nm exceeds 2000 : 1.

3.3 Cell gap tolerance

Traditional STN displays have a very tight cell gap tolerance. The 180° STN exhibits a reasonably large cell gap tolerance. Figure 5.5 shows the $V-R$ curves of the 180° STN cells with $d = 4.4$, 4.7 and 5.0 μm. The beta angle used is $\beta = 3°$. As the cell gap changes, only the bright state reflectance is slightly affected. The dark state reflectance basically remains unchanged. A large cell gap tolerance is essential to enhance the manufacturing yield.

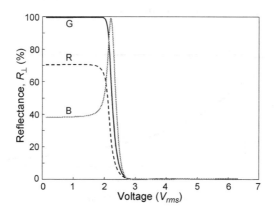

Figure 5.4 The voltage-dependent reflectance of the 180° STN. $d\Delta n = 592$ nm and $\beta = 3°$

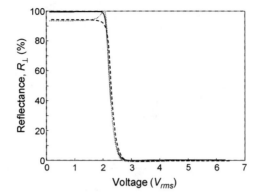

Figure 5.5 Cell gap tolerance of the 180° STN cell. Dashed, solid and gray lines are for $d=4.4$, 4.7 and 5.0 μm, respectively. $\beta=3°$

3.4 Duty ratio

From the Alt–Pleshko theory of multiplexing [8], the number of multiplexing lines (N) is determined by the selection ratio of the on- and off-state voltages as

$$\frac{V_{on}}{V_{off}} = \left(\frac{\sqrt{N}+1}{\sqrt{N}-1}\right)^{1/2} \tag{5.4}$$

A graph of Equation 5.4 is shown in Figure 5.6 for convenience. In order to have $N > 200$ for a high-information content STN display, the voltage selection ratio should be less than 1.073.

From Figure 5.3, the 180° STN cell has a bright state voltage at $V_{90} \approx 2.15\,V_{rms}$, and a dark state voltage at $V_{10} \approx 2.4\,V_{rms}$. Thus, the on/off ratio is 1.12. This corresponds to $N \simeq 90$. By choosing a high K_{33}/K_{11} ratio LC material, the duty ratio might be pushed to $1/100$. Thus, the 180° STN is attractive for medium duty ratio application. For a higher duty ratio, a larger twist angle needs to be considered.

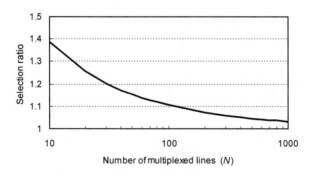

Figure 5.6 The relationship between voltage selection ratio and multiplexing lines

4 Reflective 210° STN

Let us continue to examine the reflective 210° STN cell. Again, Equation 5.1 is used to obtain optimal cell parameters. Figure 5.7 shows the $d\Delta n/\lambda$ dependent normalised reflectance at $\beta = -10$, 0, 10 and 20°. From Figure 5.7, the maximum reflectance that can be obtained is $R_\perp \simeq 96\%$ which occurs at $\beta = 10°$ and $d\Delta n \simeq 630$ nm. Let us continue to use the MLC-6297-000 LC mixture as an example; a 5 μm cell gap would satisfy this requirement.

The voltage-dependent reflectance of such a 210° STN cell is shown in Figure 5.8. This normally white mode has a bright state voltage at about 2.35 V_{rms} and a dark state voltage at about 2.50 V_{rms}. Thus, the voltage selection ratio is about 1.064. According to the chart plotted in Figure 5.6, the number of multiplexing lines is $N \simeq 250$. Similar to the 180° STN results shown in Figure 5.5, the 210° STN also has a large cell gap tolerance.

5 Reflective 240° STN

The constraint of Equation 5.1 is on the use of a linear polariser and a quarter-wave film. This simple structure can only be extended to 210° STN. For the 240° STN, the $d\Delta n/\lambda$ needs to be as large as 1.9. Moreover, the color dispersion in the voltage off state is too large and a common dark state for all three colors is difficult to obtain. Therefore, a more complicated cell design needs to be considered. Two approaches involving one or two phase compensation films on the 240° STN displays have shown promise and are described in the following sections.

5.1 Single film approach

The device configuration is similar to that shown in Figure 5.1, except for different parameters: $\phi_{LC} = 240°$, $\beta = 9°$, $d\Delta n$ of the LC is 790 nm, $d\Delta n$ of the compensation film is

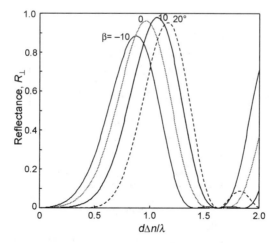

Figure 5.7 Normalised reflectivity of the 210° STN cell as a function of $d\Delta n/\lambda$ and β. From left to right, $\beta = -10$, 0, 10 and 20°, respectively

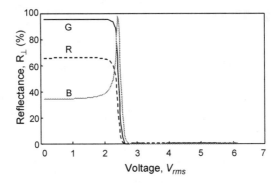

Figure 5.8 The voltage-dependent reflectance of the 210° STN cell. LC cell: $d\Delta n = 630$ nm, pretilt angle $\alpha = 7°$ and $\beta = 10°$

$R_1 = 400$ nm, and the orientation angle of the compensation film is $\phi_1 = 98°$ [9]. In these circumstances, a normally white display mode is obtained. Results on the wavelength-dependent reflectance are shown in Figure 5.9. The reflectance in the 400–600 nm region is reasonably achromatic. The dark state depends on the quality of the polariser employed. By choosing a phase matched compensation film, a contrast ratio of about 20 : 1 can be achieved.

5.2 Double film approach

The device configuration is shown in Figure 5.10. A single polariser is used. The double phase retardation films are arranged at a certain angle to reduce color dispersion [10]. However, no information on the film thickness, orientation angle or LC twist angle was given [11, 12].

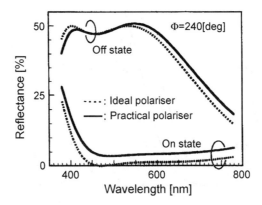

Figure 5.9 The wavelength-dependent light reflectance of a 240° STN cell. Solid lines are results using practical polarisers, and dashed lines are for the ideal polarisers (after [9], reproduced by permission of SID)

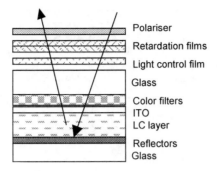

Figure 5.10 Device configuration of a Mutsushita 240° STN display employing two phase retardation films

The voltage-dependent light reflectance of such an STN panel is depicted in Figure 5.11. This normally black mode has a threshold voltage at 2.3 V_{rms} and full-on voltage at 2.4 V_{rms}. From Figure 5.3, the number of multiplexing lines that can be achieved exceeds $N=480$. A display with VGA resolution, 4096 colors, 1/240 duty ratio, 15% reflectance, 14:1 contrast ratio, 250 ms response time and 89 mW power consumption has been demonstrated [13].

5.3 Twisted film-compensated STN

The phase compensation films discussed above are all uniaxial. The double STN cell [14] provides an effective way to achieve a high contrast ratio display, except that two STN cells with reversed twist need to be used. As a result, the display cost and weight are nearly doubled. Replacing the compensation LC cell with a polymeric film would reduce the device weight substantially [15]. In particular, a twisted liquid crystal polymer offers a significant advantage for compensating temperature variations.

The phase compensation film is normally designed for a given operation temperature. As the temperature increases, the birefringence of a LC mixture decreases. In the meantime, the birefringence of a thermally cured phase compensation film, such as polycarbonate, is rather

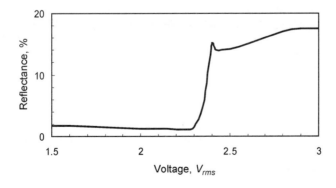

Figure 5.11 Voltage-dependent light reflectance of a Mutsushita 240° STN cell (redrawn from [13])

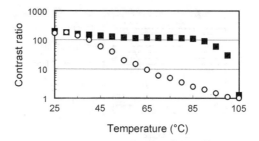

Figure 5.12 Temperature-dependent contrast ratio of an STN display compensated with a temperature-matched twisted film (squares) and a non-temperature matched twisted film (circles) (redrawn from [16])

insensitive to temperature. As a result, a good phase match can only be preserved within a limited temperature range. A phase compensation film made of twisted liquid crystal polymer (LCP), showing a similar temperature dependence as LC mixtures, has been developed [16]. Figure 5.12 shows the contrast ratios of a 240° STN cell using a temperature-matched −240° twisted LCP film and a temperature mismatched phase compensation film. The clearing temperature of the LCP is 110°C. With the mismatched film, the contrast ratio drops rapidly as the temperature increases. On the other hand, the temperature-matched LCP film preserves good contrast ratio up to 90°C.

An ideal LCP film should have a similar $d\Delta n$, reversed twist angle, and similar temperature and wavelength dependencies to the LC employed. To select a temperature-matched LCP film, the clearing point of the film should be close to that of the LC mixture. Moreover, to match the wavelength dispersions for color displays, the molecular structures and conjugation length of the film and LC mixture should be as similar as possible.

References

1   Merck KgaA, Darmstadt, Germany.
2   T. J. Scheffer, Liquid crystal display with high multiplex rate and wide viewing angle, *Proc. 3rd Intl. Display Research Conf.* pp. 400–403 (1983).
3   E. P. Raynes, The theory of supertwist transitions, *Mol. Cryst. Liq. Cryst. Letters* **4**, 1 (1986).
4   M. Akatsuka, K. Katoh, K. Sawada, and M. Nakayama, Electro-optical properties of supertwisted nematic display obtained by rubbing technique, *Proc. SID* **28**, 159 (1987).
5   T. J. Scheffer and B. Clifton, Active addressing method for high contrast video rate STN displays, *SID Tech. Digest* **23**, 228 (1992).
6   T. Sonehara and O. Okumura, *Japan Display'89*, p. 192 (1989).
7   H. S. Kwok, Parameter space representation of liquid crystal display operation modes, *J. Appl. Phys.* **80**, 3687 (1996).
8   P. M. Alt and P. Pleshko, Scanning limitations of liquid crystal displays, *IEEE Trans. Electron Devices ED-21*, p. 146 (1974).
9   I. Fukuda, E. Sakai, Y. Kotani, M. Kitamura, and T. Uchida, A new achromatic reflective STN-LCD with one polarizer and one retardation film, *J. SID* **3**, 83 (1995).
10 S. Komura, O. Itou, K. Kuwabara, K. Funahata, K. Kondo, and K. Kubo, Improvement of color purity in reflective color STN-LCDs, *Proc. The fifth Intl. Display Workshops*, pp. 217–220 (1998).

11 H. Yamaguchi, S. Fujita, N. Naito, H. Mizuno, T. Otani, T. Sekime, T. Ogawa, and N. Wakita, A reflective color STN-LCD with a single polarizer and double retardation films, *SID Tech. Digest* **28**, 647 (1997).

12 S. Fujita, H. Mizuno, M. Nishinaka, H. Yamaguchi, T. Ohtani, T. Hatanaka, and T. Ogawa, Reflective color STN-LCD technologies, *Proc. The fifth Intl. Display Workshops*, pp. 213–216 (1998).

13 S. Fujita, H. Yamaguchi, H. Mizuno, T. Ohtani, T. Sekime, T. Hatanaka, and T. Ogawa, A reflective color STN-LCD with a single polarizer and double retardation films, *J. SID* **7**, 135 (1999).

14 K. Katoh, Y. Endo, M. Akatsuka, M. Ohgawara, and K. Sawada, Application of retardation compensation; a new highly multiplexable black-white liquid crystal display with two super-twisted nematic layers, *Jpn. J. Appl. Phys.* **26**, L1784 (1987).

15 Y. Takiguchi, A. Kanemoto, H. Iimura, T. Enomoto, S. Iida, T. Toyooka, H. Ito, and H. Hara, Achromatic super twisted nematic LCD having a polymeric liquid crystal compensator, *Proc. 10th Int'l Display Research Conf.* pp.96–99 (1990).

16 M. Bosma, Optimized film-compensation of STN displays, *The 5th Int'l Display Workshops* (*Kobe, Japan*), p. 239 (1998).

6

Guest–Host Displays

1 Introduction

The guest–host (GH) display is one of the few liquid crystal technologies that does not require a polariser. Because no polariser is involved, the guest–host display has the potential to achieve 60% reflectivity and a wide viewing angle. However, without a polariser, its contrast ratio is limited to 5–10:1, similar to that of newspapers. A low contrast ratio implies a poor color rendering. Thus, the guest–host display is more suitable for graphic applications. For video applications, a contrast ratio higher than 20:1 is required.

The underlying light modulation mechanism for a guest–host display is voltage-dependent light absorption [1, 2]. In a guest–host cell, the non-absorbing LC host is doped with 1–5% of guest dichroic dyes. These rod-like dye molecules have large absorption anisotropy in the visible spectral region. When these dyes are aligned in the LC host, they absorb the incident light with its polarisation parallel to the principal molecular axis and transmit the perpendicular. Therefore, by controlling the applied voltage to the guest–host cell, the outgoing beam is modulated accordingly.

The guest–host cell can be configured in reflective and transmissive displays [3]. In a reflective display, the incident light traverses the guest–host medium twice. Thus, its cell gap can be reduced to one-half of the transmissive one, if the dye concentration remains unchanged. As a result, the response time is four times faster.

Dichroic dyes and device configuration play key roles in determining the performance of a guest–host display. In this chapter we briefly review the material properties of dichroic dyes and then discuss the device structures and their corresponding light modulation mechanisms.

2 Guest–Host Medium

The guest–host medium consists of the host LC and the guest dichroic dye. The LC alignment has a great influence on the orientation of dye molecules which, in turn, determines the display brightness and contrast ratio. For a dichroic dye, clear color, high dichroic ratio, high solubility, good electric character and good stability are the key performance parameters.

2.1 LC alignment

The dielectric anisotropy ($\Delta\varepsilon$) of the host LC determines what type of LC alignment should be used for a guest–host display. When $\Delta\varepsilon > 0$, both homogeneous and twisted cells can be considered. In this mode of operation, strong absorption occurs in the voltage-off state so that the display appears dark. When a sufficiently high voltage is applied, both LC and dye molecules are reoriented along the direction of the electric field. As a result, the absorption is reduced and the display looks bright.

In contrast, if the host LC has a negative $\Delta\varepsilon$, then the preferred alignment is homeotropic. In these circumstances, the dye molecules are aligned perpendicular to the substrate surfaces in the voltage-off state. The absorption is minimal and the display appears bright. As the applied voltage exceeds the threshold, the LC directors start to tilt and the absorption gradually increases, leading to a dark state.

Several specific guest–host display modes have been developed to accept unpolarised ambient light. These device configurations will be discussed in detail later.

2.2 Dichroic ratio

For simplicity while not losing generality, we use a transmissive cell for theoretical analysis. The normalised transmittance (ignoring the reflection and absorption losses from substrate and reflector) of a homogeneous guest–host cell can be described as follows:

$$T_{\parallel} = \exp\left(-\alpha_{\parallel} cd\right) \tag{6.1}$$

$$T_{\perp} = \exp\left(-\alpha_{\perp} cd\right) \tag{6.2}$$

$$\alpha_{\parallel} = (2S + 1)\alpha_o \tag{6.3}$$

$$\alpha_{\perp} = (1 - S)\alpha_o \tag{6.4}$$

Here, $T_{\parallel,\perp}$ and $\alpha_{\parallel,\perp}$ represent the transmittance and absorption coefficients of the GH cell for the extraordinary and ordinary rays, respectively, α_o is the absorption coefficient of the dye molecules in an isotropic host, S is the order parameter, c is the dye concentration and d is the cell gap. The dichroic ratio (DR) is a measure of the dye's absorption characteristics in a LC host. The dichroic ratio is defined as

$$DR = A_{\parallel}/A_{\perp} = (1 + 2S)/(1 - S) \tag{6.5}$$

Here, $A_{\parallel} = \alpha_{\parallel} cd$ and $A_{\perp} = \alpha_{\perp} cd$ are the absorbances parallel to and perpendicular to the LC directors, respectively. Note that the optical density $OD_{\parallel,\perp} = -\log(T_{\parallel,\perp})$ is different from absorbance by a conversion factor $OD_{\parallel,\perp} = A_{\parallel,\perp}/2.3$. However, the ratios $OD_{\parallel}/OD_{\perp}$ and $A_{\parallel}/A_{\perp}$ remain the same. From Equation 6.5, the order parameter has an important effect on the dichroic ratio of the dyes [4].

The display contrast ratio (CR) is different from the dichroic ratio of the dye. The contrast ratio is defined by the transmittance (or reflectance) ratio at the voltage on and off states, depending on whether the normally white or normally black mode is considered. For a normally white mode employing a homeotropic alignment, the bright state occurs at $V = 0$ and the dark state depends on the applied voltage. On the other hand, for a normally black mode employing a homogeneous cell, the dark state happens at $V = 0$ and the bright state depends on the applied voltage. To fully reorient the LC directors and dyes requires a relatively high voltage (about $30\,V_{rms}$) which is usually not achievable by TFT (less than $10\,V_{rms}$). Thus, the normally white mode is used to obtain high brightness with trade-off in contrast ratio, and normally a black mode is designed to trade high contrast ratio with brightness.

In addition to the dichroic ratio, the display contrast ratio is also affected by the cell gap and dye concentration. For a given cell thickness, a higher dye concentration leads to a higher contrast ratio, but the corresponding transmittance (or reflectance) is reduced and the response times are increased. Many dichroic dyes have been developed for displays in the visible region. Due to the long conjugation of dye molecules, their viscosity is usually much higher than that of the host LC. For example, adding merely 1% of black dye into ZLI-2806 increases the flow viscosity from 57 to $70\,mm^2/s$ at 20°C. The viscosity of the LC–dye system increases in proportion to the dye concentration.

2.3 Absorption wavelengths

The absorption wavelength and bandwidth of dye molecules determine the color of a guest–host display. The absorption peaks of a dye depend on its molecular structure. In general, a linearly conjugated dye has a longer absorption wavelength and a larger dischroic ratio. For a black and white display, a black dye mixed from red, green and blue dyes is needed. For a dye that absorbs strongly in red, its appearance color is blue. Among the RGB dyes, red dye has the longest molecular conjugation. Owing to the bulky structure, the red dye usually has the highest viscosity.

Figure 6.1 shows the polarised absorption spectra of a Merck D2 azo dye with the structure shown below:

Its peak absorption in E7 LC host ($T_c \simeq 60$°C) is 495 nm, the dichroic ratio is about 10, and order parameter 0.75. Due to guest–host interactions, the absorption wavelengths, dichroic ratio and order parameter of a dye could vary from host to host.

Many dyes with absorption spectra covering the whole visible range have been developed by Merck, Nippon Kayaku, Atomergic Chemicals, Mitsui Toatsu, etc. An extensive list can be found in Bahadur's book [5]. Table 6.1 lists the dichroic ratio and absorption peak of a

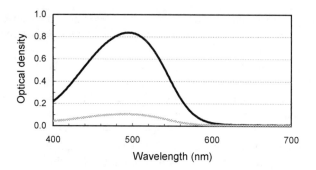

Figure 6.1   Absorption spectra of Merck D2 dye. Concentration = 1% in E7 host

Table 6.1   Dichroic ratio (DR), absorption wavelength (λ_{max}, in nm), order parameter (S) and solubility (at 25°C) of some selected Merck anthraquinone dyes in E7 LC host

Dye no.	R_1	R_2	λ_{max}	DR	S	Solubility
D5	C_4H_9		594	5.5	0.56	>5%
D16	OC_9H_{19}		596	6.6	0.63	2.2%
D27	$N(CH_3)_2$		612	6.1	0.63	0.8%
D35		C_2H_5	554	7.1	0.67	1.7%
D37		C_4H_9	556	6.8	0.66	>5%
D43		OC_5H_{11}	557	7.4	0.68	0.4%
D52		$N(CH_3)_2$	546	6.6	0.65	0.8%
D77		$CH(CH_3)_2$	558	7.4	0.68	1.6%

few Merck anthraquinone dichroic dyes, with structures shown below.

These dyes have very good photo and thermal stability, but their dichroic ratio is less than 10 : 1 and their viscosity is relatively large.

To improve the dichroic ratios, linearly conjugated dyes are desirable. Table 6.2 lists some high dichroic ratio Mitsui dyes in three LC hosts: E8, ZLI-1840 and ZLI-1565 [6]. In particular, the SI-252 dye exhibits a surprisingly high dichroic ratio (20.5) in the ZLI-1840 host. In addition to the detailed molecular interactions, the clearing temperature T_c of the host also makes an important contribution to the dichroic ratio and order parameter.

For the host LC alone, the order parameter can be approximated as [7]:

$$S = (1 - T/T_c)^{\beta} \tag{6.6}$$

Table 6.2 Dichroic ratio and absorption wavelength (λ_{max}, in nm) of some selected Mitsui dyes in three LC hosts: E8, ZLI-1840, and ZLI-1565

Dye no.	Dichroic ratio (λ_{max})		
	in E8	in ZLI-1840	in ZLI-1565
SI-209	–	10.0 (450)	9.7 (440)
SI-254	–	16.4 (472)	6.6 (462)
M-524	10.3 (577)	12.3 (576)	9.4 (575)
SI-252	17.2 (588)	20.5 (572)	13.9 (570)
M-137	10.4 (641)	11.0 (639)	9.8 (636)
M-141	10.2 (641)	9.9 (639)	–
M-483	9.2 (643)	10.4 (640)	8.7 (637)
M-412	6.0 (640)	7.3 (639)	–
M-403	11.8 (685)	10.3 (680)	–
M-406	10.5 (685)	10.6 (684)	–

Here T is the operation temperature, and $\beta \simeq 0.25$ can be treated as a constant. From Equation 6.6, a high T_c host would contribute to a large dichroic ratio. The clearing points of E8, ZLI-1840 and ZLI-1565 are 70, 90 and 85°C, respectively. Thus, the slightly lower dichroic ratio in E8 as compared to ZLI-1840 could result from the clearing point difference. However, some substantially lower dichroic ratios, e.g. SI-254 and SI-252, in the ZLI-1565 host cannot be explained by the temperature factor alone. The detailed guest–host interaction has to be taken into consideration.

From Figure 6.1, the absorption bandwidth (full width half maximum) of the D2 dye is about 120 nm. To cover the whole visible spectrum (400 to 700 nm), two or three types of dye need to be mixed together. For example, the commercially available LC-dye mixture, ZLI-3094 (Merck) consists of 1% black dye in the ZLI-2806 LC host. The host mixture is optically transparent in the whole visible region. The dielectric anisotropy of ZLI-3094 is −4.8. The polarisation-dependent absorption spectra of a 10 μm thick ZLI-3094 LC cell are shown in Figure 6.2. The measured dichroic ratio is 9.2 and the order parameter is 0.73.

As shown in Figure 6.2, the black dye does cover the whole visible spectrum although a smaller dichroic ratio is observed near both ends. The perpendicular-state absorption in Figure 6.2 represents the optical loss and should be minimised. A more linear rod-like

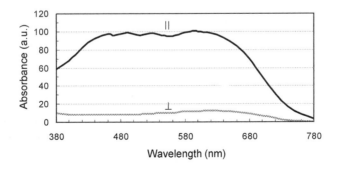

Figure 6.2 The measured absorption spectra of a 10 μm thick ZLI-3094 LC layer. The black dye concentration is 1%. Dichroic ratio DR = 9.2 and order parameter $S = 0.73$

molecule would exhibit a larger dichroic ratio. Such dye molecules are highly desirable from the viewpoint of improving the dichroic ratio and minimising absorption loss.

2.4 Solubility

The solubility of a dye is determined by its melting temperature and heat fusion enthalpy (ΔH). Owing to the relatively long conjugation of a dye molecule, its melting temperature is usually higher than 80°. Low melting point and small ΔH are equally important for improving the dye solubility.

2.5 Viscosity

The viscosity of a dye affects the response time of the display device. The key factors that determine the dye viscosity are the molecular shape and moment of inertia. A linear dye is more favorable for a large dichroic ratio and low viscosity. The anthraquinone dye, as expected, exhibits a relatively large viscosity. The dye molecules are usually bulkier than the LC so that their viscosity is much larger. For example, adding merely 1% of anthraquinone dye to a biphenyl LC host could cause a significant increase in response time.

2.6 Material stability

Since the guest–host cell is constantly driven by an electric field and exposed to ambient light, an ideal dye should possess a good electrochemical, photochemical and thermal stability [8]. In a TFT-LCD, a high resistivity LC is required to achieve a high voltage holding ratio. In the guest–host cell, if the dyes employed contain enough ions or impurities [9, 10], the overall resistivity could be reduced significantly and image flickering would occur. The most commonly employed dyes are of the azo biphenyl type. These azo dyes have adequate photo and thermal stability. For outdoor applications, an extra UV-protecting layer is always helpful.

3 Liquid Crystal Dyes

Some LC dyes exhibit solubility higher than 20%. Three series of nitro-amino dyes (biphenyl azo, tolane and diphenyldiacetylene) are discussed here.

3.1 Nitro-amino biphenyl azo dyes

These LC dyes possess a relatively low enthalpy of fusion and a low melting point (about 90°C). Thus, an excellent solubility to many LC hosts is achieved. In addition, the dielectric anisotropy $\Delta\varepsilon$ of these LC dyes is huge due to the presence of a strong on-axis dipole moment. Adding a few percent of such a dye into a LC matrix greatly enhances the birefringence Δn and lowers the threshold voltage of the guest–host mixture. However, the viscosity of the guest–host system also increases in proportion to the dye concentration.

Phase transitions

The structure of the nitro-amino biphenyl LC dyes [11] is shown below:

$$H_{2n+1}C_n—N(H)—\text{(phenyl)}—N=N—\text{(phenyl)}—NO_2$$

The phase transition temperatures of the C4, C5 and C6 homologues are listed below:

C4:   K $\underset{104.0\ °C}{\overset{109.1\ °C}{\rightleftharpoons}}$ I

C5:   K $\underset{74.0}{\overset{88.5\ °C}{\rightleftharpoons}}$ N $\underset{78.5}{\rightleftharpoons}$ I

C6:   K $\underset{69.0}{\overset{92.9\ °C}{\rightleftharpoons}}$ N $\underset{81.3}{\rightleftharpoons}$ I

The symbols K, I and N stand for the crystalline, isotropic and nematic phases, respectively. The C5 and C6 homologues show the nematic phase during the cooling process. The heat fusion enthalpies ΔH of the C4, C5 and C6 homologues are 6.55, 5.79 and 7.82 kcal/mol, respectively. Among these three homologues, C5 exhibits the lowest melting point and the smallest ΔH. Thus, it is the most attractive LC dopant for practical applications. A binary mixture designated as M56, consisting of 57 wt% C5 and 43 wt% C6 homologues shows a regular nematic phase: $66.1 < T < 82.8°C$. The $7 \leq n \leq 14$ homologues exhibit a smectic-A phase; thus, they are less desirable when formulating nematic mixtures [12].

Absorption spectra

Figure 6.3 shows the polarised absorption spectra of the C5 azo dye. In this measurement, 1% of C5 was dissolved in E7 and a 6 μm homogeneous cell was used. Thus, the results shown in Figure 6.3 represent the optical density of a 0.06 μm C5 dye at a reduced

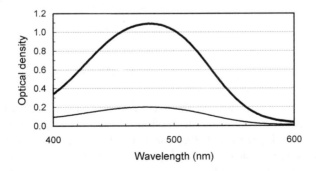

Figure 6.3 The measured absorption spectra of C5 biphenyl azo dye. Concentration = 1% in E7 host

temperature T_r ($= T/T_c$; T_c being the clearing point of the guest–host mixture). At $T = 23°C$ and $T_c = 60°C$ (E7 + 1% C5), T_r is found to be 0.868. In Figure 6.3, solid and dashed lines represent the optical density of a 0.06 μm C5 dye for the extraordinary and ordinary rays, respectively. The absorption of C5 peaks at 480 nm with an extraordinary molar extinction coefficient of 6.2×10^5. From Figure 6.3, the dichroic ratio of C5 is 5.5 at $\lambda = 480$ nm. This dichroic ratio is too low for display applications. For a guest–host display, a dichroic ratio greater than or equal to 10 is desirable. The low dichroic ratio of the azo dye is due to the bending at the double bond so that the molecule is not linear.

Dielectric anisotropy

The dielectric constants of a LC dye can be measured from the capacitance of a single parallel-aligned cell at low and high voltage regimes. The extrapolated dielectric anisotropy of M56 is about 37 at $T_r = 0.95$ (or $T = 65°C$). As T_r is decreased, e.g. at room temperature, $\Delta\varepsilon$ is estimated to be 55. The large $\Delta\varepsilon$ of these LC dyes results from their extraordinarily strong dipole moment due to the amino-nitro intra-molecular charge transfer. The NO_2 group acts as an acceptor that induces electron flow from the benzene ring. On the other hand, the $(C_nH_{2n+1})NH$ group acts as an electron donor which produces an opposite charge flow. This push-pull effect has been used frequently to design molecules for nonlinear optics applications. The large $\Delta\varepsilon$ makes nitro-amino dye useful as a dopant for reducing the operation voltage of a non-polar LC host.

3.2   Nitro-amino tolane dyes

The structure of the amino-nitro tolane dyes are shown below:

Here, n represents the number of side-chain carbon atoms. For simplicity, the tolane dyes are abbreviated as nNH-PTP-NO_2, where P stands for a phenyl ring, and T for the carbon–carbon triple bond.

Phase transitions

Four tolane dyes with $n = 4$–7 have been studied and their $[T_m, \Delta H]$ are [119.3, 6.27], [98.3, 4.69], [93.4, 8.82] and [87.5, 6.29], respectively [13]. Unlike the azo dyes, these tolane dyes exhibit no mesogenic phase. As the side chain length increases, the melting point gradually decreases. The odd-even effect on the transition temperature is not clearly observed. The C5 homologue exhibits the lowest enthalpy of fusion of the four homologues studied. Thus, its solubility should be the highest. The more linear structure of the tolane dyes means they should possess a lower viscosity and higher dichroic ratio than the corresponding azo dyes.

Dichroic ratio

The polarised absorption spectra and dichroic ratio of several dyes (1% concentration) were measured in the E63 and ZLI-4792 LC hosts (from Merck). The results are listed in Table 6.3. From Table 6.3, the absorption peak of the C5 tolane dye in E63 occurs at 417 nm with dichroic ratio 7.8. Replacing the acetylene with an azo linking group causes a significant red shift (477 nm) and a decreased dichroic ratio (6.7). This is because the molecular conjugation of the azo group slightly tilts away from the principal molecular axis. For the homologues with different side chain length, their λ_{max} remains basically unchanged. However, the dichroic ratio slightly increases and then saturates as the chain length increases. By simply replacing the NH group by a NCH$_3$, its λ_{max} shifts toward the red by about 20 nm. The trade-off is in the reduced dichroic ratio. Host ZLI-4792 is a high-resistivity low Δn mixture developed for active-matrix LC displays. It is less polar than E63 (mainly cyano-biphenyls). Thus, the λ_{max} in ZLI-4792 is shorter than that in E63. The dichroic ratio varies slightly between these two hosts.

Birefringence

The birefringence, threshold voltage V_{th}, dielectric anisotropy and visco-elastic coefficient (γ_1/K_{11}) of the guest–host systems are normally measured by the extrapolation method. The effect of the 5NH-PTP-NO$_2$ dye on the Δn and V_{th} of E63 is shown in Figure 6.4. In general,

Table 6.3 Absorption wavelength (λ_{max} in nm) and dichroic ratio (DR) of some nitro-amino tolane (PTP) and biphenyl azo (PNNP) dyes in host E63 and ZLI-4792. $T = 22°C$

Dyes	DR (λ_{max}) in E63	DR (λ_{max}) in ZLI-4792
1. 5NH-PTP-NO$_2$	7.8 (417)	7.4 (401)
2. 5NH-PNNP-NO$_2$	6.7 (477)	7.3 (451)
3. 7NH-PNNP-NO$_2$	6.9 (476)	
4. 9NH-PNNP-NO$_2$	7.6 (478)	
5. 11NH-PNNP-NO$_2$	7.6 (477)	
6. 5NCH$_3$-PNNP-NO$_2$	6.0 (498)	6.5 (485)
7. 6NCH$_3$-PNNP-NO$_2$	6.2 (498)	

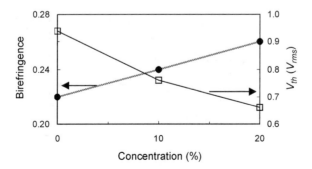

Figure 6.4 Dye concentration effect on Δn (open circles) and V_{th} (solid circles). Dye $=$ C5 tolane; LC host $=$ E63. $T = 22°C$ and $\lambda = 633$ nm

at a given wavelength and reduced temperature T_r, the Δn of a guest–host system can be approximated by $\Delta n_{gh} = x_g \Delta n_g + x_h \Delta n_h$; where $x_{g,h}$ (with $x_g + x_h = 1$) and $\Delta n_{g,h}$ are the concentration and birefringence of the guest and host, respectively. From Figure 6.4, the Δn of the tolane dye is extrapolated to be 0.42 at $T_r = 0.821$ and $\lambda = 633$ nm.

Dielectric anisotropy

The decreasing V_{th} implies that the dielectric anisotropy of 5NH-PTP-NO$_2$ dye is larger than that of E63. Nevertheless, this drop is not linear because V_{th} ($= \pi(K_{11}/\Delta\varepsilon)^{1/2}$) is not a linear function of $\Delta\varepsilon$. To estimate a dye's dielectric anisotropy, the dielectric constants of the guest–host mixture at different dye concentrations were measured. The results are shown on the right-hand side of Figure 6.5. From these data, the $\Delta\varepsilon$ of the C5 tolane dye is extrapolated to be about 60 at $T_r = 0.821$.

The large $\Delta\varepsilon$ of these tolane dyes results from their extraordinarily strong dipole moment originating from the nitro-amino intra-molecular charge transfer. The NO$_2$ group acts as an acceptor that induces electron flow from the benzene ring. On the other hand, the (C_nH_{2n+1})NH group acts as an electron donor which produces an opposite charge flow.

Viscosity

On the left-hand side of Figure 6.5, the concentration effect on the visco-elastic coefficient is depicted. Again, as the dye concentration increases, the γ_1/K_{11} of the guest–host system increases linearly. However, this increase is considered to be small compared to some commercial dyes, such as the anthraquinone type (Merck). From Figure 6.5, adding 10% of the tolane dye to the E63 LC host merely increases γ_1/K_{11} by 25%, which is about one order of magnitude smaller than that of 1% anthraquinone dye in the same host.

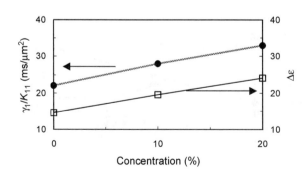

Figure 6.5 Dye concentration effect on $\Delta\varepsilon$ (right) and γ_1/K_{11} (left). Dye = C5 tolane, LC host = E63. $T = 22°C$

3.3 Nitro-amino diphenyldiacetylene dyes

Another series of nematic liquid crystal dyes is diphenyldiacetylene, as depicted below:

$$C_5H_{11}-N \overset{H}{\underset{}{|}} \diagram \equiv \equiv \diagram NO_2$$

Two homologues with $n = 5$ and 6 have been synthesised [14]. Their nematic ranges (in °C) are $127.8 \rightarrow 153.6$ and $131.0 \rightarrow 149.0$, respectively. In principle, these linearly conjugated dyes should have a longer absorption wavelength, higher birefringence, higher dielectric anisotropy and higher dichroic ratio than the corresponding tolanes. However, the UV and thermal stability of these nitro-amino diphenyldiacetylenes is insufficient, similar to the alkyl diphenyldiacetylenes discussed in Chapter 2. Therefore, these compounds are not suitable for long-term display operation.

4 Cell Configurations

Several guest–host cell configurations and dye structures have been investigated for practical applications. Extensive analyses of various operation modes and dye materials can be found in Bahadur's book [15]. In the following sections, we selectively describe the basic operation mechanisms of some commonly used guest–host displays, namely the Heilmeier–Zanoni cell, the White–Taylor cell, the Cole–Kashnow cell, double cells, three stacked cells and the amorphous chiral nematic cell.

4.1 Heilmeier–Zanoni cell

Heilmeier and Zanoni reported the first guest–host display in the late 1960s [16]. The Heilmeier–Zanoni cell utilises a polariser and an LC cell. The LC cell can be a homogeneous, homeotropic or a 90° twisted nematic cell [17, 18], depending on whether a normally black or normally white mode is planned.

Normally black mode

When a homogeneous cell is used for a guest–host display, in the voltage-off state as shown in Figure 6.6, the LC/dye directors are parallel to the polarisation axis of the incident light. In these circumstances, strong absorption occurs. This leads to the dark state of the display. As the applied voltage increases the LC directors, together with dye molecules, are reoriented along the electric field direction, resulting in a decreased absorption. This means that the reflectance is gradually increased as the applied voltage exceeds the threshold voltage. In a high voltage state, all the dye molecules are aligned nearly perpendicular to the substrates and the reflectance saturates. The major advantage of the normally black mode is the higher contrast compared with the normally white mode discussed below.

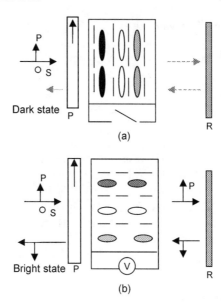

Figure 6.6 Device structure of the Heilmeier–Zanoni cell. (a) Voltage-off state, and (b) voltage-on state

Normally white mode

When a perpendicular-aligned cell is used, the LC directors and dyes are aligned perpendicular to the substrates, as shown in Figure 6.6(b). In the voltage-off state the principal axes of the dye molecules are perpendicular to the polarisation axis of the incident light. Thus, the absorption loss is minimal and the device has a maximum reflectance. When the voltage is switched on, the LC directors coupled with the dye molecules are reoriented by the electric field to be parallel to the incident light polarisation. The maximum absorption occurs and the dark state results, as illustrated in Figure 6.6(a).

By selecting a dye with a dichroic ratio of 10 : 1, the contrast ratio of the Heilmeier–Zanoni cell could exceed 50 : 1. However, due to the need for a linear polariser, the display brightness is greatly reduced.

4.2 Cole–Kashnow cell

In the Heilmeier–Zanoni cell, a polariser is used and about 60% of the incident light is absorbed after passing through the polariser twice. One way to eliminate the polariser is to add a quarter-wave film between the homogeneous guest–host cell and the reflector, as shown in Figure 6.7. This cell configuration is known as the Cole–Kashnow cell [19]. In the voltage-off state the dye molecules absorb the P wave and transmit S. The S wave is transmitted to the quarter-wave film twice and is converted to P. This P wave is again absorbed by the guest–host cell, resulting in a weak reflectance. In a high-voltage state the LC and dye molecules are reoriented by the electric field to be nearly perpendicular to the substrates. Little absorption takes place when the incident unpolarised light traverses the

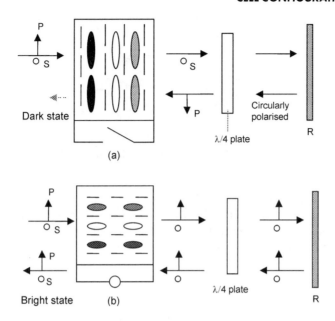

Dark state

(a)

Bright state   (b)

Figure 6.7 Device structure of the Cole–Kashnow cell. (a) Voltage-off state, and (b) voltage-on state

cell and the quarter-wave film twice. The outgoing light remains basically unpolarised and has a high optical efficiency.

The Cole–Kashnow cell can be configured in transreflective mode for dim environment applications. In the trans-reflected mode, the perfect reflector as used in Figure 6.7 is replaced by a partial reflector. Moreover, a second $1/4\lambda$ film and a linear polariser are used in front of the backlight. Unlike the transreflective mode described in Chapter 1, the transreflective Cole–Kashnow cell has the same dark and bright states as the reflective mode, so that good contrast can be preserved even if the backlight is turned on during bright ambient conditions. In the voltage-off state, both transreflective and reflective modes appear dark when a homogeneous guest–host cell is used. In the voltage-on state, both transreflective and reflective modes appear bright.

The Cole–Kashnow reflective cell exhibits two important advantages: a high reflectance and a modest contrast ratio. For a dye with a modest dichroic ratio of 10 : 1, the light reflectance can reach 50% while keeping a respectable 10 : 1 contrast ratio. A serious concern is that the rear LC substrate and quarter-wave film situated between the LC and the reflector could produce noticeable parallax, depending on the substrate and film thickness. Parallax is the occurrence of incident light and reflected light passing through different pixels. Parallax limits the resolution and viewing angle of a display device. To avoid parallax, the reflector and the quarter-wave film should be embedded in the inner side of the back substrate.

An attempt to implement the quarter-wave film inside the rear substrate has been carried out at Tohoku University [20]. Figure 6.8 shows such a device structure. First, a micro-slant reflector is deposited on the inner side of the rear glass substrate. This reflector directs the incident light at $-30°$ to the $0–60°$ cone with nearly flat top brightness of about 300% as compared to the MgO standard white. Second, a thin dielectric layer is coated on the rough

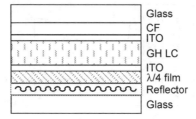

Figure 6.8 A modified version of the Cole–Kashnow cell where the $\lambda/4$ film is embedded in the inner side of the rear substrate

reflector in order to smooth its surface. Third, the $\lambda/4$ film is formed on top of the overcoat layer by a UV-curable liquid crystal polymer. Lastly, a thin ITO conductive layer is evaporated on top of the $\lambda/4$ film. Using a homeotropic-alignment guest–host cell, a reflectance of about 60% in the range 0–60°, has been achieved. The device contrast is about 5 : 1.

4.3 White–Taylor cell

By far the most commonly employed guest–host display is the White–Taylor cell [21], as sketched in Figure 6.9. The White–Taylor LC cell consists of about 2% absorbing dyes and about 1% cholesteric dopant. Owing to the presence of chiral agents, the LC directors and dye molecules form a twisted structure. The detailed twist angle is determined by the ratio of d/p, where d is the LC cell gap and p is the pitch length. For example, to make a 240° twisted White–Taylor cell, the d/p ratio is chosen to be 2/3.

Device configurations

A TFT-addressed guest–host display employing the White–Taylor cell is illustrated in Figure 6.10. The RGB color filters are deposited in the top glass substrate. Owing to the

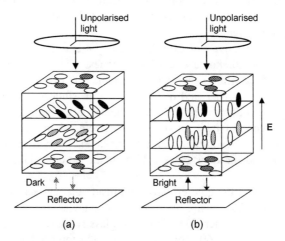

Figure 6.9 Device structure of the White–Taylor cell. (a) Voltage-off state, and (b) voltage-on state

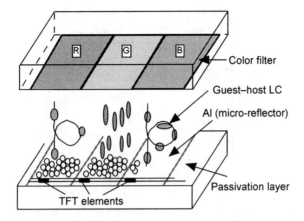

Figure 6.10 A schematic diagram of TFT-addressed White–Taylor guest–host reflective LCD

double pass of the color filters, their thickness or pigment density can be reduced by a factor of two compared to the transmissive mode. The twisted guest–host mixture is about 5 μm thick. The response time is normally in the 40–60 ms regime, depending on the exact cell gap and viscosity of the dyes employed. As illustrated in Figure 6.10, the green sub-pixel is ON, and the red and the blue are OFF. Thus, this pixel appears as green.

A reflector deposited on the inner side of the rear substrate reflects the light to the viewer. In order to avoid the specular mirror effect which exhibits a very narrow viewing angle, a micro-slant reflector (NEC, Sharp and ERSO) or a light control film (Tohoku University and ERSO) have been considered. For a reflective display, in order to achieve a high aperture ratio, the thin film transistors are preferably embedded behind the reflectors. With such a configuration the aperture ratio can be made greater than 90%. In addition to a higher display brightness and better contrast ratio, a higher aperture ratio also makes the display look like film. The pixelisation effect (also called the screen door effect) as observed in the low aperture ratio ($<70\%$) transmissive TFT-LCD is minimised.

Twist angle effect

Figure 6.11 shows computer simulation results for the twist angle-dependent reflectance and contrast ratio of White–Taylor cells with various twist angles [22]. As indicated in Figure 6.11, the voltage-on (i.e. bright) state is relatively insensitive to the twist angle. This is because in a high-voltage state virtually all the LC and dye molecules are aligned along the electric field direction. Thus, its reflectance is high and insensitive to the twist angle. However, in a voltage-off state the reflectance is strongly dependent on the twist angle in the low-twist regime. As the twist angle increases, both P and S components of the incident unpolarised light encounter more absorbing dye molecules while traversing through the cell. As a result, the reflectance decreases as the twist angle increases. For example, for a homogeneous cell (0° twist), only the P or S component is absorbed. The unabsorbed component will be reflected, resulting in a contrast ratio less than 2 : 1. To boost the contrast ratio, the twist angle needs to be increased. For example, as the twist angle increases to 360°, a 5.7 contrast ratio is obtained.

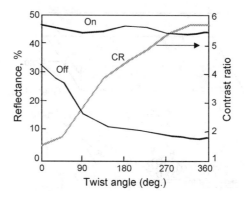

Figure 6.11   Twist angle effect on the performance of a White–Taylor cell

Hysteresis effect

It is known that a guest–host cell exhibits a hysteresis loop in its voltage-dependent transmittance curve if the twist exceeds a certain angle. Figure 6.12 shows the measured twist angle dependent hysteresis of a transmissive guest–host cell. The hysteresis becomes noticeable when the twist angle exceeds 270°. This hysteresis represents the voltage difference in the loop at 50% transmittance. Hysteresis makes precision gray scale control difficult. Thus, to avoid hysteresis, the twist angle needs to be controlled below 250°.

Gray scale control

From the contrast ratio standpoint, a larger twist angle is preferred. However, if the twist angle exceeds 240°, the EO curve becomes too sharp and gray scale control is more difficult. A standard driver IC has about 25 mV resolvable voltage. If a display is intended for eight gray levels, then the full on and off voltage difference cannot be narrower than 2 V. From the results shown in Figure 6.13, the twist angle cannot exceed 240°. Thus, the optimal twist angle is between 220° and 240° [23]. In these circumstances, the contrast ratio is about 5 : 1 and the reflectance is about 45%.

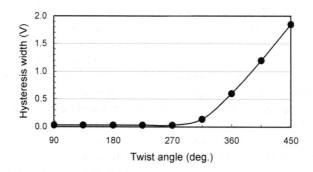

Figure 6.12   Twist angle effect on the hysteresis of a White–Taylor cell

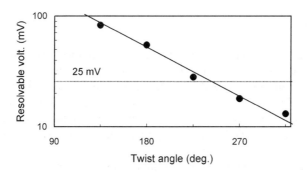

Figure 6.13   Twist angle effect on the gray scale voltage of White–Taylor cells

Birefringence effect

The birefringence Δn of the LC employed plays an important role in the White–Taylor cell, as shown in Figure 6.14. In a twisted cell, the polarisation rotation effect depends on the $d\Delta n$ of the LC layer. Let us take the 90° twisted White–Taylor cell as an example. If the LC has $\Delta n = 0$, then the only mechanism involved is absorption. The majority of the incoming P wave would be absorbed by the first few layers near the front substrates. The absorption of the subsequent layers for the P wave would be gradually decreased due to their twist angles. Conversely, the first few layers near the front substrate cause little absorption to the S wave. The major absorption occurs in those dye molecules near the back substrate (90° twisted with respect to the front substrate). Since both polarisations are absorbed, a relatively high contrast (5.5 : 1) is achieved.

As the Δn increases, the rotatory power or waveguiding effect comes into play as well. The incoming P wave is not only partially absorbed by the first few layers of the dye molecules, but it also gradually follows the LC twist. As a result, more P wave is absorbed than in the case of $\Delta n = 0$. However, due to this polarisation rotation effect, the S wave constantly faces the minimum absorption of the dye molecules, resulting in an increased reflectance, as shown in Figure 6.14. This polarisation rotation effect is greater as Δn is larger than 0.13, and thus the contrast ratio declines more noticeably.

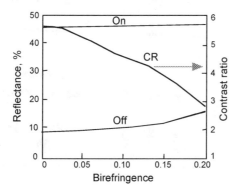

Figure 6.14   Birefringence effect on the reflectance and contrast ratio of a White–Taylor cell

4.4 Double cell

The double cell has the potential to achieve both high brightness and a high contrast ratio if the middle substrate problem can be solved. In a double cell, two homogeneous or homeotropic guest–host cells are arranged such that their molecular alignments are orthogonal to each other. Figure 6.15 illustrates the side view of double cells using either a homogeneous or a homeotropic alignment. For the reflective mode, we simply embed a reflector on the inner side of the second cell and subsequently reduce the cell thickness or dye concentration by a factor of two.

In the voltage-off state, one of the polarisation components of the incoming unpolarised light, say an S wave, is absorbed by the first cell. The transmitted P wave is then absorbed by the second guest–host cell. As a result, a good dark state is obtained. In a high-voltage state, the dye molecules in both cells are reoriented to be nearly perpendicular to the substrates. Little absorption occurs and a bright state results.

In principle, the double cell represents the most efficient way to modulate light. Unlike the twisted structure, every guest–host layer contributes directly to absorbing one component of the incoming light. High brightness and high contrast ratio can be obtained simultaneously using a minimal LC layer thickness. The thin cell gap leads to a fast response time. The major problem of the double cell approach is the need for two guest–host cells. The parallax problem resulting from two extra substrates between the LC cells could be quite severe, depending on the substrate thickness. To reduce parallax, a shared ultra-thin polyester film (0.9 μm Mylar from Du Pont or 1.3 μm Lumirrar from Toray) and planarised inner diffuser has been demonstrated [24, 25]. However, the usual rubbing techniques on this flexible thin film cannot be applied due to a lack of strength. Instead, the photo-induced molecular alignment technique has proved useful [26].

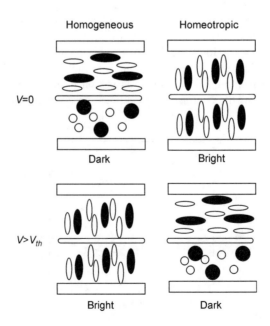

Figure 6.15   Device structure of a double guest–host cell (courtesy of Dr M. Hasegawa of Japan IBM)

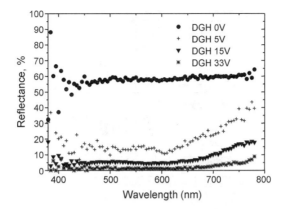

Figure 6.16 Voltage effect on the reflectance of a double guest–host display (courtesy of Dr M. Hasegawa of Japan IBM)

Figure 6.16 shows the voltage-dependent light reflectance of a double GH cell employing a homeotropic cell [27]. The LC used is MLC-6093 (Merck) and the cell gap for each layer is 2.5 μm. In the voltage-off state, the dye absorption is minimal and the reflectance reaches 60%. As the voltage increases, the dye molecules are tilted, resulting in increased absorption, i.e. decreased reflectance. At $V = 15\,V_{rms}$, a contrast ratio of about 8 : 1 is obtained. A further increase in voltage leads to a higher contrast ratio. However, such a high voltage cannot be practically achieved by the amorphous silicon TFT arrays.

4.5 Tri-layer GH cells

Another approach to achieve a high brightness color display is to stack three guest–host layers, as depicted in Figure 6.17. Unlike the conventional way of using additive colors, this approach subtracts unwanted colors from the incoming white light. The three colors selected are magenta, cyan and yellow. In a prototype demonstrated by a Toshiba group [28], the LC layer consists of 10 μm ZLI-2806 ($\Delta n = 0.0437$, $\Delta \varepsilon = -4.8$) and 0.5% Merck's ZLI-811 chiral dopant with $d/p \simeq 0.8$. Owing to the negative $\Delta \varepsilon$ LC mixture employed, the homeotropic alignment is obtained in the voltage OFF state and twisted alignment in an ON state. Therefore, the display appears white at $V = 0$ and dark as the voltage is increased to $8\,V_{rms}$. The overall reflectance is 43% with reference to the MgO standard white at 30° viewing direction. The contrast ratio reported is 5.3 : 1.

Due to a relatively thick cell gap, the rise time is 35 ms and the decay time is 115 ms. Another serious problem of this cascaded LC layer approach is parallax (or double image) at oblique viewing angles. To reduce the parallax, two 0.3 mm and two 0.5 mm-thick glass substrates are used, as shown in Figure 6.17.

To completely eliminate the intermediate substrates, a three-layer encapsulated guest–host liquid crystal display, as depicted in Figure 6.18, has been developed [29]. The encapsulated liquid crystal materials are prepared using an *in situ* method. Capsules with diameters of 1–100 μm can be obtained. These capsules are dispersed in a small amount

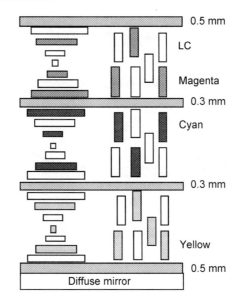

Figure 6.17 Device structure of a tri-layer guest–host display

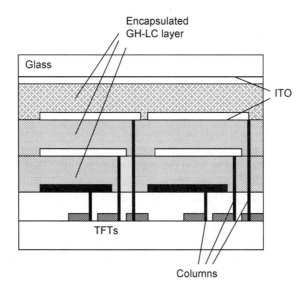

Figure 6.18 Device structure of an encapsulated three-layer guest–host display

of dispersion medium containing binder polymer. On the TFT substrate the dispersion medium is thermally evaporated to form a liquid crystal film, on which intermediate electrodes are fabricated. This process is repeated three times. A display with 60% reflectance and contrast 7 : 1 has been demonstrated.

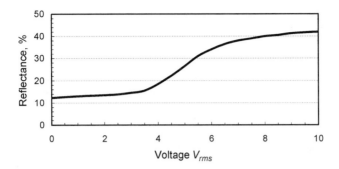

Figure 6.19 Voltage-dependent light reflectance of an a-N* guest–host display. Cell gap $d = 5.5\,\mu m$, LC: ZLI-2293. Blue dye: 1.25wt% G-256 (after [32])

4.6 Amorphous cell

In 1993 a Stanley group reported on an amorphous TN cell showing a wide and symmetric viewing angle [30, 31]. In 1995 this idea was extended to the guest–host display [32]. For a polariser-free guest–host display, a large twist angle helps improve the contrast ratio. However, the hysteresis loop increases with twist angle, as shown in Figure 6.12. Thus, for the amorphous cell, the twist angle is set at 300°. This can be controlled by doping a chiral agent (e.g. Merck S811) to the nematic LC host with $d/p = 5/6$. A typical cell gap is $d = 5.5\,\mu m$ and dye concentration is 1.25 wt%.

The fabrication process of the amorphous chiral nematic (a-N*) guest–host cell is relatively simple. The dichroic dye doped N* LC mixture is injected into a cell in the isotropic phase. The substrates are coated with polyimide films (Chisso PSIA-2101) without rubbing. The cell is then slowly cooled down to room temperature at about 0.1°C/min.

The voltage-dependent light reflectance of the a-N* guest–host cell is shown in Figure 6.19. As the applied voltage exceeds 3 V_{rms}, the reflectance starts to increase and gradually saturates at 8–10 V_{rms}. No hysteresis and stripe domains are observed. The maximum reflectance is 42%. However, the contrast is only 3.4. To enhance the contrast ratio, a dye with a larger dichroic ratio needs to be used.

5   Chroma

In addition to brightness, chroma is another critical factor for reflective color displays. The chroma C^* is defined as

$$C^* = \left(a^{*2} + b^{*2}\right)^{1/2} \tag{6.7}$$

It is standardised in the CIE 1976. In a reflective color LCD, the contrast ratio of the display and the color purity of the color filters determine the chroma. In a guest–host LCD the brightness trades with the contrast ratio, in a similar way as the color purity trades with the transmittance of a color filter. Color saturation plays a decisive role in balancing the display brightness and contrast ratio.

Table 6.4 Display characteristics of a Sharp GH and single-polariser a-Si TFT-LCDs

Display mode	Guest–host	Single-polariser
Display size	6.5″	6.5″
Resolution	640(H) × 240(V)	640(H) × 240(V)
TFT	a-Si	a-Si
No. of colors	4096	260 000
Contrast ratio	5 : 1 (0–30°)	10 : 1
Brightness	33% (ref. MgO)	30%
Response time	80 ms (on and off)	50 ms (on and off)
	150 ms (gray scales)	80 ms (gray scales)
Power consumption	200 mW	200 mW

A Sharp group has compared the chroma of three guest–host LCDs: [reflectance, contrast ratio] were [42%, 6 : 1], [46%, 5 : 1] and [57%, 3 : 1] [33]. The results indicate that the display with 46% reflectance and a 5 : 1 contrast ratio has the best color saturation at red and blue. Thus, it is the preferred choice among the three candidates considered.

These authors further compared the chroma of a GH and a single-polariser LCD: [reflectance, contrast ratio] were [46%, 5 : 1] and [38%, 10 : 1]. When the reflectance is high, the GH display has a higher chroma than the single-polariser display. On the other hand, when the reflectance is low, the single-polariser display has a better color saturation due to its higher contrast ratio. Therefore, the guest–host LCD is more suitable for graphic displays, while the single-polariser type is more suitable for video pictures such as still cameras and video movies.

Table 6.4 lists some performance characteristics of two Sharp reflective LCDs using the same high aperture ratio (about 90%) TFT and aluminum micro-reflectors. The only difference is in the LC operation modes. The left column is for the guest–host and the right column is for the single-polariser LCD. Due to its higher contrast ratio, more colors can be distinguished in the single-polariser LCD. Moreover, the single-polariser LCD has a faster response time because no dye is added.

If a guest–host LCD with 60% reflectance and a 10 : 1 contrast ratio can be developed, it will find widespread application for mobile displays. The modified Cole–Kashnow cell with an embedded $\lambda/4$ film, the double cell with a thin polyester film for eliminating parallax, and the tri-layer encapsulated guest–host LC discussed in this chapter all possess great potential to realise such a dream.

References

1 G. H. Heilmeier and L. A. Zanoni, Guest–host interactions in nematic liquid crystals: a new electro-optic effect, *Appl. Phys. Lett.* **13**, 91 (1968).

2 V. G. Chiyrinov, Liquid Crystal Devices: Physics and Applications, (Artech House, 1999), Chapter 2.

3 A. C. Lowe, Assessment of nematic guest–host systems for application to integrated liquid crystal displays, *Mol. Cryst. Liq. Cryst.* **66**, 295 (1981).

4 A. C. Lowe and R. J. Cox, Order parameter and the performance of nematic guest–host displays, *Mol. Cryst. Liq. Cryst.* **66**, 309 (1981).

5   B. Bahadur, Liquid Crystals Applications and Uses (World Scientific, Singapore, 1992), Vol. 3, Ch. 11.

6   Mitsui Toatsu Chemicals, Inc. Tokyo, Japan.

7   I. Haller, Thermodynamic and static properties of liquid crystals, *Prog. Solid State Chem.* **10**, 103 (1975).

8   A. C. Lowe, Assessment of nematic guest–host systems for application to integrated liquid crystal displays, *Mol. Cryst. Liq. Cryst.* **66**, 295 (1981).

9   H. V. Ivashchenko and V. G. Rumyantsev, Dyes in liquid crystals, *Mol. Cryst. Liq. Cryst.* **150**, 1 (1987).

10 M. Bremer, S. Naemura and K. Tarumi, Model of ion solvation in liquid crystal cells, *Jpn. J. Appl. Phys.* **37**, L88 (1998).

11 S. T. Wu, J. D. Margerum, M. S. Ho, and B. M. Fung, Liquid crystal dyes with high solubility and large dielectric anisotropy, *Appl. Phys. Lett.* **64**, 2191 (1994).

12 S. T. Wu, J. D. Margerum, M. S. Ho, B. M. Fung, C. S. Hsu, S. M. Chen, and K. T. Tsai, High birefringence liquid crystals, *Mol. Cryst. Liq. Cryst.* **261**, 79 (1995).

13 S. T. Wu, E. Sherman, J. D. Margerum, K. Funkhouser, and B. M. Fung, High solubility and low viscosity dyes for guest–host displays, *Asia Display'95*, 567 (1995).

14 M. E. Neubert, Kent State University (private communication).

15 B. Bahadur, Liquid Crystals Applications and Uses, (World Scientific, Singapore, 1992), Vol. 3, Ch. 11.

16 G. H. Heilmeier, J. A. Castellano and L. A. Zanoni, Guest–host interactions in nematic liquid crystals, *Mol. Cryst. Liq. Cryst.* **8**, 293 (1969).

17 T. Uchida, H. Seki, C. Shishido, and M. Wada, Guest–host interactions in nematic liquid crystal cells with twisted and tilted alignments, *Mol. Cryst. Liq. Cryst.* **54**, 161 (1979).

18 H. L. Ong, Electro-optical properties of guest–host nematic liquid crystal displays, *J. Appl. Phys.* **63**, 1247 (1988).

19 H. S. Cole and R. A. Kashnow, A new reflective dichroic liquid crystal display device, *Appl. Phys. Lett.* **30**, 619 (1977).

20 N. Sugiura and T. Uchida, Designing bright reflective full-color LCDs using an optimized reflector, *SID Tech. Digest* **28**, 1011 (1997).

21 D. L. White and G. N. Taylor, New absorptive mode reflective liquid-crystal display device, *Appl. Phys. Lett.* **45**, 4718 (1974).

22 Y. Itoh, *et al.*, AMLCD96 Digest p. 409 (1996).

23 H. Ikeno, H. Kanoh, N. Ikeda, K. Yanai, H. Hayama, and S. Kaneko, A reflective guest–host LCD with 4096-color display capability, *SID Tech. Digest* **28**, 1015 (1997).

24 A. C. Lowe and M. Hasegawa, High Reflectivity Double Cell Nematic Guest–Host Display, *Proc 17th International Display Research Conference*, 250 (1997).

25 M. Hasegawa, C. Hellermark, A. Nishikai, Y. Taira, and A. C. Lowe, Reflective stacked crossed guest–host display with a planarized inner diffuser, *SID Tech. Digest* **31**, 128 (2000).

26 Y. Iimura and S. Kobayashi, Prospects of the photo-alignment technique for LCD fabrication, *SID Tech. Digest* **28**, 311 (1997).

27 M. Hasegawa, K. Takeda, Y. Sakaguchi, Y. Taira, J. Egelhaaf, E. Leuder, and A. C. Lowe, A 320-dpi 4-in reflective stacked crossed guest–host display, *SID Tech. Digest* **30**, 962 (1999).

28 K. Sunohara, K. Naito, M. Tanaka, Y. Nakai, N. Kamiura, and K. Taira, A reflective color LCD using three-layer GH-mode, *SID Tech. Digest* **27**, 103 (1996).

29 K. Sunohara, K. Naito, S. Shimizu, M. Akiyama, M. Tanaka, Y. Nakai, A. Sugahara, K. Taira, H. Iwanaga, T. Ohtake, A. Hotta, S. Enomoto, and H. Yamada, Reflective color LCD composed of stacked films of encapsulated liquid crystal, *SID Tech. Digest* **29**, 762 (1998).

30 Y. Toko, T. Sugiyama, K. Katoh, Y. Limura, and S. Kobayashi, Amorphous twisted nematic-liquid-crystal displays fabricated by nonrubbing showing wide and uniform viewing-angle characteristics accompanying excellent voltage holding ratio, *J. Appl. Phys.* **74**, 2071 (1993).

31 T. Sugiyama, Y. Toko, T. Hashimoto, K. Katoh, Y. Limura, and S. Kobayashi, Analytical simulation of electro-optical performance of amorphous twisted nematic liquid crystal displays, *Jpn. J. Appl. Phys.* **32**, 5621 (1993).

32 H. Koimai, Y. Limura, S. Kobayashi, T. Hashimoto, T. Sugiyama, and K. Katoh, Polarizer-free reflective amorphous chiral nematic guest–host LCDs, *SID Tech. Digest* **26**, 699 (1995).

33 K. Nakamura, H. Nakamura and N. Kimura, Development of high reflective TFT, *Sharp Technology Journal* **69**, 33 (1997).

7

Liquid Crystal/Polymer Composites

1 Introduction

Liquid crystal/polymer composites are a relatively new class of materials which have been used in displays, light shutters and switchable windows [1–6]. They consist of low molecular weight liquid crystals and polymers, and can be divided into two subsystems: polymer dispersed liquid crystals (PDLC) and polymer stabilised liquid crystals (PSLC). In a PDLC, the liquid crystal exists in the form of micron-sized droplets which are dispersed in the polymer binder. The concentration of the polymer is comparable to that of the liquid crystal. The polymer forms a continuous medium. The liquid crystal droplets are isolated from one another. A scanning electron microscope (SEM) picture of a PDLC sample is shown in Figure 7.1. In a PSLC, the polymer forms a sponge-like structure. The concentration of the liquid crystal is much higher than that of the polymer. The liquid crystal forms a continuous medium. A SEM picture of a PSLC is shown in Figure 7.2.

In PDLCs and PSLCs the materials are usually operated between a transparent state and a scattering state. The transparent state has the appearance of a transparent glass and transmits the incident light. The scattering state has a milky appearance and scatters the incident light. As an example, a polymer stabilised cholesteric texture (PSCT) normal-mode switchable window is shown in Figure 7.3. Although the liquid crystal is birefringent, an incident light with a given incident angle and polarisation encounters only the refractive index of the liquid crystal in the polarisation direction. In the transparent state, the liquid crystal has such an orientation that the material exhibits a single refractive index. The system is an optically uniform medium and the incident light propagates through it with only a small loss. In the scattering state, the liquid crystal has a random poly-domain structure. The incident light encounters non-uniformity in the refractive index and is scattered away from the original propagation direction.

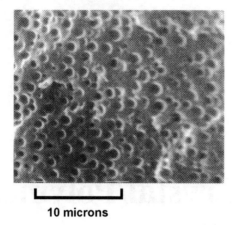

10 microns

Figure 7.1 SEM photograph of a PDLC. It was taken after the PDLC sample was fractured and the liquid crystal was extracted. The dark circles correspond to the liquid crystal droplets

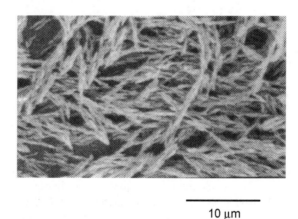

10 μm

Figure 7.2 SEM photograph of PSLC. It was taken after the PSLC sample was opened and the liquid crystal was extracted. The bright filaments correspond to the polymer network.

Liquid crystal/polymer composite technology has the advantage over regular liquid crystal technologies, such as twisted nematic (TN) and electrically controlled birefringence (ECB), in that it does not need polarisers and therefore the transmittance is very high (about 90%) in the transparent state. It is suitable for use in applications where high transmittance is required. Its compatibility with plastic substrates is another advantage, because the non-uniform bi-refringence of plastic substrates does not cause problems. For PDLCs, the polymer binder is self-adhesive, making it possible to fabricate large displays using a roll-to-roll process.

2 Polymer Dispersed Liquid Crystals

In PDLCs liquid crystals are confined in droplets. Their electro-optical properties are determined by many factors such as the anchoring on the droplet surface, the droplet size

(a)

(b)

Figure 7.3 Picture of a PSCT switchable window, showing the on and off states. Photograph courtesy of Reveo, Inc.

and externally applied fields, as well as the liquid crystal. They have many unique properties because of the confinement and the large surface-to-volume ratio.

2.1 Liquid crystal droplet configurations in PDLC

The liquid crystal dispersed in the polymer of a PDLC can be in one of many liquid crystal phases such as nematic, cholesteric, smectic-A and smectic-C* [4]. The most common PDLC is a polymer-dispersed nematic liquid crystal which is usually referred to as a polymer-dispersed liquid crystal with the word nematic omitted. Here we only discuss polymer-dispersed nematic liquid crystals. In a nematic liquid crystal, the elongated molecules are aligned in a common direction which is defined as the director n. Inside a nematic droplet, the director configuration is determined by the droplet shape and size, the anchoring condition on the droplet surface and the externally applied field, as well as the elastic constants of the liquid crystal. According to the director configuration inside the droplet, PDLC droplets can be divided into four major types, as shown in Figure 7.4. When the anchoring condition is tangential, there are two types of droplet. One is the bipolar droplet as shown schematically in Figure 7.4(a) [7–9]. The other one is the toroidal droplet as shown

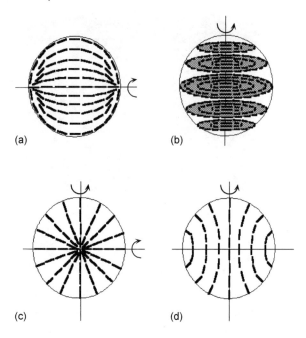

Figure 7.4 Schematics showing liquid crystal director configurations confined in droplets in PDLCs: (a) bipolar droplet, (b) toroidal droplet, (c) radial droplet and (d) axial droplet

schematically in Figure 7.4(b) [1, 10]. When the anchoring condition is perpendicular, there are also two types of droplet. One is the radial droplet as shown schematically in Figure 7.4(c) [11]. The other is the axial droplet as shown schematically shown in Figure 7.4(d) [8,11]. When a droplet is bigger than 5 μm in diameter, it is possible to identify the droplet configuration using an optical microscope.

In the bipolar droplet with strong anchoring, there are two point defects at the ends of a diameter of the droplet. This diameter is called the bipolar axis. The droplet director N is defined as a unit vector along the bipolar axis. The director on a plane containing the bipolar axis is shown in Figure 7.4(a). The liquid crystal director is tangential to the circle along the circumference and parallel to the bipolar axis along the diameter. At other places inside the circle the director is oriented in such a way that the total free energy is minimised [12]. There is a rotational symmetry of the director around the bipolar axis. A typical optical microphotograph of a sample with bipolar droplets is shown in Figure 7.5(a). The elastic deformations involved are splay and bend. When there is no externally applied field, the orientation of N is arbitrary for perfect spherical droplets. In practice, the droplets are usually deformed. The deviation of the droplet from a spherical shape results in a certain orientation of N [13,9]. Preferred deformed bipolar droplets can be made by applying stress or external fields during the formation of the droplet [14]. When a sufficiently high external electric field is applied, the liquid crystal ($\Delta\varepsilon > 0$) is reoriented with the bipolar axis parallel to the field.

In the toroidal droplet, there is a line defect along a diameter of the droplet [10]. The director on a plane cut perpendicularly to the defect line is tangential to the concentric

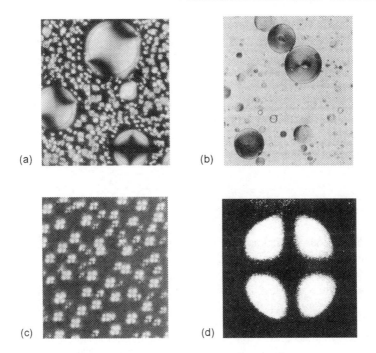

Figure 7.5 Microphotographs of PDLCs. (a) Bipolar droplet, photograph courtesy of Professor J.W. Doane, (b) toroidal droplet [10], (c) radial droplet and (d) axial droplet

circles. There is a rotational symmetry around the defect line. The bend elastic deformation is the only one involved. A toroidal droplet exists when the bend elastic constant is smaller than the splay elastic constant; otherwise the droplet takes the bipolar configuration. Toroidal droplets rarely exist because for most liquid crystals the bend elastic constant is larger than the splay elastic constant. Nevertheless, toroidal droplets have been observed as shown in the optical microphotograph (Figure 7.5(b)). When the droplet is sufficiently large, the director near the defect line will escape in the direction parallel to the line so that the total free energy is reduced. When a sufficiently high external field is applied, the director near the defect line will also escape with the defect line parallel to the applied field.

In the radial droplet, the director everywhere is along the radial direction, and there is a point defect in the center of the droplet [8, 11]. The director configuration on a plane cut through the droplet center is shown in Figure 7.4(c). There is rotation symmetry around any diameter. An optical microphotograph of a sample with radial droplets is shown in Figure 7.5(c). The dark cross at the center is narrow. The splay elastic deformation is the only one present in radial droplets.

In the axial droplet, there is a line defect along a circumference [8]. The director on a plane cut perpendicularly to the line defect and containing a diameter of the droplet is shown in Figure 7.4(d). There is a rotational symmetry around the diameter perpendicular to the line defect. An optical microphotograph of a sample with axial droplets is shown in Figure 7.5(d). In an axial droplet the dark cross at the center is wider than in a radial droplet, because the liquid crystal director is more parallel to the symmetry axis near the center in the axial droplet. Both splay and bend elastic deformations exist in axial droplets. The splay elastic energy of the axial droplet is lower than that of the radial droplet if all material parameters and droplet sizes are identical. The axial droplet is more stable than the radial droplet if the anchoring is weak, or if the droplet is small or if an external field is applied. If the liquid crystal is in the radial configuration at zero electric field, when a sufficiently high electric field is applied it can be switched to the axial configuration with the symmetry axis parallel to the field.

Besides the four droplet configurations discussed above, other droplet configurations could exist under appropriate conditions. For example, there is a twisted bipolar droplet where the splay and bend elastic energies are reduced by introducing twist deformation [15]. The point defect in the radial droplet could escape from the center in an effort to reduce the total free energy.

2.2 Preparation of PDLC

There are four methods for making PDLCs. The first one is microencapsulation which starts with an inhomogeneous mixture of a liquid crystal and an aqueous polymer solution [16–18]. In this method the liquid crystal is first encapsulated to form droplets in the aqueous solution; then the water is allowed to evaporate. The other three methods start with a homogeneous mixture of a polymer (or monomer) and a liquid crystal, and then the mixture is induced to phase-separate to form polymer-dispersed liquid crystal droplets. Three methods are identified according to how the phase separation between the liquid crystal and the polymer is realised: thermally induced phase separation (TIPS), polymerisation-induced phase separation (PIPS) and solvent-induced phase separation (SIPS).

Encapsulation

PDLCs formed by encapsulation are also called emulsion-based PDLCs or NCAP (nematic curvilinearly aligned phase) [1, 16]. In this method a nematic liquid crystal, water and a water soluble polymer, such as polyvinyl alcohol, are put into a container. Water acts as a solvent in this preparation method. The polymer is dissolved by the water to form a viscous solution. This aqueous solution does not dissolve the liquid crystal. When this system is stirred by a propeller blade at a sufficiently high speed, micron-sized liquid crystal droplets form. Smaller liquid crystal droplets form at higher stirring speeds. Then the emulsion is coated on a substrate and the water is allowed to evaporate. After the evaporation of the water, a second substrate is laminated to form PDLC devices.

Solvent-induced phase separation

A multi-component system can exist either in a homogeneous form in which the components are mixed on the molecular scale, or in a phase separated (heterogeneous) form in which there are two or more distinct phases [19]. The form of the mixture is determined by its free energy. The form with the lowest free energy is realised. There are two contributions to the free energy: one is the inter-molecular interaction energy u, and the other is the mixing entropy s. The free energy f is given by

$$f = u - Ts \qquad (7.1)$$

Here, the interaction energy u is defined as the interaction energy when the system is in the homogeneous form subtracted from the interaction energy when the system is in a completely phase separated form; u can be either positive or negative and is usually positive. The contribution of the mixing entropy is negative and its absolute value is maximised in the homogeneous phase. The mixing entropy always tends to mix the components. When f is positive, the system phase separates. When f is negative, the system is homogeneously mixed or partially mixed.

PDLCs can be made by mixing a liquid crystal, a thermoplastic and a common solvent [20]. The liquid crystal and the thermoplastic do not dissolve each other, while the solvent dissolves both the liquid crystal and the thermoplastic. When the concentration of the solvent is sufficiently high, the interaction energy between the solvent and the liquid crystal, which is negative, and the interaction energy between the solvent and the thermoplastic, which is also negative, are dominant. The interaction energy of the homogeneous form is negative or positive and small. The free energy is negative and they are homogeneously mixed. When the solvent evaporates sufficiently, the interaction energy between the liquid crystal and the thermoplastic, which is positive, becomes dominant. The interaction energy of the homogeneous form becomes positive and large. The free energy becomes positive and the system phase separates. When the phase separation takes place, provided the viscosity of the system is high, liquid crystal droplets form and do not coalesce. The droplet size depends on the solvent evaporation rate, with smaller droplets obtained at a faster evaporation rate. For example, 5% nematic liquid crystal E7 (Merck), 5% PMMA (poly methyl methacrylate) and 90% chloroform are put into a closed bottle to mix. Then the homogeneous mixture, which is clear in appearance, is cast on a glass plate. The glass plate is put into a chamber with inject and vent holes. Air is blown into the chamber and then vents out at a controllable rate. As the chloroform evaporates, the material changes to opaque when liquid crystal droplets begin to form. Then another glass plate is put on top of the first glass plate to sandwich the PDLC. In practice, the solvent-induced phase separation method is rarely used, because it is difficult to control the solvent evaporation rate.

Thermally induced phase separation

PDLCs can be made by making a binary mixture of a liquid crystal and a thermoplastic (polymer). If the interaction energy between the liquid crystal molecule and the thermoplastic molecule is lower than the average interaction energy between two liquid crystal molecules and between two thermoplastic molecules, the mixture always exists in a homogeneous form. If the interaction energy between the liquid crystal molecule and the thermoplastic molecule is higher than the average interaction energy between two liquid crystal molecules and between two thermoplastic molecules, depending on the temperature, it may exist in a homogeneous form or phase separate [19]. At a high temperature, the mixing entropy term is large and dominant; the homogeneous form has a lower free energy and is realised. At a low temperature, the mixing entropy term is small, and the mixture phase separates. The phase separation is achieved by lowering the temperature and the method is called thermally induced phase separation. Under appropriate conditions (spinodal decomposition), when the phase separation takes place, the liquid crystal forms isolated droplets.

In practice, it is difficult to mix a thermoplastic and a liquid crystal directly (at room temperature). A solvent is usually used to mix the thermoplastic and the liquid crystal. For example, 5% of nematic liquid crystal E7 and 5% of thermoplastic PMMA (poly methyl methacrylate) and 90% of chloroform are put into a bottle to mix. The material looks clear. Some glass fiber spacers are also added to the mixture. Then the mixture is cast on a glass plate. The mixture is left in an open space to allow the chloroform to evaporate slowly. When the chloroform evaporates sufficiently, the material left on the glass plate begins to phase separate and it changes from clear to opaque. The droplets formed in this system are bipolar. After the chloroform has evaporated, a cover glass plate is put on top of the mixture of the liquid crystal and thermoplastic. The glass fiber spacers control the cell thickness. When the cell is heated to 100°C, the liquid crystal and the thermoplastic are in a uniform mixture; the cell changes to clear. When the temperature is lowered to room temperature, the material phase separates again and becomes opaque. The droplet size can be controlled by the cooling rate, with smaller droplets formed at faster cooling rates. PIMB (poly isobutyl methacrylate) has a perpendicular anchoring and produces radial droplets.

Thermally induced phase separation is rarely used in manufacturing large size displays because a uniform thickness is difficult to achieve. It is, however, very useful in scientific investigations. This PDLC can be thermally cycled many times. Different droplet sizes can be obtained in one sample using different cooling rates.

Polymerisation induced phase separation

When a polymer and a liquid crystal are mixed, the mixing entropy is smaller than that of the system consisting of the corresponding monomer and the liquid crystal, because the positions of the constituent monomers in one chain of the polymer are no longer arbitrary but constrained. Once the position of the first (end) monomer in a polymer chain is fixed, the second monomer can only occupy the nearest neighboring position with respect to the first monomer. According to the Flory–Huggins theory, the mixing entropy of the system is given by [21,22].

$$s = -K_B \left[(1 - \phi_p) \ln (1 - \phi_p) + \frac{1}{x} \phi_p \ln \phi_p \right] \tag{7.2}$$

where K_B is the Boltzmann constant, and ϕ_p and x are the volume fraction and degree of polymerisation of the polymer. For a fixed polymer volume fraction, the higher the degree of polymerisation, the smaller the mixing entropy. When we start with a system consisting of a liquid crystal and a monomer (or prepolymer), the degree of polymerisation of the monomer is 1; the mixing entropy is large. The free energy of the system in homogeneous form is negative, and the liquid crystal and the monomer are in a homogeneous solution. As the monomers are polymerised at a fixed temperature, the degree of polymerisation increases, the mixing entropy decreases and the free energy increases. When the degree of polymerisation reaches a critical value, the liquid crystal phase separates from the polymer to form droplets.

Polymerisation can be thermally-initiated or photo-initiated. In thermally-initiated polymerisation the monomers are typically combinations of epoxy resins and the curing agent thiol. For example, Epon 828 (Shell Chemical), Capcure 3800 (Miller Stephenson Company) and E7 (EM Chemicals) are mixed in a 1:1:1 ratio [20]. The mixture is stirred thoroughly at room temperature. The stirring usually creates air bubbles which can be removed using a centrifuge. The mixture is then sandwiched between two substrates with electrodes and cured at an elevated temperature. Smaller droplets are formed at a higher temperature or a higher concentration of epoxy resins because of the higher reaction rate. In choosing monomers, attention should be paid to matching their refractive index to the ordinary refractive index of the liquid crystal. The curing speed is also important.

In photo-polymerisation, monomers with acrylate or methyacrylate end groups are used. Some photo-initiators are also needed [23]. Upon absorbing a photon, the photo-initiator becomes a free radical which reacts with the acrylate group. The opened double bond reacts with another acrylate group. The chain reaction propagates until the opened double bond reacts with another free radical, and then the polymerisation stops. As an example, Norland 66 (which is a combination of acrylate monomers and photo-initiators) and E7 (EM Chemicals) are mixed in the ratio $1:1$. The mixture is sandwiched between two substrates with electrodes and then cured under irradiation from UV light of a few mW/cm^2. Smaller droplets are formed under higher UV irradiation.

2.3 Operation of PDLC

We now consider how PDLCs operate. As an example, we consider a PDLC with bipolar droplets. At zero field, the droplet director N is oriented randomly throughout the cell, as shown in Figure 7.6(a). Consider a droplet with its droplet director making an angle θ with the normal of the cell. Normally incident light, with linear polarisation in the plane defined by the propagation direction of the light and the bipolar axis of the droplet, encounters a refractive index approximately given by

$$n(\theta) = \frac{n_{//}n_{\perp}}{(n_{//}^2 \cos^2 \theta + n_{\perp}^2 \sin^2 \theta)^{1/2}} \tag{7.3}$$

where $n_{//}$ and $n_{\perp}$ are the refractive indices for light polarised along and perpendicular to the liquid crystal director, respectively. The (isotropic) polymer is chosen such that its refractive index n_p is the same as $n_{\perp}$. The light encounters different refractive indices when it propagates through the polymer and the liquid crystal droplet; that is, the PDLC is a non-uniform

(a)   (b)

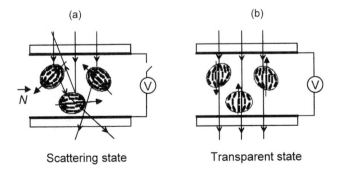

Scattering state   Transparent state

Figure 7.6 Orientation of liquid crystal droplets in the off and on states in the PDLC

medium. Therefore the light is scattered when it goes through the PDLC. When a sufficiently high electric voltage is applied across the PDLC cell, the droplets are reoriented with their droplet director N parallel to the normal of the cell as shown in Figure 7.6(b). $\theta = 0$ for all the droplets. Now, when the normal incident light propagates through the droplets, it encounters the refractive index $n_\perp$ which is the same as the encountered refractive index when it propagates through the polymer. The PDLC is a uniform medium for the light. Therefore the light goes through the PDLC without scattering. The PDLC discussed here is a normal-mode light shutter in the sense that it is opaque in the field-off state and transparent in the field-on state.

A typical voltage–transmittance curve of a PDLC cell is shown in Figure 7.7 [6]. The PDLC is made from a mixture of an epoxy and a liquid crystal and the cell thickness is 24 μm. At 0 V the material is in the scattering state and the transmittance is low. As the applied voltage is increased, the droplet director is aligned toward the cell normal direction and the transmittance increases. The drive voltage (at which the transmittance reaches 90% of the maximum value) is about 25 V. The maximum transmittance is about 90% (compared with the transmittance of an empty cell). The voltage–transmittance curve of PDLCs is not steep and therefore it is difficult to use PDLCs to make multiplexed displays on a passive matrix. It should be pointed out that the transmittance of PDLCs in the scattering state is dependent on the collection angle of the detector [24, 25]. Without specifying the collection angle, it is meaningless to talk about the transmittance and contrast. We will discuss this topic more in section 4 of this chapter.

2.4 Drive voltage of a PDLC

Many factors affect the drive voltage of a PDLC. The main factors are the droplet size and shape [9, 13], the surface anchoring of the liquid crystal on the droplet surface, the dielectric anisotropy, elastic constants and the resistivity of the liquid crystal [3, 26]. The total free energy of the system consists of its elastic energy, electric energy and the surface anchoring energy. In the equilibrium state, the liquid crystal is in the state of minimum free energy. At zero field, the liquid crystal configuration is determined by the droplet size and shape and the anchoring condition. At a sufficiently high field, the liquid crystal ($\Delta\varepsilon > 0$) is aligned parallel

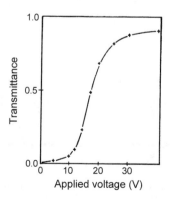

Figure 7.7 The voltage–transmittance curve of the PDLC [6]

to the field. The decrease in the electric energy compensates for the increase of the elastic energy and surface energy.

We discuss qualitatively the effects of the droplet size on the drive voltage. Let us consider a PDLC with a droplet size D. The elastic energy density f_{elas} is proportional to $(1/D)^2$, and the total elastic energy F_{elas} of a droplet is proportional to $D^3 \cdot (1/D)^2 = D$. Because of the different orientations of the liquid crystal in the bulk, the difference of the total elastic energy between the on-state and off-state is $\Delta F_{elas} = AD$, where A is a constant. The total surface energy F_s is proportional to the total area of the droplet surface, i.e. D^2. Because of the different orientations of the liquid crystal at the droplet surface, the difference of the total surface energy between the on-state and off-state is $\Delta F_s = BD^2$, where B is a constant. The total electric energy F_{elec} is proportional to $-\Delta\varepsilon(V/h)^2 D^3$, where h is the cell thickness. Because of the different orientation of the liquid crystal, the difference in the total electric energy between the on-state and off-state is

$$\Delta F_{elec} = C\Delta\varepsilon \left(\frac{V}{h}\right)^2 D^3$$

where C is a constant. At the drive voltage V_d (the critical voltage), the change of the total free energy is approximately zero, that is

$$\Delta F = \Delta F_{elas} + \Delta F_s + \Delta_{elec} = AD + BD^2 - C\Delta\varepsilon \left(\frac{V_d}{h}\right)^2 D^3 = 0 \qquad (7.4)$$

In the case of strong anchoring, the liquid crystal on the droplet surface is anchored along the easy direction in both the off-state and on-state; the change of the surface energy is zero, that is, $B = 0$. The drive voltage is then given by

$$V_d = \left(\frac{A}{C\Delta\varepsilon}\right)^{1/2} \frac{h}{D} \qquad (7.5)$$

Therefore $V_d D = $ constant. In the case of weak anchoring, the elastic energy is zero at a sacrifice of the surface energy in both off-state and on-state; that is, $A = 0$. The drive voltage is given by

$$V_d = \left(\frac{B}{C\Delta\varepsilon}\right)^{1/2} \frac{h}{D^{1/2}} \qquad (7.6)$$

Therefore $V_d D^{1/2} = $ constant. The anchoring condition may be established during the formation of the liquid crystal droplets.

In PDLCs with strong anchoring, the drive voltage is usually high. Such an example is the PDLC made from PVA and ZLI2061 (from Merck) using the NCAP method [27]. The cell thickness is 13 μm. The drive voltage V_d is approximately linearly proportional to $1/D$, as shown in Figure 7.8.

In PDLCs with weak anchoring, the drive voltage is usually low. Such an example is the PDLC made from E7 (from Merck) and NOA65 (Norland Optical Adhesive) by photo-polymerisation induced phase separation [28]. The cell thickness is 12 μm. The drive

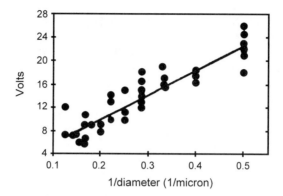

Figure 7.8 Drive voltage of the PDLC with different droplet sizes [27]

voltages for various droplet sizes are listed in Table 7.1. The product of the drive voltage and the square root of the droplet radius is approximately a constant.

Droplet shape is another factor affecting the drive voltage. When the droplet shape deviates from spherical, the elastic energy in the on-state is usually increased, thus the drive voltage is increased [9, 13]. Oblate droplet shapes can be obtained by using polymers with large thermal expansion coefficients. When forming the droplets at high temperatures and then decreasing to room temperature, oblate droplets form due to the contraction of the polymer in the cell normal direction. Prolate droplets can be obtained by shearing during the droplet formation.

In PDLCs, ions may accumulate on the droplet surface [29], changing the effective electric field inside the droplet, and thus changing the drive voltage. When the conductivity of a dielectric medium is non-zero due to ions, the effective dielectric constant is given by $\varepsilon_{eff} = \varepsilon + i(\sigma/\omega)$, where $i = \sqrt{-1}$, ε is the real part of the dielectric constant, σ is the conductivity and ω is the frequency of the applied field. When the resistivity $\rho\,(=(1/\sigma))$ is low or the frequency is low, the imaginary part of the dielectric constant is dominant. The drive voltage is proportional to $(\rho_p + 2\rho_{lc})/(3\rho_{lc})$, where ρ_p and ρ_{lc} are the resistivity of the polymer and the liquid crystal, respectively. Therefore the drive voltage is lower for the system with a smaller ρ_p/ρ_{lc} [3]. The drive voltage also changes when different frequencies are used [26].

It is difficult to derive a general rule to obtain low drive voltage PDLCs. Many factors are involved in preparing PDLCs. It is important to choose a suitable polymer and liquid crystal. For a particular system of polymer and liquid crystal, the important factors are the concentrations of the materials, and the curing conditions such as polymerisation speed, temperature, thermal expansion and shearing. The point at which the phase separation takes place in the temperature–concentration phase diagram during the droplet formation is also important. A more uniform droplet size is obtained in spinodal decomposition. The optimisation has to be determined experimentally.

Table 7.1 Drive voltages of the PDLC with various droplet radii [28]

Droplet radius R (μm)	0.3	0.6	0.8	1.1	1.3
Drive voltage V_d(V)	16	9.1	7.6	7.0	6.2
$V_d R$	4.8	5.5	6.1	7.7	8.1
$V_d R^{1/2}$	8.7	7.1	6.8	7.3	7.1

2.5   Dichroic dye doped PDLC

Dichroic dyes can be incorporated into PDLCs [30, 31]. The dye molecules are usually elongated and can be dissolved in the liquid crystal. The dye and the liquid crystal phase separate from the polymer binder. They are inside the droplet and can be switched. The dye used must be a positive type in the sense that the absorption transition dipole is along the long molecular axis. When the polarisation of the incident light is parallel to the long axis of the dye molecules, the light is absorbed [32]. When the polarisation of the incident light is perpendicular to the long axis of the dye molecules, the light is not absorbed. In the field-off state, the dye molecules are randomly oriented with the droplets, as shown in Figure 7.9(a). When the cell is sufficiently thick, there are droplets oriented in every direction. The un-polarised incident light is absorbed. In the field-on state, the dye molecules are aligned in the cell normal direction with the liquid crystal, as shown in Figure 7.9(b). They are always perpendicular to the polarisation of normally incident light. Therefore the light ideally passes through the cell without absorption. In practice, there is some absorption even in the field-on state, because of the thermal fluctuation of the dye molecules and the anchoring of the curved surface of the droplet.

Figure 7.10 shows the typical transmittance–voltage curve of a dye doped PDLC [33]. At 0 V the liquid crystal droplets are randomly oriented throughout the cell and the absorbance is high due to the absorption of the dye and the scattering due to the refractive index mismatch between the liquid crystal and the polymer. As the applied voltage is increased, the droplets become more and more aligned in the cell normal direction, and the absorbance decreases. The maximum transmittance is about 0.70, which is lower than that of pure liquid crystal PDLCs, due to the non-perfect order of the dye molecules.

There are a few points worth emphasising. First, it is desirable that the solubility of the dye in the polymer should be very low. The dye molecules dissolved in the polymer do not change their orientation under the applied field, and therefore they tend to decrease the contrast of the PDLC. Second, an oblate droplet shape is desirable, because inside such a droplet the dye molecules are oriented more in the plane parallel to the cell surface and they absorb the incident light more strongly in the field-off state [30]. Third, in the field-off state the scattering of the material increases the optical path length of the light inside the cell, and therefore enhances the absorption. Fourth, dye doped PDLCs do not need polarisers, because of the random orientation of the droplets in the field-off state, which is an advantage over

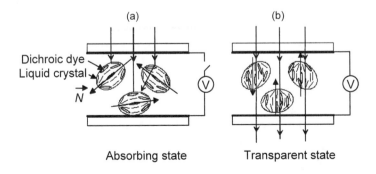

Figure 7.9   Schematic diagram showing how the dichroic dye doped PDLC works

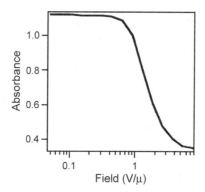

Figure 7.10   Absorbance of the dye doped PDLC versus the applied field [33]

nematic dichroic dye displays. Fifth, dye doped PDLCs have gray levels, because the droplets are gradually aligned by increasing the applied field, which is an advantage over cholesteric dichroic dye displays.

3 Polymer Stabilised Liquid Crystals

In polymer stabilised liquid crystals (PSLC), the polymer concentration is usually less than 10%. The monomer used may be mesogenic with rigid core structures similar to those of liquid crystal molecules [2, 34–37]. Before polymerisation, the mixture of the monomer and the liquid crystal is in a liquid crystal phase. The polymerisation environment is anisotropic due to the aligning effect of the liquid crystal on the monomer and the anisotropic diffusion of the monomer in the liquid crystal. The anisotropic fiber-like polymer networks which usually form, mimic the structure of the liquid crystal during the polymerisation [38 39]. Furthermore, external fields and surface alignment techniques can be applied in the polymerisation process to create various polymer network structures. Therefore PSCLs are a rich field and many fascinating structures can be achieved.

3.1 Preparation of PSLCs

Polymer-stabilised liquid crystals are usually made from mixtures of liquid crystals and monomers. The concentration of the monomer is usually less than 10%. The monomer can be directly dissolved in the liquid crystal. Although any type of polymerisation method can be used, photo-initiated polymerisation is fast and is used most often. The monomer is usually an acrylate or a methyl-acrylate because of their fast reaction rate. In order to form stable polymer networks, a bi-functional monomer or a mixture of bi-functional and mono-functional monomers must be used. A small amount of photo-initiator is added to the mixture. The concentration of the photo-initiator is typically 1–5% of the monomer. When irradiated by UV light, the photo-initiator produces free radicals which react with the double bonds of the monomer and initiate the chain reaction of polymerisation.

When the mixture of the liquid crystal, the monomer and the photo-initiator is irradiated by UV light, the monomer is polymerised to form a polymer network. The UV intensity is usually a few mW/cm^2 and the irradiation time is around half an hour. SEM [38, 40], neutron scattering, confocal microscopy [41], birefringence study and the Freederickzs transition technique have all been used to study the morphology of polymer networks in polymer stabilised liquid crystals [39, 42]. The results suggest a bundle structure for the polymer networks [39, 42]. The lateral size of the bundle, as shown in Figure 7.2, is a few tenths of a micron. The bundle consists of polymer fibrils with a lateral size of a few nanometers and liquid crystals. The morphology of the polymer network is affected by the following factors: the structure of the monomer, the UV intensity, the photo-initiator type and concentration, and the temperature. Polymer bundles with larger lateral sizes are formed at low a UV intensity, low photo-initiator concentration and a higher temperature [39]. For example, 96.7% nematic liquid crystal E7, 3% monomer BAB6 {4,4'-bis[6-(acryloyloxy)-hexy]-1,1'-biphenylene} and 0.3% BME (benzoin methyl ether) are mixed. The mixture is in nematic phase at room temperature. The viscosity of the mixture is comparable to that of the nematic liquid crystal and can be easily put into cells in a vacuum chamber. The cells are then irradiated by UV light for the monomer to form a polymer network. During polarisation, external electric fields can be applied to control the orientation of the mixture.

Monomers used in PSLCs preferably have rigid cores and flexible tails. They form anisotropic fibril-like networks. If the monomer does not have flexible tails, it forms bead-like structures which are not stable under perturbations such as externally applied fields [38]. If the monomer does not have a rigid core and is flexible, it can still form anisotropic networks.

3.2   Working modes of PSLCs

Polymer networks formed in liquid crystals are anisotropic and affect the orientation of a liquid crystal. They tend to align the liquid crystal in the direction of the fibrils. They are used to stabilise desirable liquid crystal configurations and to control the electro-optical properties of liquid crystal devices. Polymer networks have been used to improve the performance (drive voltage and response times) of conventional liquid crystal devices such as TN, IPS displays [43–45]. In this chapter we will, however, concentrate on scattering displays from polymer stabilised liquid crystals.

Polymer stabilised nematic liquid crystals

The Philips group developed a polymer stabilised nematic liquid crystal light shutter [34, 46–48]. It is made from a mixture of a nematic liquid crystal and a diacrylate liquid crystal monomer. The mixture is put into cells with homogeneous anchoring and then photopolymerised in the nematic phase. The concentration of the monomer is a few percent. The system is also called anisotropic gel. The nematic liquid crystal used has a positive dielectric anisotropy. Figure 7.11 shows schematically how the shutter works. In Figure 7.11(a) there is no applied voltage across the cell; the liquid crystal is homogeneously aligned in the x direction. The polymer network is anisotropic like the liquid crystal and its concentration is low. When light passes through the material, it encounters a uniform refractive index, and therefore it passes through the material without scattering. In Figure 7.11(b) there is a voltage applied across the cell. The applied field tends to align the liquid

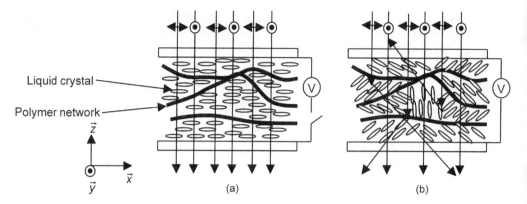

Figure 7.11 Schematic diagram showing how the polymer-stabilised nematic liquid crystal light shutter works

crystal in the z direction while the polymer network tries to keep the liquid crystal in the x direction. As a result of the competition between the applied field and the polymer network, the liquid crystal is switched into a multi-domain structure. The liquid crystal directors in the domains are random in the x–z plane. When light polarised in the x direction passes through the cell, it encounters different refractive indices in different domains and therefore is scattered. When light polarised in the y direction passes through the cell, it always encounters the ordinary refractive index of the liquid crystal because the liquid crystal is oriented in the x–z plane, and therefore it can pass through the cell without scattering. The polymer stabilised nematic liquid crystal is a reverse-mode light shutter in the sense that it is transparent in the field-off state and opaque in the field-on state.

The response of the polymer stabilised nematic liquid crystal light shutter is shown in Figure 7.12, where the voltage is the peak-to-peak value [46]. At 0 V, the material is in the homogeneous state and the transmittance is high (about 100%). When the applied voltage is increased above a threshold, the material is switched to the multi-domain structure and the transmittance decreases. The minimum transmittance is about 50% because the incident

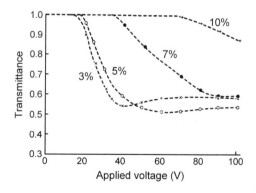

Figure 7.12 Transmittance versus applied voltage curves of the polymer-stabilised nematic liquid crystal light shutter. The concentration of the polymer network is varied from 3% to 10% [46]

light is unpolarised and only light polarised in one direction is scattered. As the concentration of the polymer network is varied, the electro-optical properties of the shutter change. The drive voltage is higher for the shutter with a higher polymer network concentration because the interaction between the liquid crystal and the polymer network is stronger. The turn-off time is decreased from 15 ms to 5 ms when the concentration of the polymer network is increased from 3% to 7%. The concentration of the polymer network also affects the size of the liquid crystal domains. When the concentration is increased to 10%, the domains become too small and do not strongly scatter visible light.

In order to overcome the problem that only one polarisation component is scattered by a polymer-stabilised homogeneously aligned nematic liquid crystal, Hikmet introduced the polymer-stabilised homeotropically aligned nematic liquid crystal [49]. The liquid crystal has a negative dielectric anisotropy ($\Delta\varepsilon < 0$). It is mixed with a small amount of diacrylate monomer and put into cells with homeotropic alignment layers. The cells are irradiated by UV light for photopolymerisation. The resulting polymer network is perpendicular to the cell surface.

At zero field the liquid crystal is in the uniform homeotropic state shown in Figure 7.13(a); the material is a homogeneous medium. The light propagates through the material without scattering. When an electric field is applied, the liquid crystal molecules are tilted away from the field direction because of their negative dielectric anisotropy. The material is switched to a multi-domain structure, as shown in Figure 7.13(b). At the boundary between the domains, the refractive indices change abruptly. The material is optically non-uniform and therefore light is scattered when passing through it. When the liquid crystal molecules are tilted away from the field direction, the director in the domains is random in the x–y plane. Therefore light polarised in both the x and y directions is scattered. The polymer-stabilised homeotropic nematic liquid crystal is a reverse-mode light shutter in the sense that it is transparent in the field-off state and opaque in the field-on state.

The voltage–transmission curve of an 8 μm polymer-stabilised homeotropic nematic liquid crystal light shutter is shown in Figure 7.14 [49]. At 0 V the material is in the uniform homeotropic state and its transmittance is nearly 100%. When the applied voltage is increased above a threshold, the transmittance begins to decrease. The minimum transmittance is below 10%. From this figure, two effects can be seen. First, as the polymer concentration

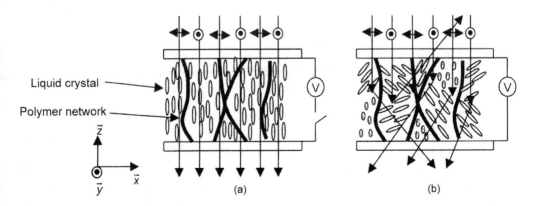

Figure 7.13 Schematic diagram showing how the polymer-stabilised homeotropical nematic liquid crystal light shutter works

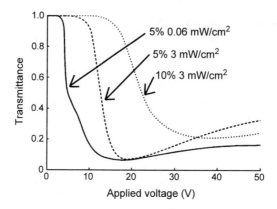

Figure 7.14 Transmittance versus applied voltage curves of the polymer-stabilised homeotropic nematic liquid crystal light shutter [49]

is increased, the aligning effect of the polymer network (tending to keep the liquid crystal in the normal direction of the cell) becomes stronger, and therefore the drive voltage is increased. Second, when the curing UV intensity is increased, polymer bundles with smaller lateral size form and the distance between them becomes shorter [39]; the aligning effect of the polymer network also becomes stronger, and therefore the drive voltage is increased.

Polymer stabilised cholesteric liquid crystals

Cholesteric liquid crystals have a helical structure in which the liquid crystal director twists around a perpendicular axis (the helical axis). They exhibit three major textures, determined by the direction of the helical axis and the applied fields. When a cholesteric liquid crystal is in the planar texture, the helical axis is perpendicular to the cell surface, and the material reflects light around the wavelength $\bar{n}P$, where $\bar{n}$ is the average refractive index and P is the pitch of the liquid crystal. When the cholesteric liquid crystal is in the focal conic texture, the helical axis is more or less random throughout the cell, and the material is usually optically scattering. When a sufficiently high field is applied across the cell (along the cell normal direction) and the cholesteric liquid crystal has a positive dielectric anisotropy, it can be switched to the homeotropic texture where the helical structure is unwound, the liquid crystal director is aligned in the cell normal direction, and the material is transparent. (More detailed discussions on cholesteric liquid crystals are given in the chapter on cholesteric reflective displays.)

Yang, Doane and Chien developed the polymer-stabilised cholesteric texture (PSCT) [36, 50, 51]. The PSCT normal-mode material is made from a mixture of cholesteric liquid crystal and a small amount of biacrylate monomer. The pitch of the liquid crystal is a few microns (0.5–5 μm). No special cell surface treatment is needed. The mixture is in the cholesteric phase and is sandwiched between two parallel glass plates with conducting coating. When the cell is exposed to UV irradiation for photo-polymerisation, the material is in the homeotropic texture in the presence of a sufficiently high electric field. The polymer network formed is perpendicular to the cell surface, as shown in Figure 7.15 [58].

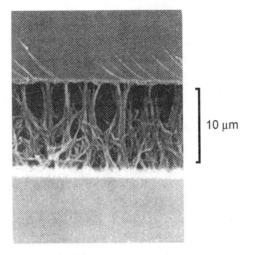

Figure 7.15   SEM photograph of polymer network in PSCT normal-mode material [38]

When a PSCT normal-mode light shutter is in the zero field state, the liquid crystal tends to have a helical structure while the polymer network tends to keep the liquid crystal director parallel to it. Thus the material has a poly-domain structure, as shown in Figure 7.16(a) and thus the focal conic texture is stabilised. In this state, it is optically scattering. When a sufficiently high electric field is applied across the cell, the liquid crystal ($\Delta\varepsilon > 0$) is switched to the homeotropic texture, as shown in Figure 7.16(b), and therefore it becomes transparent. Because the concentration of the polymer is low and both the liquid crystal and the polymer are aligned in the cell normal direction, the PSCT normal-mode light shutter is transparent at any viewing angle.

A typical voltage–transmission curve of the PSCT normal-mode light shutter is shown in Figure 7.17. At zero field, the material is in the stabilised focal conic texture and is scattering, and the transmittance is low. As the applied voltage is increased above a threshold, its transmittance begins to increase. When the applied voltage is sufficiently high, the material

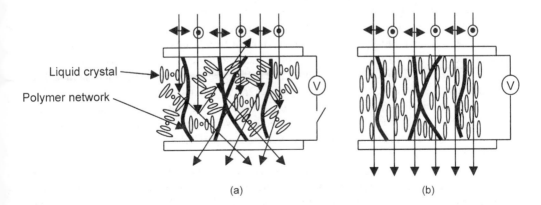

Liquid crystal

Polymer network

(a)   (b)

Figure 7.16   Schematic diagram showing how the polymer-stabilised cholesteric texture normal-mode light shutter works

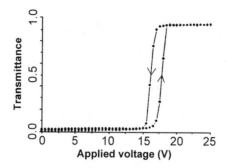

Figure 7.17   Voltage–transmittance curve of a PSCT normal-mode light shutter

is switched to the homeotropic texture with the high transmittance at around 90%. The light loss is mainly caused by the reflection from the two glass–air interfaces. There is a hysteresis in the transition between the focal conic texture and the homeotropic texture, which also exists in pure cholesteric liquid crystals.

The electro-optical properties of the PSCT normal-mode light shutter are affected by the pitch of the liquid crystal, the polymer concentration and the curing conditions. In order to scatter visible light strongly, the focal conic domain size has to be around the wavelength of the light. The main factors affecting the domain size are the pitch, the polymer concentration and the curing UV intensity. The drive voltage is mainly determined by the pitch and the dielectric anisotropy of the liquid crystal. A faster response can be achieved with shorter pitch cholesteric liquid crystals.

Although the hysteresis in the focal conic-homeotropic transition makes it unsuitable for gray scale operation, it is very useful in bistable operation. The material is bistable at a bias voltage in the hysteresis loop. For pure cholesteric liquid crystals (without polymer), in order to obtain a sufficiently large hysteresis for bistable operation, homeotropic alignment layers are used [52, 53]. In this case, the hysteresis depends on the cell thickness. The hysteresis decreases with increasing cell thickness, which imposes a trade-off between bistability and contrast. This problem is overcome by using polymer networks as in the PSCT normal-mode material. The hysteresis can be enhanced greatly using polymer networks and a large cell thickness can be employed [54–57]. Figure 7.18 shows the voltage–transmittance curves of several PSCT normal-mode light shutters with various polymer concentrations [55].

The PSCT reverse-mode light shutter is also made from a mixture of a cholesteric liquid crystal ($\Delta\varepsilon > 0$) and a small amount of a biacrylate monomer [36, 50, 58]. The pitch of the liquid crystal is a few microns (3–15 µm). The cell has homogeneous alignment layers. The mixture is in the cholesteric phase and is put into the cell in a vacuum. Due to the alignment layers, the material is in the planar texture with the helical axis perpendicular to the cell surface at zero field. When the cell is irradiated by UV light to polymerise the monomers, it is in the planar texture. The polymer network formed is parallel to the cell surface, as shown in Figure 7.19 [38].

At zero field, the material is in the planar texture as shown in Figure 7.20(a). Because the pitch is in the infrared region, the material is transparent to visible light. When an external field is applied across the cell, the field tends to align the liquid crystal in the cell normal direction while the polymer network tends to keep the liquid crystal in the planar texture. As a result of the competition between these two factors, the liquid crystal is switched to the

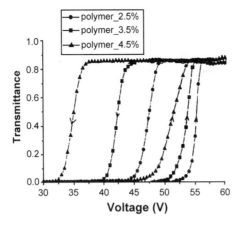

Figure 7.18 The voltage–transmittance curves of the PSCT normal-mode shutters with three different polymer concentrations [55]

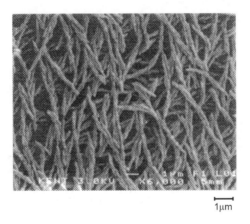

1μm

Figure 7.19 SEM photograph of the polymer network in PSCT reverse-mode light shutter [38]

poly-domain focal conic texture as shown in Figure 7.20(b), and the material becomes scattering.

The voltage–transmittance curve of a PSCT reverse-mode light shutter is shown in Figure 7.21. At zero field the material is in the planar texture and has a high transmittance (about 90%) for visible light. When the applied voltage is increased above a threshold, the liquid crystal begins to tilt away from the cell plane and transfers into the poly-domain focal conic texture, and the transmittance decreases. When the applied voltage is sufficiently high, all the liquid crystal is switched to the focal conic texture and the transmittance is decreased to the minimum. The main factors controlling the electro-optical properties of a PSCT reverse-mode light shutter are the cholesteric pitch and the polymer concentration. The cholesteric pitch has to be sufficiently short so that the scattering of the focal conic texture is independent of the incident light polarisation. The polymer concentration must be sufficiently high (the polymer network is stronger with higher concentration) so that the polymer network is

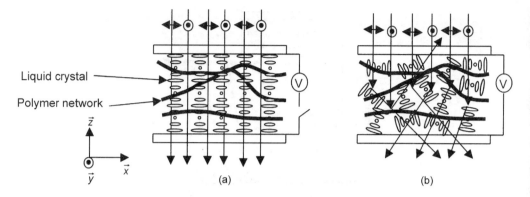

Figure 7.20 Schematic diagram showing how the polymer-stabilised cholesteric texture reverse-mode light shutter works

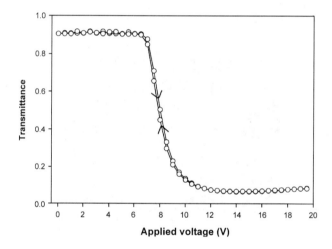

Figure 7.21 The voltage–transmittance curve of a PSCT reverse-mode light shutter [58]

strong and would not be damaged (reoriented) under the applied field. Also, precautions should be taken not to apply too high a voltage in order to avoid damaging the polymer network. Strong polymer networks are obtained with low UV curing intensities.

Aerosil-frame stabilised nematic liquid crystal

Kreuzer *et al.* created a dispersion of nematic liquid crystal and silica particles [59–62]. The volume fraction of the silica particles is about 2 to 3%. The silica particles with nm size form aggregates via $\equiv$Si–O–Si$\equiv$moieties. The aggregates are linked to form an aerosil-frame by hydrogen bonds between the $\equiv$Si–OH groups on the surface of the aggregates. The aerosil-frame creates the poly-domain structure shown in Figure 7.22(a), and the material is scattering. When a sufficiently high field is applied, the liquid crystal ($\Delta\varepsilon>0$) is aligned

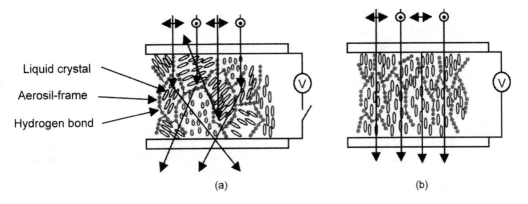

Figure 7.22 Schematic diagram showing how the aerosil-frame stabilised nematic liquid crystal light shutter works

homeotropically as shown in Figure 7.22(b), and the material becomes transparent. The uniformly aligned liquid crystal exerts a torque on the aerosil-frame. A unique feature of the aerosil-frame is that the hydrogen bonds can be unlocked under mechanical interaction, and the aerosil-frame is broken. Then the $\equiv$Si–OH groups on the surface of the aggregates re-group to form hydrogen bonds again and the aggregates form a new frame which favors the homeotropic orientation of the liquid crystal. When the applied field is turned off, the material remains in the homeotropic state.

The scattering loss and voltage curve of the aerosil-frame stabilised nematic liquid crystal material is shown in Figure 7.23 [59]. At 0 V the material is in the poly-domain scattering state; the light scattering loss is high. As the applied voltage is increased, the liquid crystal is aligned towards the homeotropic state, and the material becomes less scattering. The symbol $\times$ represents the scattering loss data measured with the voltage applied. The symbol $\circ$ represents the scattering loss data measured when the applied voltage is turned off. When the

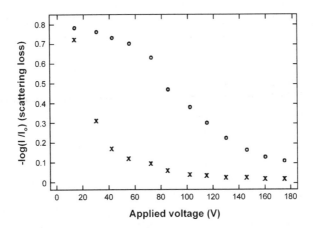

Figure 7.23 Scattering loss versus applied voltage for the aerosil-frame stabilised nematic liquid crystal light shutter [59]

applied voltage is low, for example at 50 V, the liquid crystal is aligned homeotropically and the scattering loss is low when the voltage is applied. At this voltage, the aerosil-frame has not yet been reoriented. When the voltage is turned off, the liquid crystal relaxes back to the original structure and the scattering loss is high. When the applied voltage is high, for example at 160 V, the liquid crystal is aligned homeotropically and the aerosil-frame is reoriented, and the scattering loss is low. When the voltage is turned off, the material remains in the homeotropic state and the scattering loss remains low.

The aerosil-frame stabilised nematic liquid crystal material has a memory effect and is multistable. Once the material is switched to the transparent homeotropic state, it remains there. Two schemes can be employed to rewrite a display made from the material. In the first scheme, a regular nematic liquid crystal with a positive dielectric anisotropy is used. The display is first switched to the transparent homeotropic state by applying a high voltage. It is then selectively addressed to the scattering state using laser addressing. The laser irradiation heats the material to the isotropic phase locally. When the material is cooled quickly to room temperature, the material ends in the scattering state. Thus images can be generated. In the second scheme, a dual frequency nematic liquid crystal is used. When a low-frequency voltage is applied, the liquid crystal has a positive dielectric anisotropy and is therefore switched to the transparent state. When a high-frequency voltage is applied, the liquid crystal has a negative dielectric anisotropy and therefore is switched to the scattering state.

Holographic polymer-dispersed liquid crystal

In PDLCs formed by polymerisation-induced phase separation, spatial variations in structure can be achieved when non-uniform polymerisation conditions are introduced [63]. An example of this is the holographically formed PDLC [64–70]. A mixture of a liquid crystal and a photo-polymerisable monomer is sandwiched between two glass substrates. A laser light is used to initiate the polymerisation. In the polymerisation, the cell is irradiated by the laser beam from both sides, as shown in Figure 7.24. The two incident light beams interfere with each other inside the cell and form the intensity pattern shown on the right-hand side of the figure. In the region where the light intensity is high, more free radicals are produced, and monomers are polymerised to form polymers. The liquid crystal is squeezed out of the region. Thus alternating polymer rich and liquid crystal rich layers are formed. The quantity d (the thickness of one layer of polymer plus the thickness of one layer of liquid crystal) is determined by the wavelength and incident angle of the laser beam.

The liquid crystal and polymer are chosen such that the ordinary refractive index n_o of the liquid crystal is equal to the refractive index n_p of the polymer. At zero field, the liquid crystal has a random orientation structure, as shown in Figure 7.25(a) and the refractive index of the cell oscillates periodically. If the incident light with an angle of incidence θ satisfies the Bragg condition $\lambda = 2d \cos \theta$, it will be reflected. When a sufficiently high external electric field is applied across the cell, the liquid crystal ($\Delta \varepsilon > 0$) will be aligned perpendicular to the layers, as shown in Figure 7.25(b). The incident light encounters the same refractive index in the polymer rich layer and liquid crystal rich layer, and passes through the material without reflection. Thus the holographic PDLC can be used for reflective displays.

The spectral response of a holographic PDLC to applied electric fields is shown in Figure 7.26 where white incident light is used [68]. At 0 V, due to the periodic undulation of the

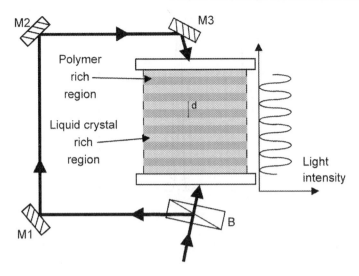

Figure 7.24 Schematic diagram showing how the holographic PDLC is formed. B: beam splitter, M: mirror

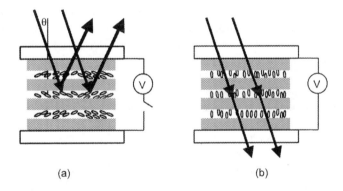

Figure 7.25 Schematic diagram showing how the holographic PDLC is used for a reflective display

refractive index, the cell has an intense narrow reflection peak. When the applied voltage is increased, the liquid crystal is aligned toward the layer normal direction. The amplitude of the oscillation of the refractive index decreases and the reflection of the cell decreases. There are two problems with the holographic PDLC when used for reflective displays. First, the drive voltage is high. The drive voltage is approximately equal to the field threshold of the Freedericksz transition of the liquid crystal layer multiplied by the cell thickness. If the anchoring of the liquid crystal at the polymer surface is strong, the field threshold is inversely proportional to the thickness of the liquid crystal layer, which is less than 0.5 μm. The second problem is that stacked layers have to be used in order to make full-color displays, which will be expensive.

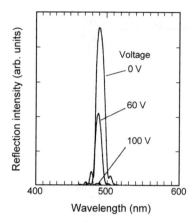

Figure 7.26 Reflection spectra of the holographic PDLC under various applied voltages [68]

4 Scattering Properties of Liquid Crystal/Polymer Composites

Light scattering by liquid crystal/polymer composites is caused by the spatial variation of refractive index in the materials. It is similar to the scattering of clouds which consist of water droplets. As discussed, in PDLCs the scattering is due to the refractive index mismatch between the liquid crystal and the polymer. In PSLCs, the scattering is due to the refractive index mismatch between the liquid crystal and the polymer network, as well as that between liquid crystal domains. A precise calculation of the scattering of PDLCs is very difficult because of the birefringence of the liquid crystal, the variation in droplet size and the irregularity of the droplet shape. A precise calculation of the scattering of PSLCs is almost impossible because of the lack of understanding of the liquid crystal domain structure. Here we will offer only some qualitative discussion on the theory of the scattering and present available experimental results. The scattering of PDLCs and PSLCs is very similar in that most of the incident light is scattered in the forward direction. Much more work needs to be done on the scattering of PSLCs. Here our discussion is focussed on that of PDLCs.

4.1 Basic properties

In PDLCs the liquid crystal and the polymer are chosen in such a way that the refractive index of the polymer n_p is equal to the ordinary refractive index n_o of the liquid crystal. As shown in Figure 7.6, when a PDLC is in the field-off state, the orientation of the droplets is random through the sample; the incident light encounters different refractive indices in the polymer binder and the liquid crystal droplets, and therefore it is scattered. When the PDLC is in the field-on state, the droplets are aligned in the cell normal direction; the normal incident light encounters the same refractive index in the polymer binder and the liquid crystal, and therefore is transmitted. For obliquely incident light, as shown in Figure 7.27, if the polarisation of the incident light is perpendicular to the plane defined by the liquid crystal director and the wavevector of the incident light, the light encounters the refractive index n_o and therefore is

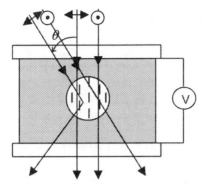

Figure 7.27 Schematic diagram showing the scattering of the PDLC in the field-on state

not scattered. If the polarisation of the incident light is in the plane defined by the liquid crystal director and the wavevector of the incident light, the light encounters the refractive index $n(\theta) = n_o n_e / (n_o^2 \sin^2 \theta + n_e^2 \cos^2 \theta)^{1/2}$ which is different from the refractive index n_p of the polymer, and therefore is scattered. The transmittance of the light with this polarisation at various incident angles α is shown in Figure 7.28 [7]. The transmittance decreases by half when the incident angle is increased to 30°. If the obliquely incident light is unpolarised, the component with the polarisation in the plane defined by the droplet director and the light propagation direction is scattered, which makes the PDLC hazy. This limitation on the viewing angle can be eliminated when a linear polariser is laminated on the PDLC with the trade-off that the on-state transmittance is decreased by half.

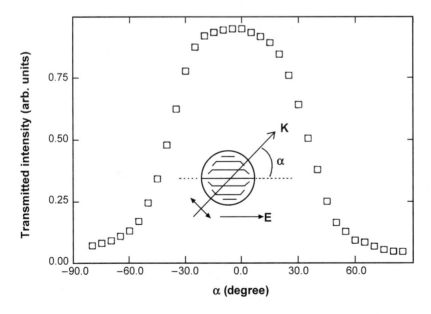

Figure 7.28 Angular dependence of the transmittance of the PDLC in on-state for the incident light linearly polarised in the incident plane [7]

4.2 Rayleigh–Gans scattering theory

When a light is propagating in a medium, the electric field of the light induces a dipole moment at each point, which oscillates with the same frequency as the light. The oscillating dipole moments radiate light in all directions [71], and the net electric field at any point is the vector sum of the fields produced by all the dipole radiators, as schematically shown in Figure 7.29. The incident light is collimated and the electric field at the source point r is

$$E_{in}(r, t) = E_o \exp(iK_o r - i\omega t) \tag{7.7}$$

The total induced dipole moment in the volume element d^3r is given by

$$dP(r, t) = \alpha \vec{E}_{in} = \alpha \vec{E}_o \exp(iK_o r - i\omega t)\ d^3r \tag{7.8}$$

where α is the polarisability and K_o is the wavevector of the incident light. α is related to the dieletric tensor by $\alpha(r) = \varepsilon_o[\varepsilon(r) - I]$, where I is the unit matrix. K_o is related to the frequency ω by $K_o = K_o k_o = (\omega/C)\hat{k}_o$, where C is the speed of light in a vacuum and $\hat{k}_o$ is a unit vector along the incident direction. The wavevector of the scattered light is $K' = K_o \hat{k}'$, as shown in Figure 7.30. $K_S = 2K_o \sin(\theta/2)$, where θ is the scattering angle.

The scattered field radiated by the dipole moment in the volume element d^3r is given by

$$
\begin{aligned}
dE_s(R, t) &= \left. \frac{\hat{k}' \times [\hat{k}' \times d\ddot{P}(r, t')]}{4\pi\varepsilon_o C^2 |R - r|} \right|_{t'=t-(1/C)|R-r|} \\
&= \left. \frac{\hat{k}' \times \{\hat{k}' \times [-\omega^2\alpha(r)E_{in}(r, t')d^3r]\}}{4\pi\varepsilon_o C^2 |R - r|} \right|_{t'=t-(1/C)|R-r|} \\
&= \frac{-\omega^2 \hat{k}' \times \{\hat{k}' \times [\alpha(r)E_o]\}}{4\pi\varepsilon_o C^2 |R - r|} \exp\left[iK_o r - i\omega\left(t - \frac{1}{C}|R - r|\right)\right] d^3r \\
&= \frac{-\omega^2 \hat{k}' \times \{\hat{k}' \times [\alpha(r)E_o]\}}{4\pi\varepsilon_o C^2 |R - r|} \exp[-i\omega t + i[K_o r - K'(R - r)]] d^3r \\
&= \frac{-\omega^2 \hat{k}' \times \{\hat{k}' \times [\alpha(r)E_o]\}}{4\pi\varepsilon_o C^2 |R - r|} \exp(-iK_S r) \exp(-\omega t + iK'R) d^3r
\end{aligned} \tag{7.9}
$$

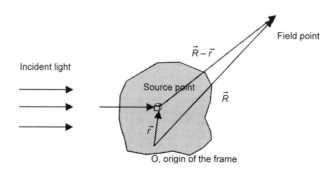

Figure 7.29 Schematic diagram showing the scattering of a medium

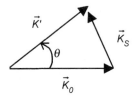

Figure 7.30 Schematic diagram showing the wavevectors of the incident and scattered light

For a far field point, $R \gg r$, $|R-r| \simeq R$. Equation 7.9 becomes

$$dE_s(R, t) = \frac{-K_o^2 \hat{k}' \times \{\hat{k}' \times [\alpha(r)E_o]\}}{4\pi\varepsilon_o R} \exp\left(-iK_s r\right) \exp\left(-i\omega t + iK'R\right) d^3r \qquad (7.10)$$

The total scattered field is given by

$$E_s(R, t) = -\frac{K_o^2}{4\pi\varepsilon_o R} \exp\left(-i\omega t + iK'R\right) \int \hat{k}' \times \{\hat{k}' \times [\alpha(r)E_o]\} \exp\left(-iK_s r\right) d^3r$$

$$= -\frac{K_o^2}{4\pi\varepsilon_o R} V \exp\left(-i\omega t + iK'R\right) \hat{k}' \times \{\hat{k}' \times [\alpha(K_s)E_o]\} \qquad (7.11)$$

where

$$\alpha(K_S) = \frac{1}{V} \int \alpha(r) \exp\left(-iK_s r\right) d^3r$$

is the Fourier component of the polarisability and V is the volume of the scattering medium. The intensity of the scattered light

$$I_S = |E_s|^2 \propto \left|\frac{1}{4\pi\varepsilon_o}\alpha(K_S)\right|^2 = \left|\frac{1}{4\pi}\varepsilon(K_S)\right|^2 \qquad (7.12)$$

For a qualitative discussion, in a PDLC with droplet size D, the Fourier transformation of the dielectric tensor is peaked at the wavevector $(2\pi/D)$. Roughly speaking, the scattering angle is given by

$$2K_o \sin\left(\frac{\theta}{2}\right) = 2\frac{2\pi}{\lambda} \sin\left(\frac{\theta}{2}\right) = \frac{2\pi}{D}$$

That is, $\sin (\theta/2) = (\lambda/2D)$. We can make a few remarks on the scattering. First, in order to scatter the light, the droplet size should be comparable to the wavelength of the light. Second, larger droplets scatter the light more in the forward direction (small scattering angle) than smaller droplets.

We define the coordinate for the incident light in such a way that the z axis is parallel to the incident direction and the x axis is in the plane defined by K_o and K'; and we define the coordinate for the scattered light in such a way that the z' axis is parallel to the scattering direction and the x' axis is also in the plane defined by K_o and K', as shown in Figure 7.31. The incident field E_o makes an angle α with the x axis. In matrix form,

$$E_o = \begin{pmatrix} E_{lo} \\ E_{ro} \end{pmatrix} = E_o \begin{pmatrix} \cos\alpha \\ \sin\alpha \end{pmatrix}$$

defined in the frame xyz. The scattered field is

$$E_s = \begin{pmatrix} E_{ls} \\ E_{rs} \end{pmatrix}$$

defined in frame $x'y'z'$. Rewrite Equation 7.11 in the matrix form

$$
\begin{aligned}
E_s &= -\frac{K_o^2}{4\pi\varepsilon_o R} V \exp\left(-i\omega t + iK'R\right) \hat{k} \times \left\{ \hat{k}' \times \left[\alpha(K_s)E_o\right] \right\} \\
&= \frac{1}{iK_o R} \exp\left(-i\omega t + iK'R\right) S \cdot E_o
\end{aligned}
\tag{7.13}
$$

where

$$S = S(\theta, \alpha) = \begin{pmatrix} S_{ll} & S_{lr} \\ S_{rl} & S_{rr} \end{pmatrix}$$

is the scattering matrix. The two components of the differential scattering cross section are

$$\left(\frac{d\sigma}{d\Omega}\right)_l = \frac{|E_{ls}|^2}{|E_o|^2} R^2 = \frac{1}{K_o^2} |S_{ll}\cos\alpha + S_{lr}\sin\alpha|^2 \tag{7.14}$$

$$\left(\frac{d\sigma}{d\Omega}\right)_r = \frac{|E_{rs}|^2}{|E_o|^2} R^2 = \frac{1}{K_o^2} |S_{lr}\cos\alpha + S_{rr}\sin\alpha|^2 \tag{7.15}$$

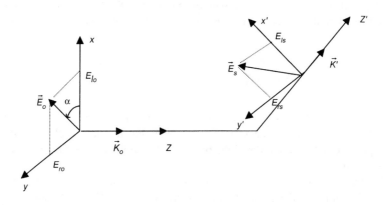

Figure 7.31 The coordinates for the incident and scattered light

The total differential scattering section is

$$\frac{d\sigma}{d\Omega} = \left(\frac{d\sigma}{d\Omega}\right)_l + \left(\frac{d\sigma}{d\Omega}\right)_r$$

Let us first consider the scattering of an isotropic spherical droplet with the refractive index n and radius a.

$$\alpha(K_S)E_o = \alpha(K_S)E_o$$

$$
\begin{aligned}
E_{ls} &= \hat{x}'\left(-i\frac{1}{4\pi\varepsilon_o}K_o^3 V\alpha(K_s)\hat{k}' \times \left[\hat{k}' \times E_o\right]\right) \\
&= -i\frac{1}{4\pi\varepsilon_o}K_o^3 V\alpha(K_s)\hat{x}'\left[\hat{k}'\left(\hat{k}'E_o\right) - E_o\right] \\
&= i\frac{1}{4\pi\varepsilon_o}K_o^3 V\alpha(K_s)\hat{x}'E_o \\
&= i\frac{1}{4\pi\varepsilon_o}K_o^3 V\alpha(K_s)(\cos\ \theta\hat{x} + \cos\ \theta\hat{z})(E_{lo}\hat{x} + E_{ro}\hat{y}) \\
&= i\frac{1}{4\pi\varepsilon_o}K_o^3 V\alpha(K_s)\ \cos\ \theta E_{lo} \\
E_{rs} &= \hat{y}'\left(-i\frac{1}{4\pi\varepsilon_o}K_o^3 V\alpha(K_s)\hat{k}' \times \left[\hat{k}' \times E_o\right]\right) \\
&= i\frac{1}{4\pi\varepsilon_o}K_o^3 V\alpha(K_s)\hat{y}'E_o \\
&= i\frac{1}{4\pi\varepsilon_o}K_o^3 V\alpha(K_s)\hat{y}(E_{lo}\hat{x} + E_{ro}\hat{y})
\end{aligned}
\tag{7.16}
$$

$$= i\frac{1}{4\pi\varepsilon_o}K_o^3 V\alpha(K_s)E_{ro} \tag{7.17}$$

Therefore the scattering matrix is

$$S = i\frac{1}{4\pi\varepsilon_o}K_o^3 V\alpha(K_s)\begin{pmatrix} \cos\ \theta & 0 \\ 0 & 1 \end{pmatrix} \tag{7.18}$$

Now we calculate the Fourier component of the polarisability.

$$
\begin{aligned}
\alpha(K_s) &= \frac{1}{V} \int\limits_{whole\ space} \alpha(r)\ \exp\left(-iK_s r\right) d^3 r \\
&= \frac{1}{V} \int\limits_{inside\ droplet} \alpha(r)\ \exp\left(-iK_s r\right) d^3 r + \frac{1}{V} \int\limits_{outside\ droplet} \alpha(r)\ \exp\left(-iK_s r\right) d^3_r \\
&= \frac{1}{V} \int\limits_{whole\ space} \alpha_{out}\ \exp\left(-iK_s r\right) d^3 r + \frac{1}{V} \int\limits_{inside\ droplet} (\alpha_{in} - \alpha_{out})\ \exp\left(-iK_s r\right) d^3 r
\end{aligned}
$$

The first integral is zero. $\alpha_{in} = \varepsilon_o(\varepsilon - 1) = \varepsilon_o(n^2 - 1)$. The medium outside the droplet is also isotropic with refractive index $n_o \cdot \alpha_{out} = \varepsilon_o(n_o^2 - 1)$. Hence

$$\alpha(K_s) = \frac{1}{V}\varepsilon_o\left(n^2 - n_o^2\right) \int\limits_{inside\ droplet} \exp\left(-iK_s r\right)d^3r \tag{7.19}$$

Define

$$Q(\theta) = \frac{1}{V} \int\limits_{inside\ droplet} \exp\left(-iK_s r\right)d^3r$$

For the integration, we use a polar coordinate with K_s in the polar direction.

$$
\begin{aligned}
Q(\theta) &= \frac{2\pi}{V}\int_0^\pi d\beta\left[\int_0^a \exp\left(-iK_s r\ \cos\ \beta\right)\ \sin\ \beta r^2 dr\right] \\
&= \frac{2\pi}{V}\int_0^a r^2 dr\left[\int_0^\pi \exp\left(-iK_s r\ \cos\ \beta\right)\ \sin\ \beta d\beta\right] \\
&= \frac{2\pi}{V}\int_0^a \frac{2}{K_s}\ \sin\ (K_s r)\ rdr \tag{7.20} \\
&= \frac{4\pi}{VK_s^3}\int_0^{aK_s} u\ \sin\ u du \\
&= \frac{4\pi}{VK_s^3}\left[\sin\ (aK_s) - aK_s\ \cos\ (aK_s)\right]
\end{aligned}
$$

The differential scattering cross section for unpolarised incident light is given by

$$
\begin{aligned}
\frac{d\sigma}{d\Omega} &= \frac{K_o^6}{K_o^2}\frac{1}{16\pi^2}(n^2 - n_o^2)\left(\cos^2\theta < \cos^2\alpha > + < \sin^2\alpha >\right) \\
&\quad \times \left(\frac{4\pi}{K_s^3}\left[\sin\ (aK_s) - aK_s \cos\ (aK_s)\right]\right)^2
\end{aligned} \tag{7.21}
$$

We know that $K_s = 2K_o \sin(\theta/2)$ and $\langle\sin^2\alpha\rangle = \langle\cos^2\alpha\rangle = 1/2$. Let $A = aK_o$, then

$$\frac{d\sigma}{d\Omega} = \pi a^2(n^2 - n_o^2)^2\frac{(\cos^2\theta + 1)}{128\pi A^2}\left(\frac{\sin\left[2A \sin(\theta/2)\right] - 2A \sin(\theta/2)\cos\left[2A \sin(\theta/2)\right]}{\sin^3(\theta/2)}\right)^2 \tag{7.22}$$

The light scattered in the forward direction is given by

$$
\begin{aligned}
\sigma_{forward} &= 2\pi \int_0^{\pi/2} \frac{d\sigma}{d\Omega} \sin\theta d\theta \\
&= \pi a^2 (n^2 - n_o^2)^2 \int_0^{\pi/2} \frac{(\cos^2\theta + 1)}{64 A^2} \\
&\quad \times \left(\frac{\sin[2A\sin(\theta/2)] - 2A\sin(\theta/2)\cos[2A\sin(\theta/2)]}{\sin^3(\theta/2)} \right)^2 \sin\theta d\theta
\end{aligned}
\tag{7.23}
$$

The light scattered in the backward direction is given by

$$
\begin{aligned}
\sigma_{forward} &= 2\pi \int_{\pi/2}^{\pi} \frac{d\sigma}{d\Omega} \sin\theta d\theta \\
&= \pi a^2 (n^2 - n_o^2)^2 \int_{\pi/2}^{\pi} \frac{(\cos^2\theta + 1)}{64 A^2} \\
&\quad \times \left(\frac{\sin[2A\sin(\theta/2)] - 2A\sin(\theta/2)\cos[2A\sin(\theta/2)]}{\sin^3(\theta/2)} \right)^2 \sin\theta d\theta
\end{aligned}
\tag{7.24}
$$

The scattering cross sections of the materials with $n_o = 1.5$ and $n = 1.7$ are plotted in Figure 7.32 where the unit of the vertical axis is πa^2. When the droplet size a is smaller than the wavelength λ $(A = 2\pi(a/\lambda)$, the forward and backward scattering cross sections are about the same. When the droplet size is larger than the wavelength, most of the scattered light is in the forward direction.

The scattering of liquid crystal droplets can be calculated in the same way, except that dielectric tensors have to be used [12]. The calculation is more complicated and is not presented here. Readers interested in the detailed calculation of the scattering of PDLCs are referred to the papers published by Žumer et al. [12, 72, 73]. The formulation presented in this section is called Rayleigh–Gans scattering, which uses the three following assumptions: $|n/n_o - 1| \ll 1$, so that refraction at the droplet interface can be neglected;

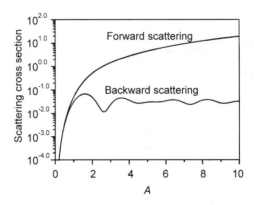

Figure 7.32 The scattering cross sections versus A

$2K_o a|n-n_o| \ll 1$; and $\sigma \ll 1$, so that there is no multiple scattering inside the medium and the incident light intensity at any point inside the medium is the same.

4.3 Anomalous diffraction scattering theory

In PDLCs the wavelength of visible light is around 500 nm, the droplet diameter is about 1 μm and $|n-n_o| \simeq 0.2$; therefore $2K_o a|n-n_o| \simeq 1$. Hence the Rayleigh–Gans scattering theory does not describe the scattering of PDLCs very well. A better approach for PDLCs is the anomalous diffraction theory [74, 75]. Here the assumptions are: $|n/n_o-1| \ll 1$, so that refraction at the droplet interface can be neglected; and $K_o a \gg 1$, so that the light ray can be traced. In this theory, the light ray is assumed to go through the liquid crystal droplet without scattering, as shown in Figure 7.33. The phase and polarisation states of the ray are calculated using the Jones matrix method. Consider a plane beyond the droplet which is perpendicular to the incident light propagation direction; the phase is no longer a constant on the plane but is modified by the liquid crystal droplet. Take all the points on the plane as the centers of secondary spherical waves, as in Huygen's principle. The scattered field is calculated by summing over all the fields of the secondary waves. One of the main results of the theory is that the scattered light is concentrated in a cone of linear angle less than 30° in the forward direction.

4.4 Multiple scattering and scattering profile

So far we have only considered single scattering, that is, a light ray is scattered only once by a droplet. This single scattering is adequate for PDLC films with a cell gap of less than 5 μm. For PDLC films thicker than this, multiple scattering becomes important and has profound effects on the scattering profile.

Kelly *et al.* used Monte Carlo simulation to study multiple scattering of PDLCs [76]. In their study, wave optics were used to study the single scattering by one droplet, and the

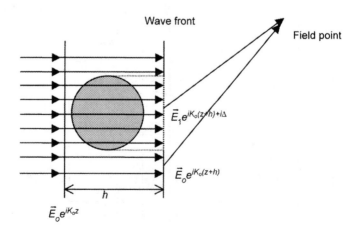

Figure 7.33 Schematic diagram showing how the scattered field is calculated in the anomalous diffraction Theory

photon was used to study multiple scattering by multiple droplets. Consider a PDLC film with thickness h, as shown in Figure 7.34. Divide the film into $M = h/D$ layers, where D is the droplet diameter. The probability that a photon is scattered when passing through one layer is q which is equal to the droplet density multiplied by the layer thickness and also by the scattering cross section of one droplet. The probability that a photon is scattered n times when passing through the PDLC film is given by the binomial distribution function:

$$P(n, M) = \frac{M!}{n!(M-n)!} q^n (1-q)^{M-n} \qquad (7.25)$$

The average number of times that a photon is scattered when passing through the film is

$$\bar{n} = qM = qh/D \qquad (7.26)$$

When $M \gg 1$, $q \ll 1$ and $\bar{n}$ is finite, the distribution function becomes the Poisson distribution:

$$P(n, \bar{n}) = \frac{\bar{n}^n}{n!} \exp(-\bar{n}) \qquad (7.27)$$

The probability that a photon is not scattered when passing through the film is given by

$$P(0, \bar{n}) = \frac{\bar{n}^0}{0!} \exp(-\bar{n}) = \exp(-\bar{n}) \qquad (7.28)$$

The transmittance (percentage of light not scattered) of the PDLC film is given by

$$T = \frac{I_{out}}{I_{in}} = \frac{P(0, \bar{n})}{\sum_{n=0}^{M} P(n, \bar{n})} = \frac{P(0, \bar{n})}{1} = \exp(-\bar{n}) \qquad (7.29)$$

Hence $\bar{n} = -\ln(T)$. T can be experimentally measured.

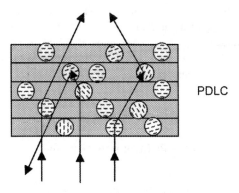

Figure 7.34 Schematic diagram showing multiple scattering in a PDLC

The scattering profile $(d\sigma/d\Omega)$ of the single scattering is obtained by theoretical calculation using the anomalous diffraction theory or, experimentally, by measuring the scattering profile of a thin PDLC film:

$$\frac{d\sigma}{d\Omega} = f_1(\theta, \phi) \tag{7.30}$$

where θ and ϕ are the polar angle and the azimuthal angle, respectively, and the polar axis is along the incident light direction. $f_1(\theta, \phi)$ is the probability that a photon is scattered in the direction described by (θ, ϕ). In multiple scattering, the scattering profile is the same for each individual scattering in the local frame where the polar axis is parallel to the incident light direction in that scattering event. For a photon scattered n times, in the first scattering, it is scattered in the direction (θ_1, ϕ_1) with the probability $f_1(\theta_1, \phi_1)$; in the second scattering, it is scattered in the direction (θ_2, ϕ_2) in the local frame with the probability $f_2(\theta_2, \phi_2)$; ...; in the nth scattering, it is scattered in the direction (θ_n, ϕ_n) in the local frame with the probability $f_1(\theta_n, \phi_n)$. The propagation direction of the photon before the first scattering is $\hat{k}_o = (0, 0, 1)^P$, where the superscript P indicates the transpose. The propagation direction of the photon after the nth scattering is given by

$$\hat{k}_n = \begin{pmatrix} k_{nx} \\ k_{ny} \\ k_{nz} \end{pmatrix} = \left[\prod_{i=1}^{n} M_i(\theta_i, \phi_i) \right] \hat{k}_o \tag{7.31}$$

where $M(\theta, \phi)$ is the standard three-dimensional rotation matrix

$$M(\theta, \phi) = \begin{pmatrix} \cos\theta\,\cos\phi & -\sin\phi & \sin\theta\,\cos\phi \\ \cos\theta\,\sin\phi & \cos\phi & \sin\theta\,\sin\phi \\ -\sin\theta & 0 & \cos\theta \end{pmatrix} \tag{7.32}$$

Because for a given final direction $\hat{k}_n$ (when n is large) there are many different ways of scattering to obtain $\hat{k}_n$, it is difficult to calculate the scattering profile after scattering n times. Nevertheless, it can be calculated using Monte Carlo simulation. The scattering space (θ, ϕ) is discretised into Q units with location described by (θ_j, ϕ_l) $(j = 1, 2, \ldots, M_j;\ l = 1, 2, \ldots, M_l;\ Q = M_j M_l)$.

In the first scattering, the scattering angle (θ_1, ϕ_1) is generated according to $f_1(\theta_1, \phi_1)$. In the second scattering, the scattering angle (θ_2, ϕ_2) is generated according to $f_1(\theta_2, \phi_2)$. In the nth scattering, the scattering angle (θ_n, ϕ_n) is generated according to $f_1(\theta_n, \phi_n)$.

Repeat the process for N $(N \gg 1)$ photons; $N_{j,l}$ photons end in the unit at (θ_j, ϕ_l). Then the scattering profile of a photon after scattering n times is given by

$$f_n(\theta_j, \phi_l) = \frac{N_{j,l}}{N} \tag{7.33}$$

Finally, the scattering profile of the PDLC film is given by

$$f(\theta_j, \phi_l) = \sum_{n=1}^{M} P(n, \bar{n}) f_n(\theta_j, \phi_l) \tag{7.34}$$

The experimental and calculated results are shown in Figure 7.35, where the droplet size is 1.2 μm and the cell thickness is 25 μm [76]. The scattering is independent of the azimuthal angle. The modeled scattering profile of single scattering is obtained using the anomalous diffraction scattering theory of a PDLC with 1.2 μm droplet size. It does not agree with the experimental result. For the 25 μm film, multiple scattering is important. After taking the multiple scattering into account, the modeled result agrees well with the experimental result. The FWHM (full width at half maximum) of the scattering profile peak is about 30°.

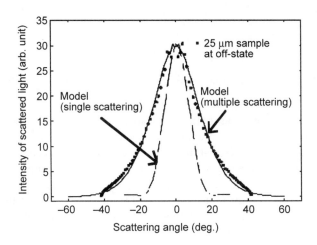

Figure 7.35   Scattering profiles of the PDLC film [76]

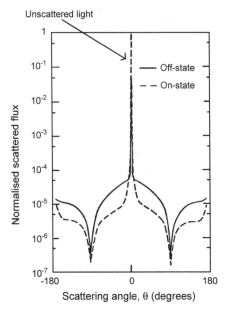

Figure 7.36   Scattering profile of the PDLC at all angles [77]

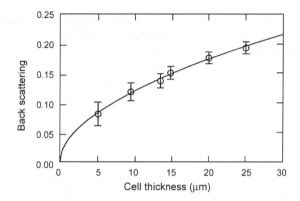

Figure 7.37 Cell thickness dependence of backward scattering of the PDLC films [78]

As mentioned before, the measured transmittance of PDLCs and PSLCs depends on the collection angle of the detection system. In the scattering state, the larger the collection angle, the more light is detected, and the higher the measured transmittance. In discussing transmittance and contrast, the collection angle must be specified. This is very important in designing projection displays made from PDLCs and PSLCs [24, 25].

In a reflective display application, it is desirable that backward scattering is high. In PDLCs, unfortunately, most of the scattering is forward. An experimentally measured scattering profile in all angles is shown in Figure 7.36 [77]. The sharp peak around 0° corresponds to the unscattered light. In the scattering state, about 25% of the light is scattered backward (including reflection from the glass–air interfaces). The cell thickness dependence of the backward scattering is shown in Figure 7.37 [78]. As the cell thickness is increased, the backward scattering is increased, with the trade-off of increased drive voltage. For switchable window applications, the performance is optimised with a droplet size of about 1 μm. Backward scattering can also be increased by using smaller droplets, which has the trade-off that the drive voltage is higher and the scattering becomes more wavelength-dependent.

5 PDLC Reflective Displays

The major applications of PDLCs and PSLCs are switchable windows and projection displays, because of their high on-state transmittance and compatibility with plastic substrates. They have also been used for reflective displays [78–80]. Here we discuss only reflective displays from PDLCs. PSLCs can be used in the same ways [81]. There are three types of display design, depending on the light condition and display background, as presented below.

5.1 PDLC reflective display with a black background

The display structure is schematically shown in Figure 7.38 [78]. Behind the PDLC film, there is a black film serving as the absorbing layer. When a pixel of the display is

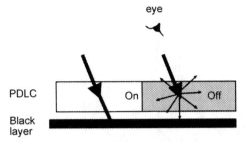

Figure 7.38 Schematic diagram showing the structure of the PDLC reflective display with a black background

in the field-on state, the material is transparent. The incident light passes through the PDLC without scattering, reaches the black film and is absorbed. Therefore the pixel has a black appearance. When a pixel is in the field-off state, the material is scattering. The incident light is scattered in all directions. Some of the scattered light reaches the reader's eye. Therefore the pixel has a white appearance.

In this design, the incident light can be either collimated or isotropic (incident at all angles). The contrast ratio can be higher than 10. The viewing angle is large without compensation films. The problem is the reflectivity, which usually is less than 20% for a reasonable cell thickness. The transmittance–voltage curve of PDLCs is not steep, as shown in Figure 7.7, so that an active matrix has to be used in high content information displays. Because of the limited voltage that can be provided by an active matrix, the cell thickness usually has to be less than 20 μm.

There have been many efforts to improve the reflectivity of PDLCs. The most successful approach is that using brightness enhancing films (BEFs) [79, 80]. BEFs were originally designed to improve the brightness of TN displays by producing collimated incident light. The structure of a BEF is shown in Figure 7.39(a). As an example, a BEF is made from a polymer with a refractive index of 1.59. The critical angle for total internal reflection on the bottom surface of the BEF is

$$\alpha = \sin^{-1}\left(\frac{1.0}{1.59}\right) = 38.97°$$

The corresponding incident angle on the top surface of the BEF is

$$\beta = \sin^{-1}\left[\frac{1.59\sin\,(45° - \alpha)}{1.0}\right] = 9.6°$$

The incident angle with respect to the vertical direction is $\gamma = -(45° - 9.6°) = -35.4°$. As shown in Figure 7.39(b), when the incident angle is in the range from $-35.4°$ to $45°$, it can pass through the BEF. When the incident light is in the range from $-90°$ to $-35.4°$ and from $45°$ to $90°$, it is totally reflected back. For incident light on the right side of the saw teeth of the BEF, a similar argument holds.

The structure of the PDLC reflective display with BEFs is shown in Figure 7.40. Some of the light scattered at relatively large angles in the forward direction by the material in the

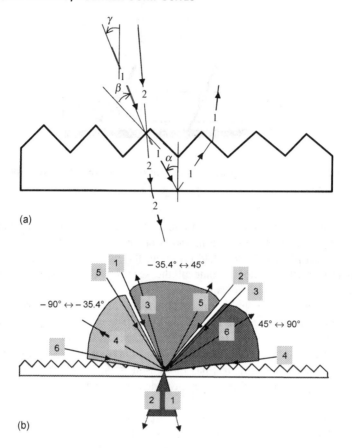

Figure 7.39 The structure of the BEF

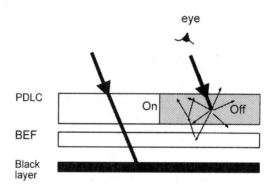

Figure 7.40 Structure of the PDLC reflective display with the BEF

field-off pixel is reflected by the BEF. Thus the light reflected by the display is enhanced. The reflection spectra of a PDLC reflective display with BEFs are shown in Figure 7.41 [80]. The reflection of the BEF itself is low, about 2%. The reflection of the PDLC is about

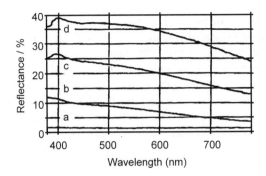

Figure 7.41 Reflection spectra of the PDLC reflective display with BEFs. (a) BEF, (b) PDLC (c) PDLC + 1 BEF, and (d) PDLC + 2 BEF [80]

10%. The addition of one BEF increases the reflection by about 10%. The drawback of adding the BEF is that the viewing angle is decreased.

5.2 PDLC reflective display with a mirror

The display structure is schematically shown in Figure 7.42. In this display the incident light must be collimated. For the on-state pixel, the incident light passes through the PDLC material, reaches the mirror and is reflected at the specular angle. The reader orients the display in such a way that the reflected light at the specular angle cannot reach the eye. The pixel has a dark apperance. For the pixel in the off-state, the incident light is scattered. Although most of the light is scattered in the forward direction, it reaches the mirror and is reflected. The reflected light is no longer at the specular angle. It is scattered again by the PDLC material on reflection. Some of the scattered light reaches the reader's eye. The pixel has a white appearance. In this design, the PDLC acts as a switchable diffuser. The light reflected in the backward direction is 100%.

If the incident light is partially collimated, the display with the above structure has a low contrast ratio and does not work well. The performance can be improved by using a

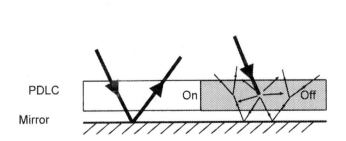

Figure 7.42 Structure of the PDLC reflective display with a mirror

dielectric mirror, as shown in Figure 7.43 [82]. The material for the dielectric mirror is deposited on the lower substrate. It is a short wavelength pass filter. It reflects light whose wavelength λ and incident angle θ satisfy the condition $\lambda/\cos\theta > \lambda_{co}$. The incident angle θ is the angle between the incident light and the normal of the dielectric mirror. λ_{co} is the cut-off wavelength and is usually chosen to be around 560 nm. The reader looks at the display in the normal direction. For the pixel in the on-state, the PDLC material is transparent. For light with a small incident angle and short wavelength, it passes through the PDLC material and dielectric mirror, and is absorbed by the black layer. For the pixel in the off-state, the PDLC material scatters the incident light. Although most of the light is scattered in the forward direction, it has large incident angles at the dielectric mirror and therefore is reflected by the mirror. This display has high reflection and a large viewing angle, as shown in Figure 7.44 [82]. The reflection is independent of the azimuthal angle. If the black layer is replaced by a white diffusion reflector, the reflection is increased, but the contrast ratio is decreased. The reflection shown in the figure is normalised to a standard diffusion reflector.

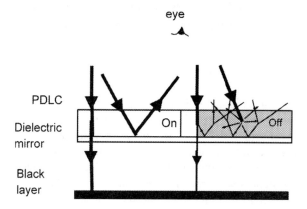

Figure 7.43   Structure of the PDLC reflective display with the dielectric mirror

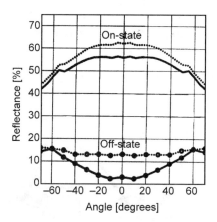

Figure 7.44   Reflection spectrum of the PDLC reflective display with a dielectric mirror. The solid line is for the display with a black layer. The dashed line is for the display with a white diffusion reflector [82]

5.3 Dichroic dye doped PDLC reflective display

The structure of a dichroic dye doped PDLC is depicted in Figure 7.45 [30, 83–86]. Behind the PDLC film there is a color reflector. The doped dichroic dye can be either black (absorbing all visible light) or color (absorbing the light in the band of the reflector), but black dyes are preferred. When a pixel of the display is in the field-off state, the material is absorbing and scattering. The incident light is absorbed, except that a small amount of light is scattered backward, and the pixel has a natural black appearance. When a pixel of the display is in the field-on state, the material is transparent. The incident light reaches the color reflector and the portion in the reflection band is reflected, and the pixel has a colored appearance. Fluorescent reflectors may be used, which have a more attractive appearance. The incident light can be either collimated or diffused.

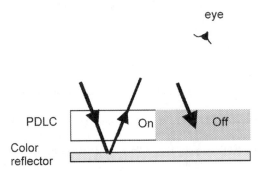

Figure 7.45 Schematic diagram showing the structure of a dichroic dye doped PDLC reflective display

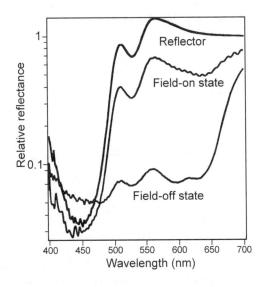

Figure 7.46 Reflection spectra of the reflector and the dichroic dye doped PDLC reflective display [30]

The typical reflection spectra of a dichroic dye doped PDLC are shown in Figure 7.46 [30]. The reflectance of the field-on state is high and the contrast is good. The viewing angle is large. Another advantage of the display is the compatibility of PDLC with plastic substrates. Fabrication of large displays is easy and low cost. It is also possible to laminate touch switches to the display.

References

1   P. S. Drzaic, "Liquid crystal dispersions" (World Scientific, NJ, 1995).

2   G. P. Crawford and S. Žumer, "Liquid crystals in complex geometries" (Taylor & Francis, London, 1996).

3   J. W. Doane, Chapter 14, Polymer dispersed liquid crystal displays, "Liquid crystals, applications and uses," Vol. 1, ed. by B. Bahadur (World Scientific, Singapore, 1990).

4   H.-S. Kitzerow, Polymer-dispersed liquid crystals, from the nematic curvilinear aligned phase to ferroelectric films, *Liq. Cryst.* **16**, 1 (1994).

5   G. P. Crawford, J. W. Doane, and S. Žumer, Chapter 9, Polymer dispersed liquid crystals: nematic droplets and related systems, "Handbook of liquid crystal research," ed. by P. J. Collings and J. S. Patel (Oxford University Press, New York, 1997).

6   J. W. Doane, N. A. Vaz, B.-G. Wu, and S. Žumer, Field controlled light scattering from nematic microdroplets, *Appl. Phys. Lett.* **48**, 269 (1996).

7   J. W. Doane, A. Golemme, J. W. West, J. B. Whitehead, Jr., and B.-G. Wu, Polymer dispersed liquid crystals for display application, *Mol. Cryst. Liq. Cryst.* **165**, 511 (1988).

8   R. Ondris-Crawford, E. P. Boyko, B. G. Wagner, J. H. Erdman, S. Žumer, and J. W. Doane, Microscope textures of nematic droplets in polymer dispersed liquid crystals, *J. Appl. Phys.* **69**, 6380 (1991).

9   H. Lin, H. Ding, and J. R. Kelly, The mechanism of switching a PDLC film, *Mol. Cryst. Liq. Cryst.* **262**, 99 (1995).

10   P. S. Drzaic, A new director alignment for droplets of nematic liquid crystal with low bend-to-splay ratio, *Mol. Cryst. Liq. Cryst.* **154**, 289 (1988).

11   J. H. Erdman, S. Žumer, and J. W. Doane, Configuration transition in a nematic liquid crystal confined to a small cavity, *Phys. Rev. Lett.* **64**, 1907 (1990).

12   S. Žumer and J. W. Doane, Light scattering from small nematic droplet, *Phys. Rev. A* **34**, 3373 (1986).

13   B. G. Wu, J. H. Erdman, and J. D. Doane, Response times and voltages for PDLC light shutters, *Liq. Cryst.* **5**, 1453 (1989).

14   J. D. Margerum, A. M. Lackner, J. H. Erdman, and E. Sherman, Addressing factors for polymer dispersed liquid crystal displays, *SPIE Proc.* **1455**, 27 (1991).

15   G. E. Volovik and O. D. Lavrentovich, Topological dynamics of defects: boojums in nematic droplets, *Sov. Phys. JEPT* **58**, 1159 (1983).

16   J. L. Fergason, Polymer encapsulated nematic liquid crystals for scattering and light control applications, *SID Intnl. Symp. Digest Tech. Papers*, **16**, 68 (1985).

17   P. S. Drzaic, Polymer dispersed nematic liquid crystals for large area displays and light valves, *J. Appl. Phys.* **60**, 2142 (1986).

18   P. S. Drzaic, Reorientation dynamics of polymer dispersed nematic liquid crystal films, *Liq. Cryst.* **3**, 1543 (1988).

19   C. Kittel and H. Kroemer, "Thermal Physics" (W. H. Freeman and Company, California, 1980).

20   J. L. West, Phase separation of liquid crystals in polymers, *Mol. Cryst. Liq. Cryst.* **157**, 427 (1988).

21   P. J. Flory, Thermodynamics of high polymer solutions, *J. Chem. Phys.* **10**, 51 (1942).

22 M. L. Huggins, Thermodynamic properties of solutions of long-chain compounds, *Ann. New York Acad. Sci.* **43**, 1 (1942).

23 Y. Hirai, S. Niiyama, H. Kumai, and T. Gunjima, Phase diagram and phase separation in LC/prepolymer mixture, *Proc. SPIE* **1257**, 2 (1990).

24 T. Tomita and P. Jones, Projection displays using nematic dispersions, *SID Intl. Symp. Digest Tech. Papers*, **23**, 579 (1992).

25 Y. Ooi, M. Sekine, S. Niiyama, Y. Hirai, M. Kunigita, T. Wakabayashi, M. Yuki, and T. Gunjima, LCPC projection display system for HDTV, *Proc. Japan Display 92*, 113 (1992).

26 D. Seekola and J. Kelly, A comparative study of the dielectric and optical response of PDLC films, *Proc. SPIE* **1455**, 19 (1991).

27 P. S. Drzaic, Polymer dispersed nematic liquid crystal for large area displays and light valves, *J. Appl. Phys.* **60**, 2142 (1986).

28 W. Wu, Single and multiple light scattering studies of PDLC films in the presence of electric fields, Dissertation, Kent State University, 1999. Private communication with J. R. Kelly.

29 J. H. Erdman, J. W. Doane, S. Žumer, and G. Chidichimo, Electro-optic response of PDLC light shutters, *Proc. SPIE* **1080**, 32 (1998).

30 P. S. Drzaic, Nematic droplet/polymer films for high-contrast coloured reflective displays, *Display*, 2–13 (1991).

31 J. L. West and R. Ondris-Crawford, Characterization of polymer dispersed liquid crystal shutters by ultraviolet/visible and infrared absorption spectroscopy, *J. Appl. Phys.* **70**, 3785 (1991).

32 B. Bahadur, Dichroic liquid crystal displays, "Liquid crystals: Applications and uses," Vol. 3, ed. by B. Bahadur (World Scientific, Singapore, 1990).

33 P. S. Drzaic, Droplet size and shape effects in nematic droplet/polymer films, *Proc. SPIE*, **1257**, 29 (1990).

34 R. A. M. Hikmet, Anisotropic gels obtained by photopolymerization in the liquid crystal state, "Liquid crystals in complex geometries," ed. by G. P. Crawford and S. Zumer (Taylor & Francis, London, 53–82, 1996).

35 D. J. Broer, Networks formed by photoinitiated chain cross-linking, "Liquid crystals in complex geometries," ed. by G. P. Crawford and S. Zumer (Taylor & Francis, London, 239–255, 1996).

36 D.-K. Yang, L.-C. Chien, and Y. K. Fung, Polymer stabilized cholesteric textures: materials and applications, "Liquid crystals in complex geometries," ed. by G. P. Crawford and S. Žumer (Taylor & Francis, London, 103–143, 1996).

37 D. J. Broer, R. G. Gossink, and R. A. M. Hikmet, Oriented polymer networks obtained by photopolymerization of liquid crystal-crystalline monomers, Die Angewandte Makromolekulare Chemie, **183**, 45 (1990).

38 Y. K. Fung, D.-K. Yang, Y. Sun, L. C. Chien, S. Žumer, and J. W. Doane, Polymer networks formed in liquid crystals, *Liq. Cryst.* **19**, 797–901 (1995).

39 R. Q. Ma and D.-K. Yang, Freedericksz Transition in Polymer stabilised Nematic Liquid Crystals, *Phys. Rev. E.* **61**, 1576 (2000).

40 I. Dierking, L. L. Kosbar, A. C. Lowe, and G. A. Held, Two-stage switching behavior of polymer stabilized cholesteric textures, *J. Appl. Phys.* **81**, 3007 (1997).

41 G. A. Held, L. L. Kosbar, I. Dierking, A. C. Lowe, G. Grinstein, V. Lee, and R. D. Miller, Confocal microscopy study of texture transitions in a polymer stabilised cholesteric liquid crystal, *Phys. Rev. Lett.* **79**, 3443 (1997).

42 Y. K. Fung, A. Borstnik, S. Zumer, D.-K. Yang, and J. W. Doane, Pretransitional nematic ordering in liquid crystals with dispersed polymer networks, *Phys. Rev. E* **55**, 1637 (1997).

43 P. J. Bos, J. Rahman, and J. W. Doane, *SID Intl. Symp. Digest Tech. Papers*, **24**, 887 (1993).

44 M. Escuti, C. C. Bowley, and G. P. Crawford, A model of the fast-switching polymer-stabilized IPS configuration, *SID Intl. Symp. Digest Tech. Papers*, **30**, 32 (1999).

45 R. A. M. Hikmet, H. M. J. Boots, and M. Michielsen, Ferroelectric liquid crystal gels network stabilized ferroelectric display, *Liq. Cryst.* **19**, 65 (1995).

46 R. A. M. Hikmet, Electrically induced light scattering from anisotropic gels, *J. Appl. Phys.* **68**, 4406 (1990).

47 R. A. M. Hikmet, Anisotropic gels and plasticised networks formed by liquid crystal molecules, *Liq. Cryst.* **9**, 405 (1991).

48 R. A. M. Hikmet and H. M. J. Boots, Domain structure and switching behavior of anisotropic gels, *Phys. Rev. E* **51**, 5824 (1995).

49 R. A. M. Hikmet, Electrically induced light scattering from anisotropic gels with negative dielectric anisotropy, *Mol. Cryst. Liq. Cryst.* **213**, 117 (1992).

50 D. K. Yang, L. C. Chien, and J. W. Doane, Cholesteric liquid crystal/polymer gel dispersion for haze-free light shutter, *Appl. Phys. Lett.* **60**, 3102 (1992).

51 M. Pfeiffer, Y. Sun, D.-K. Yang, J. W. Doane, W. Sautter, E. Lueder, and Z. Yaniv, Design of PSCT materials for MIM addressing, *SID Intl. Symp. Digest Tech. Papers*, **25**, 837 (1994).

52 A. Mochizuki, T. Yoshihara, M. Iwasaki, Y. Yamagishi, Y. Koike, M. Haraguchi, and Y. Kaneko, A 1120 × 768 pixel four color double-layer liquid-crystal projection display, *SID Intl. Symp. Digest Tech. Papers*, **21**, 155 (1990).

53 Y. Yabe, H. Yamada, T. Hoshi, T. Yoshihara, A. Mochizuki, and Y. Yoneda, A 5-Mpixel overhead projection display utilizing a nematic-cholesteric phase-transition liquid crystal, *SID Intl. Symp. Digest Tech. Papers*, **22**, 261 (1991).

54 Y. Fung, D. K. Yang, and J. W. Doane, Projection display from polymer stabilized cholesteric textures, *Proc. Eurodisplay'93*, 157 (1993).

55 R. Sun, W. Jang, and D.-K. Yang, Optimization of polymer stabilised cholesteric texture materials for high-brightness projection displays, *SID Intl. Symp. Digest Tech. Papers*, **30**, 652 (1999).

56 W. G. Jang, R. Sun, R. J. Twieg, and D.-K. Yang, Dichroic dye-doped bistable polymer-stabilized cholesteric-texture light valve, *J. SID*, **8**, 73 (2000).

57 R.-P. Sun and D.-K. Yang, Fast drive scheme for polymer stabilised cholesteric texture display, *Proc. Intl. Display Research Conf.*, 34 (2000).

58 R. Q. Ma and D.-K. Yang, Polymer stabilized cholesteric texture reverse-mode light shutter: Cell design, *J. SID*, **6**, 125 (1998).

59 M. Kreuzer and T. Tschudi, Erasable optical storage in bistable liquid crystal cells, *Mol. Cryst. Liq. Cryst.* **223**, 219 (1992).

60 M. Kreuzer, H. Gottschling, and T. Tschudi, Structure from formation and self-organization phenomena in bistable optical elements, *Mol. Cryst. Liq. Cryst.* **207**, 219 (1991).

61 M. Kreuzer, W. Balzer, and T. Tschudi, Formation of spatial structures in bistable optical elements containing nematic liquid crystals, *Mol. Cryst. Liq. Cryst.* **198**, 231 (1991).

62 M. Kreuzer, H. Leigeber, R. Maurer, and A. Miller, Bacteriorhodopsin, a dynamic high resolution optical recording material for parallel optical information processing, *Proc. Japan Display'92*, 175 (1992).

63 R. L. Sutherland, Bragg scattering in permanent nonlinear-particle composite gratings, *J. Opt. Soc. Am. B*, **8**, 1516 (1991).

64 R. L. Sutherland, V. P. Tondiglia, and L. V. Natarajan, Electrically switchable volume gratings in polymer-dispersed liquid crystal, *Appl. Phys. Lett.* **64**, 1074 (1994).

65 T. J. Bunning, L. V. Natarajan, and V. P. Tondiglia, *et al.*, Holographic polymer-dispersed liquid crystals (H-PDLCs), *Annu. Rev. Mater. Sci.* **30**, 83 (2000).

66 T. J. Bunning, L. V. Natarajan, and V. P. Tondiglia *et al.*, Morphology of reflection holograms formed in situ using polymer-dispersed liquid crystals, *Polymer*, **14**, 3147 (1996).

67 G. P. Crawford, T. G. Fiske, and L. D. Silverstein, Reflective color LCDs based on H-PDLC and PSCT technologies, *SID Intl. Symp. Digest Tech. Papers*, **27**, 99 (1996).

68 K. Tanaka, K. Kato, S. Tsuru, and S. Sakai, Holographically formed liquid-crystal/polymer device for reflective color display, *J. SID* **2**, 37 (1994).

69 M. J. Escuti, P. Kosssyrev, C. C. Bowley, S. Danworaphong, G. P. Crawford, T. G. Fiske, J. Coledrove, L. D. Silverstein, A. Lewis, and H. Yuan, Diffuse H-PDLC reflective displays: an enhanced viewing-angle approach, *SID Intl. Symp. Digest Tech. Papers*, **31**, 766 (2000).

70 C. C. Bowley, A. K. Fontecchio, and G. P. Crawford, Electro-optica investigations of H-PDLCS: the effect of monomer functionality on display performance, *SID Intl. Symp. Digest Tech. Papers*, **30**, 958 (1999).

71 H. C. van de Hulst, "Light scattering by smal particles" (Dover Publications, New York, 1981).

72 G. P. Montgomery, Jr., Angle-dependent scattering of polarized light by polymer dispersed liquid-crystal films, *J. Opt. Am. B*, **5**, 774 (1988).

73 G. P. Montgomery, Jr. and N. Vaz, Light-scattering analysis of the temperature-dependent transmittance of a polymer-dispersed liquid-crystal film in its isotropic phase, *Phys. Rev. A* **40**, 6580 (1989).

74 S. Zumer, Light scattering from nematic droplets: anomalous-diffraction approach, *Phys. Rev. A*, **37**, 4006 (1988).

75 J. R. Kelly, W. Wu, and P. Palffy-Muhoray, Wavelength dependence of scattering in PDLC film: droplet size effect," *Mol. Cryst. Liq. Crys.* **223**, 251 (1992).

76 J. R. Kelly and W. Wu, Multiple-scattering effects in polymer-dispersed liquid-crystals, *Liq. Cryst.* **14**, 1683 (1993).

77 N. A. Vaz, G. W. Smith, and G. P. Montgomery, Jr., A light control film composed of liquid crystal droplets dispersed in a uv-curable polymer, *Mol. Cryst. Liq. Cryst.* **146**, 1 (1987).

78 P. Nolan, M. Tillin, D. Coates, E. Ginter, E. Lueder, and T. Kallfass, Reflective mode PDLC displays-paper white display, *Proc. EuroDisplay'93*, 397 (1993).

79 J. D. LeGrange, T. M. Miller, P. Wiltzius, K. R. Amudson, J. Boo, A. van Blaaderer, M. Srinivasarao, and A. Kmetz, Brightness enhancement of reflective polymer-dispersed LCDs, *SID Intl. Symp. Digest Tech. Papers*, **26**, 275 (1995).

80 A. Kanemoto, Y. Matsuki, and Y. Takiguchi, Back scattering enhancement in polymer dispersed liquid crystal display with prism array sheet, *Proc. Intnl. Display Research Conf.*, 183 (1994).

81 W. Sautter, R. Bunz, R. Harjung, and E. Lueder, A reflective MIM-addressed PSCT display suited for video applications, *Proc. EuroDisplay'96*, 523 (1996).

82 H. J. Cornelissen, J. H. M. Neijzen, F. A. M. A. Paulissen, and J. M. Schlangen, Reflective direct-view LCDs using polymer dispersed liquid crystal (PDLC) and dielectric reflectors, *Proc. Intl. Display Research Conf. 97*, 144 (1997).

83 P. S. Drzaic, Light budget and optimization strategies for display applications of dichroic nematic droplet/polymer films, *Proc. SPIE* **1455**, 255 (1991).

84 P. S. Drzaic, R. Wiley, J. McCoy, and A. Guillaime, High-brightness reflective displays using nematic droplet/polymer films, *SID Intnl. Symp. Digest Tech. Papers*, **21**, 210 (1990).

85 P. S. Drzaic, R. Wiley, J. McCoy, and A. Guillaime, High-brightness and color contrast displays constructed from nematic droplet/polymer films incorporating pleochroic dyes, *Proc. SPIE* **1080**, 41 (1989).

86 P. S. Drzaic, A. M. Gonzales, P. Jones, and W. Montoya, Dichroic-based displays from nematic dispersions, *SID Intl. Symp. Digest Tech. Papers*, **23**, 571 (1992).

8

Cholesteric Reflective Displays

1 Introduction

The cholesteric (Ch) phase is a liquid crystal phase exhibited by chiral molecules or mixtures containing chiral components. Chiral molecules do not have reflection symmetry. A cholesteric liquid crystal is similar to a nematic liquid crystal: it has long-range orientational order but no long-range positional order. It has, however, one property that is different from the nematic liquid crystal, in that it has a helical structure, as shown in Figure 8.1 [1]. In a plane perpendicular to the helical axis, the average direction of the long axes of the molecules is along a common direction represented by the liquid crystal director n as in the nematic liquid crystal. Along the helical axis, the liquid crystal directors on two neighboring planes are twisted slightly with respect to one another. The distance along the helical axis for the director to rotate 2π is called the pitch and is denoted by P_o. The period is $P_o/2$, because n and $-n$ are equivalent.

As a matter of fact, the first observed liquid crystal phase was the cholesteric phase [2]. The material was cholesteryl benzoate, and therefore the phase is called the cholesteric phase. Nowadays, the cholesteric liquid crystals used in applications are biphenyl or triphenyl molecules, which have the typical molecular structure of nematic liquid crystals, with chiral centers [3]. For this reason cholesteric liquid crystals are also called chiral nematic liquid crystals.

The elastic free energy of a cholesteric liquid crystal is given by

$$f = \frac{1}{2}K_{11}(\nabla \cdot n)^2 + \frac{1}{2}K_{22}(n \cdot \nabla \times n + q_o)^2 + \frac{1}{2}K_{33}(n \times \nabla \times n)^2 \tag{8.1}$$

The first term is for splay deformation, the second term is for twist deformation and the third term is for bend deformation. $n \cdot \nabla \times n$ is a pseudo-scalar and it changes sign under the reflection symmetry operation. Since nematic liquid crystals have reflection symmetry, their

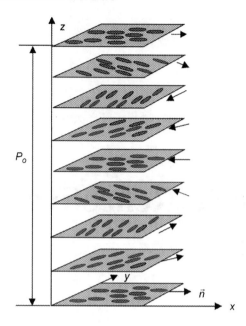

Figure 8.1 Schematic diagram showing the cholesteric structure

free energy does not change under the reflection symmetry operation, and it is required that $q_o = 0$. In the ground state, $\nabla \cdot \boldsymbol{n} = 0$, $\boldsymbol{n} \cdot \nabla \times \boldsymbol{n} = 0$ and $\boldsymbol{n} \times \nabla \times \boldsymbol{n} = 0$; that is, the director $\boldsymbol{n}$ is unidirectional in space. Cholesteric liquid crystals do not have reflection symmetry; the free energy may change under the reflection symmetry operation and therefore $q_o = 0$ is not required. In the ground state, $\boldsymbol{n} \cdot \nabla \times \boldsymbol{n} = -q_o \neq 0$, which corresponds to the structure shown in Figure 8.1. q_o is the chirality of the liquid crystal and is related to the pitch P_o by $q_o = 2\pi/P_o$. The sign of q_o is meaningful: a positive sign corresponds to a right-hand twist and a negative sign corresponds to a left-hand twist.

For a cholesteric liquid crystal with a given pitch, the state of the liquid crystal is characterised by the direction of the helical axis as shown in Figure 8.2. In the planar texture, the helical axis is perpendicular to the cell surface as shown in Figure 8.2(a). The material reflects light centered at the wavelength given by $\lambda_o = nP_o$, where n is the average refractive index. If λ_o is in the visible light region, the cell has a bright colored appearance. A microphotograph of the planar texture is shown in Figure 8.3(a). In the planar texture, there are usually disclinations called oily streaks, the dark lines shown in Figure 8.3(a), which are bent cholesteric layers [4–9]. In the focal conic texture, the helical axis is more or less parallel to the cell surface, as shown in Figure 8.2(b). When the pitch is short, the cholesteric liquid crystal can be regarded as a layered structure. The focal conic texture looks like that of a smectic-A, as shown in Figure 8.3(b). It is a multiple domain structure and the material is scattering. When the pitch is long, the texture is called the fingerprint texture, because it looks like a fingerprint. For a cholesteric liquid crystal with positive dielectric anisotropy, the pitch can be elongated by applying external fields [10]. There is no sharp transition between the focal conic texture and the fingerprint texture; therefore we do not distinguish between these two textures in this chapter. When the applied field is larger than a critical field E_c, the helical structure is unwound with the liquid crystal director

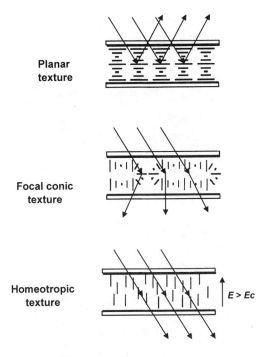

Figure 8.2 The structures of the cholesteric textures

aligned in the cell normal direction as shown in Figure 8.2(c). This texture is called the homeotropic texture. With the appropriate surface anchoring condition or dispersed polymer, both the planar texture and the focal conic texture can be stable at zero field.

2 Optical Properties of Cholesteric Liquid Crystals

Cholesteric liquid crystals exhibit two stable states at zero field: the reflecting planar texture and the non-reflecting (scattering) focal conic texture [11–13]. Their optical properties, such as color and reflectivity, are important in display applications.

2.1 Basic optical properties of cholesteric liquid crystals

When a cholesteric liquid crystal is in the planar texture, there is a periodic structure of the refractive index in the cell normal direction. The refractive index oscillates between the ordinary refractive index n_o and the extraordinary refractive index n_e. The period is $P_o/2$ because $\boldsymbol{n}$ and $-\boldsymbol{n}$ are equivalent. The liquid crystal exhibits Bragg reflection at the wavelength $\lambda_o = 2n(P_o/2) = nP_o$ for normally incident light [1]. The reflection is strong and multiple reflection inside the liquid crystal is important. The reflection peak is broad and has a bandwidth given by $\Delta n P_o$, where $\Delta n = n_e - n_o$ is the birefringence. Circularly polarised light with the same handedness as the helical structure is reflected strongly because of the constructive interference of the light reflected from different positions, while circularly

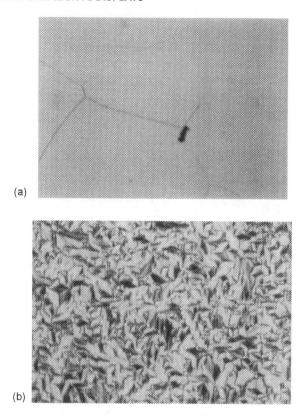

(a)

(b)

Figure 8.3 (a) Microphotograph of the planar texture. (b) Microphotograph of the focal conic texture

polarised light with the opposite handedness to the helical structure is not reflected because of the destructive interference of the light reflected from different positions. If the (normally) incident light is unpolarised, then the maximum reflection from the cholesteric liquid crystal is 50%. 100% reflection can be achieved by stacking a left-handed cholesteric liquid crystal and a right-handed cholesteric liquid crystal.

2.2 Reflection spectrum of cholesteric liquid crystals

For normally incident light, the reflection of a cholesteric liquid crystal can be calculated by solving Maxwell's equation exactly [14], or by using the Jones matrix or the Berreman 4×4 method [4, 15, 16]. For obliquely incident light, the Berreman 4×4 method is the best. Here we briefly discuss the Berreman 4×4 method [17–20]. For an optical medium, such as a cholesteric liquid crystal in the planar texture, whose dielectric tensor is only a function of the coordinate z

$$\varepsilon(z) = \begin{bmatrix} \varepsilon_{11}(z) & \varepsilon_{12}(z) & \varepsilon_{13}(z) \\ \varepsilon_{21}(z) & \varepsilon_{22}(z) & \varepsilon_{23}(z) \\ \varepsilon_{31}(z) & \varepsilon_{32}(z) & \varepsilon_{33}(z) \end{bmatrix} \tag{8.2}$$

For light incident in the xz plane with an incident angle α with respect to the z axis, as shown in Figure 8.4, the fields of the optical wave are

$$\boldsymbol{E} = \boldsymbol{E}(z) \exp\left(-ik_x x + i\omega t\right) \tag{8.3}$$

$$\boldsymbol{H} = \boldsymbol{H}(z) \exp\left(-ik_x x + i\omega t\right) \tag{8.4}$$

Maxwell's equations for the optical wave are

$$\nabla \cdot \boldsymbol{D} = \nabla \cdot (\varepsilon_o \varepsilon \cdot \boldsymbol{E}) = 0 \tag{8.5}$$

$$\nabla \cdot \boldsymbol{B} = \nabla \cdot (\mu_o \boldsymbol{H}) = 0 \tag{8.6}$$

$$\nabla \times \boldsymbol{E} = -\frac{\partial \boldsymbol{B}}{\partial t} = -i\mu_o \omega \boldsymbol{H} \tag{8.7}$$

$$\nabla \times \boldsymbol{H} = \frac{\partial \boldsymbol{D}}{\partial t} = -i\varepsilon_o \omega \varepsilon \cdot \boldsymbol{E} \tag{8.8}$$

Because of Equation 8.5, it is required that

$$k_x = k_o \sin \alpha = \frac{2\pi}{\lambda} \sin \alpha = \text{const.} \tag{8.9}$$

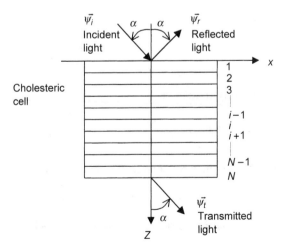

Figure 8.4 The coordinate system used to describe the light propagation in the Berreman 4 × 4 method

We consider a special case where $\varepsilon_{13} = \varepsilon_{23} = \varepsilon_{31} = \varepsilon_{32} = 0$. From Maxwell's equations, it can be obtained that

$$
\frac{\partial \psi}{\partial z} = ik_o
\begin{bmatrix}
0 & 0 & 0 & \eta_o \left(\frac{k_x^2}{k_o^2} \cdot \frac{1}{\varepsilon_{33}} - 1 \right) \\
0 & 0 & \eta_o & 0 \\
\dfrac{\varepsilon_{21}}{\eta_o} & \dfrac{1}{\eta_o} \left(\varepsilon_{22} - \dfrac{k_x^2}{k_o^2} \right) & 0 & 0 \\
\dfrac{-\varepsilon_{121}}{\eta_o} & \dfrac{-\varepsilon_{12}}{\eta_o} & 0 & 0
\end{bmatrix}
\psi = Q(z) \cdot \psi
\qquad (8.10)
$$

where $\eta_o = (\mu_o / \varepsilon_o)^{1/2} = 377\,\Omega$ is the impedance of a vacuum. $\psi = (E_x, E_y, H_x, H_y)^T$, where T denotes the transpose matrix, is the Berreman vector and Q is the Berreman matrix.

$$
Q = ik_o
\begin{bmatrix}
0 & 0 & 0 & \eta_o \left(\frac{k_x^2}{k_o^2} \cdot \frac{1}{\varepsilon_{33}} - 1 \right) \\
0 & 0 & \eta_o & 0 \\
\dfrac{\varepsilon_{21}}{\eta_o} & \dfrac{1}{\eta_o} \left(\varepsilon_{22} - \dfrac{k_x^2}{k_o^2} \right) & 0 & 0 \\
\dfrac{-\varepsilon_{121}}{\eta_o} & \dfrac{-\varepsilon_{12}}{\eta_o} & 0 & 0
\end{bmatrix}
=
\begin{bmatrix}
0 & 0 & 0 & Q_{14} \\
0 & 0 & Q_{23} & 0 \\
Q_{31} & Q_{32} & 0 & 0 \\
Q_{41} & Q_{42} & 0 & 0
\end{bmatrix}
$$

$$(8.11)$$

Outside the cholesteric cell, the medium is a vacuum. The Berreman vector has only two independent components. For the incident light, the Berreman vector is

$$
\psi_i = \left(E_{xi}, E_{yi}, \frac{-\cos\alpha}{\eta_o} E_{yi}, \frac{1}{\eta_o \cos\alpha} E_{xi} \right)
\qquad (8.12)
$$

For the reflected light, the Berreman vector is

$$
\psi_r = \left(E_{xr}, E_{yr}, \frac{\cos\alpha}{\eta_o} E_{yr}, \frac{-1}{\eta_o \cos\alpha} E_{xr} \right)
\qquad (8.13)
$$

For the transmitted light, the Berreman vector is

$$
\psi_t = \left(E_{xt}, E_{yt}, \frac{-\cos\alpha}{\eta_o} E_{yt}, \frac{1}{\eta_o \cos\alpha} E_{xt} \right)
\qquad (8.14)
$$

Note that if the incident light is linearly polarised in the x–z plane with amplitude E_o, then

$$
E_{xi} = E_o \cos\alpha
\qquad (8.15)
$$

For the cholesteric liquid crystal in the planar texture with the helical axis parallel to the z axis, we have

$$
\varepsilon_{11} = \frac{1}{2}(\varepsilon_{/\!/} + \varepsilon_{\perp}) + \frac{1}{2}(\varepsilon_{/\!/} - \varepsilon_{\perp}) \cos\left(\frac{4\pi}{P_o} z \right)
\qquad (8.16)
$$

$$\varepsilon_{22} = \frac{1}{2}(\varepsilon_{//} + \varepsilon_{\perp}) - \frac{1}{2}(\varepsilon_{//} - \varepsilon_{\perp})\cos\left(\frac{4\pi}{P_o}z\right) \tag{8.17}$$

$$\varepsilon_{33} = \varepsilon_{\perp} \tag{8.18}$$

$$\varepsilon_{12} = \varepsilon_{21} = \frac{1}{2}(\varepsilon_{//} - \varepsilon_{\perp})\sin\left(\frac{4\pi}{P_o}z\right) \tag{8.19}$$

For a small Δz, from Equation 8.10 we obtain that

$$\psi(z + \Delta z) = \exp\left(\boldsymbol{Q}(z)\Delta z\right) \cdot \psi(z) \tag{8.20}$$

We use the Cayley–Hamilton theory, which gives

$$\boldsymbol{B}(z) = \exp\left(\boldsymbol{Q}(z)\Delta z\right) = \gamma_1\boldsymbol{I} + \gamma_2(\Delta z\boldsymbol{Q}) + \gamma_3(\Delta z\boldsymbol{Q})^2 + \gamma_4(\Delta z\boldsymbol{Q})^3 \tag{8.21}$$

where $\boldsymbol{I}$ is the unit matrix and γ_i are determined by the equations

$$\exp q_i = \gamma_1 + \gamma_2 q_i + \gamma_3 q_i^2 + \gamma_4 q_i^3, \quad i = 1, 2, 3, 4 \tag{8.22}$$

q_i are the eigenvalues of $\Delta z\boldsymbol{Q}$ and are given by

$$q_i = \pm\Delta z\left\{\frac{1}{2}(Q_{23}Q_{32} + Q_{41}Q_{14})\right.$$
$$\left.\pm\frac{1}{2}\left[(Q_{23}Q_{32} + Q_{41}Q_{14})^2 - 4Q_{14}Q_{23}(Q_{32}Q_{41} + Q_{31}Q_{42})\right]^{1/2}\right\}^{1/2} \tag{8.23}$$

Inside a glass plate with refractive index n_g, $\varepsilon_{11} = \varepsilon_{22} = \varepsilon_{33} = n_g^2$ and $\varepsilon_{21} = \varepsilon_{12} = 0$. q_i is degenerate and has only two values q_1 and q_2. Then

$$\exp\left(\boldsymbol{Q}(z)\Delta z\right) = \gamma_1\boldsymbol{I} + \gamma_2(\Delta z\boldsymbol{Q}) \tag{8.24}$$

where γ_i are calculated from the equations

$$\exp q_i = \gamma_1 + \gamma_2 q_i, \quad i = 1, 2 \tag{8.25}$$

We divide the cholesteric cell (including the glass plates, electrodes and alignment layers) into N slabs with thicknesses $\Delta z_i(i = 1, 2, \ldots, N-1, N)$. The Berreman vectors at the boundaries between the slabs are:

$$\psi(0) = \psi_i + \psi_r$$
$$\psi(1) = \boldsymbol{B}(z_1) \cdot \psi(0)$$
$$\psi(2) = \boldsymbol{B}(z_2) \cdot \psi(1) = \boldsymbol{B}(z_2) \cdot \boldsymbol{B}(z_1) \cdot \psi(0)$$
$$\vdots$$
$$\psi_t = \psi(N) = \boldsymbol{B}(z_N) \cdot \psi(N-1) = \boldsymbol{B}(z_N) \cdot \boldsymbol{B}(z_{N-1}) \cdot \ldots \cdot \boldsymbol{B}(z_1) \cdot \psi(0)$$
$$= \boldsymbol{P} \cdot (\psi_i + \psi_r) \tag{8.26}$$

$\boldsymbol{P} = \boldsymbol{B}(z_N) \cdot \boldsymbol{B}(z_{N-1}) \cdot \ldots \cdot \boldsymbol{B}(z_1)$ can be calculated numerically. In components, Equation 8.26 contains four equations. ψ_i is given. ψ_r and ψ_t have two unknown variables each and can be found by solving Equation 8.26. The reflectance can be calculated by

$$R = \left[\left(E_{xr}/\cos \alpha \right)^2 + E_{yr}^2 \right] \bigg/ \left[\left(E_{xi}/\cos \alpha \right)^2 + E_{yi}^2 \right] \tag{8.27}$$

Therefore the transmittance can be calculated by

$$T = \left[\left(E_{xt}/\cos \alpha \right)^2 + E_{yt}^2 \right] \bigg/ \left[\left(E_{xi}/\cos \alpha \right)^2 + E_{yi}^2 \right] \tag{8.28}$$

For a cholesteric liquid crystal with parameters $P_o = 330\,\text{nm}$, $n_o = 1.516$, $n_e = 1.744$, the simulated reflection spectra of the liquid crystal with various cell thicknesses are shown in Figure 8.5, where the liquid crystal is in the perfect planar texture without defects and the

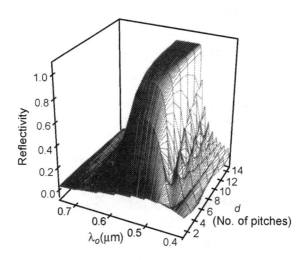

Figure 8.5 The reflection spectra of the cholesteric liquid crystal in the planar texture in cells with various cell thicknesses [19]

incident light is circularly polarised with the same handedness as the liquid crystal [19]. The reflection is saturated when the cell thickness is increased to $10P_o$.

The reflection of a cholesteric liquid crystal in the planar texture depends on the polarisation state of the incident light and the properties of the detection system. As an example, we consider a cholesteric liquid crystal with the following parameters: $P_o = 338$ nm, cell thickness $h = 5070$ nm, $n_o = 1.494$ and $n_e = 1.616$. The incident angle is $22.5°$. The cell has polyimide homogeneous alignment layers. The reflection spectra are shown in Figure 8.6 [20]. In Figure 8.6(a) and (b), the incident light is linearly polarised perpendicular to the incident plane and the detection is also linearly polarised in the same direction. In Figure 8.6(a) the polarisation of the incident light is parallel to the liquid crystal director in the entrance plane, while in Figure 8.6(b) the polarisation is perpendicular to the liquid crystal director in the entrance plane. The spectra are very different because of the interference between the light reflected from the liquid crystal and the light reflected from the interfaces between the glass substrate, the ITO electrode, the alignment layer and the liquid crystal. In Figure 8.6(a) the components interfere destructively and therefore there is a dip in the middle of the reflection band. In Figure 8.6(b) they interfere constructively and therefore the reflection is higher in the middle of the reflection band. In Figure 8.6(c) crossed polarisers are used. The light reflected from the interfaces cannot go through the analyser and so is not detected. The linearly polarised incident light can be decomposed into two circularly polarised components, and one of them is reflected. The reflected circularly polarised light can be decomposed into two linearly polarised components, and one of them passes through the analyser. Therefore the maximum reflection is 25%. In Figure 8.6(d) the incident light is unpolarised and all the reflected light is detected. The reflection in the band is slightly higher than 50% because of the light reflected from the interfaces. The fringes are due to the finite thickness of the liquid crystal. It exists even when the substrates have

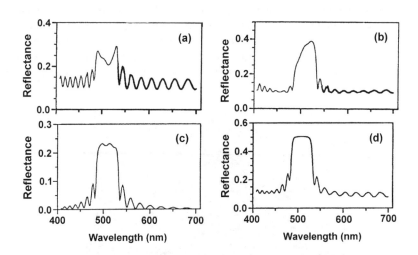

Figure 8.6 The reflection spectra of the cholesteric liquid crystal. (a) Incident light: σ-polarisation and parallel to the liquid crystal director on the entrance plane; detection: σ-polarisation. (b) Incident light: σ-polarisation and perpendicular to the liquid crystal director on the entrance plane; detection: σ-polarisation. (c) Incident light: σ-polarisation; detection: π-polarisation. (d) Incident light: unpolarised; detection: unpolarised [20]

a refractive index matched to that of the liquid crystal, but disappears for infinitely thick samples. These simulated results agree very well with experimental results [20].

2.3 Viewing angle of cholesteric displays

When light is obliquely incident at the angle θ on the cholesteric liquid crystal, the central wavelength of the reflection band is given by $\lambda = nP_o \cos \theta$. When θ is increased, the reflection band is shifted to a shorter wavelength, the reflection band becomes broader and the peak reflection becomes higher. The shift of the position of the reflection band is undesirable in display applications, because the color of the reflected light changes with the viewing angle. For the perfect planar texture, there is another problem that for incident light at an angle θ, the reflected light is only observed at the corresponding specular angle. These problems can be partially solved by dispersing a small amount of polymer in the liquid crystal or by using an alignment layer which gives weak homogeneous anchoring or home-otropic anchoring. The dispersed polymer and the alignment layer produce defects and create a poly-domain structure, as shown in Figure 8.7, which is different from the perfect planar texture shown in Figure 8.3(a). The dark spots correspond to defects. In this structure, the helical axis in the domains is no longer exactly parallel to the cell normal but distributed around the normal. In this imperfect planar texture, for incident light at a given angle, light rays reflected from different domains are in different directions, as shown in Figure 8.8(a). Under room light condition, light reflected from different domains can be observed at one location, as shown in Figure 8.8(b). Because the observed light is a mixture of different colors, the colors observed at different viewing angles are not very different. This poly-domain structure improves the viewing angle of the cholesteric display [21]. Furthermore the dispersed polymer and the alignment layers stabilise the focal conic texture at zero field. The display stabilised by polymer is called the polymer-stabilised cholesteric display and the one stabilised by the alignment layer is called the surface-stabilised cholesteric display [22–25].

The deviation of the helical axis from the cell normal direction in the poly-domain planar texture depends on the amount of the dispersed polymer. In the regular polymer stabilised cholesteric display, the polymer concentration is low and the deviation is small. The

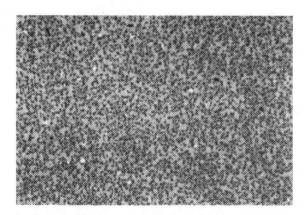

Figure 8.7 Microphotograph of the imperfect planar texture

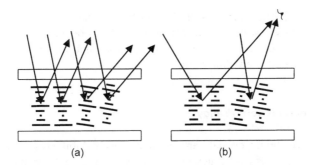

Figure 8.8 Schematic diagrams showing the reflection from the poly-domain planar texture

reflection spectrum of the planar texture is not very broad, as shown in Figure 8.9(a) [24]. The color of the reflected light is pure. The reflection spectra of surface-stabilised cholesteric displays are similar. The reflection of the focal conic texture is low. If the polymer concentration is sufficiently high, the deviation becomes large. When nP_o has the wavelength of red light, the reflection spectrum of the planar texture becomes very broad, as shown in

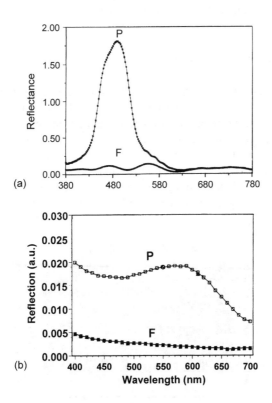

Figure 8.9 The reflection spectra of the cholesteric displays. P: planar texture, F: focal conic texture. (a) The regular polymer stabilised cholesteric display [24]. (b) The polymer stabilised black–white display [26]

Figure 8.9(b) [26, 27]. The planar texture has a white appearance. The display is called the polymer-stabilised black–white cholesteric display. The drawback of this display is that the scattering of the focal conic texture is stronger than that of the focal conic texture of the regular polymer stabilised cholesteric displays.

2.4 Gray scale of cholesteric displays

Although cholesteric liquid crystals are bistable, they exhibit gray scale memory states because of their multi-domain structure [24, 28–30]. Starting from the imperfect planar texture, some domains can be switched to the focal conic texture at lower voltages than other domains. Once a domain has been switched to the focal conic texture, it remains there even after the applied voltage is turned off. Figure 8.10 shows microphotographs of the gray scale states of a cholesteric display. From right to left, the states are achieved by applying voltage pulses with increasing amplitude, and the reflectance decreases. The domain size is around $10\,\mu m$ and the domain structure cannot be observed by the naked eye. The typical pixel size is about $100\,\mu m$. It is misleading to call the cholesteric display multistable. A cholesteric domain has only two stable states at zero field: it is either in the planar texture or in the focal conic texture. In a cholesteric display, it is observed that the domains in the planar texture have the same optical properties, independent of the states of other domains.

2.5 Cell design of cholesteric displays

When a cholesteric liquid crystal is in the planar texture, it reflects a single color light; when it is in the focal conic texture, it is scattering. In order to achieve high contrast, it is desirable that the backward scattering of the focal conic texture be minimised. As discussed in the liquid crystal/polymer composite chapter, the back scattering can be reduced by creating large focal conic domains. It is also desirable that the light from the back of the display be controlled. This is done by coating a color absorption layer on the back plate of the display, as shown in Figure 8.11. A black display is made with a black absorption layer [23, 25]. The display has a bright color appearance when the liquid crystal is in the planar texture and a

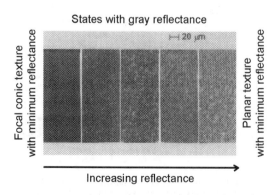

Figure 8.10 The microphotographs of the gray scale states of the cholesteric display

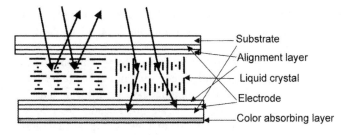

Planar texture   Focal conic texture

Figure 8.11 Schematic diagram of the cell structure of the cholesteric display

black appearance when the liquid crystal is in the focal conic texture. A white display is made with a color absorption layer [31]. As an example, the liquid crystal reflects yellow color light and the absorption layer reflects blue color. When the liquid crystal is in the planar texture, yellow light is reflected from the liquid crystal and blue light is reflected from the absorption layer, and therefore the display has a white appearance. When the liquid crystal is in the focal conic texture, only blue light is reflected from the absorption layer, and the display has a blue appearance. Higher contrast can be achieved by putting the absorption layer inside the cell on top of the electrode of the back plate.

The reflection of a cholesteric liquid crystal display is cell thickness-dependent. For perfect planar texture, $3\,\mu m$ is sufficiently thick to obtain the saturated reflection from a liquid crystal with $\Delta n \geq 0.2$. In cholesteric displays, because of the defects produced by the dispersed polymer or alignment layers, usually a $5\,\mu m$ cell is needed to obtain a saturated reflection.

In polymer-stabilised bistable cholesteric displays, it is preferable to mix a cholesteric liquid crystal with a few percent of a monomer. The viscosity of the mixture is low and it is put into cells in a vacuum. Then the cells are irradiated by UV light for photopolymerisation of the monomer. A blacker focal conic texture can usually be obtained by polymerising the monomer in the homeotropic texture in the presence of an applied voltage [23].

In surface-stabilised bistable cholesteric displays, either weak homogeneous alignment layers or homeotropic alignment layers can be used [25, 32]. When homeotropic alignment layers are used, the focal conic texture appears darker and the response times are shorter in some cases [33], but the reflection of the planar texture is lower.

Liquid crystals with large Δn and $\Delta \varepsilon$ are always desirable. When Δn is large, a thin cell can be used, and therefore the drive voltage is low. When $\Delta \varepsilon$ is large, the drive voltage is low. In display applications, the cholesteric liquid crystal is a mixture of a nematic liquid crystal and chiral dopants. The components should be chosen carefully such that the cholesteric phase temperature range is wide and the pitch does not shift with temperature. Liquid crystals with S_A phase should not used, because they usually exhibit a large pitch change with temperature [34].

2.6 Color of cholesteric displays

In a cholesteric display with one cholesteric liquid crystal, only a single color can be displayed. In order to make multiple color displays, cholesteric liquid crystals with various pitches must be used. This can be done either by stacking multiple layers of cholesteric liquid crystals with different pitches or by using one layer of cholesteric liquid crystals partitioned in plane.

Multiple color displays from one layer can be made by pixelation of colors. The displays have three alternating types of stripe with three different pitches reflecting blue, green and red light. Partition or other means of preventing inter-stripe diffusion must be used. Polymer walls, especially field-induced polymer walls, are good candidates [35, 36]. The different pitches can be achieved by two methods. In the first method, three cholesteric liquid crystals with different pitches are put into empty cells with partitions. The second method is photo color tuning [37–39]. A photo-sensitive chiral dopant is added to the liquid crystal. The dopant undergoes a chemical reaction under UV irradiation and thus its chirality changes, and the pitch of the liquid crystal changes. After the mixture has been put into display cells, the cells are irradiated by UV light with photo masks. By varying the irradiation time, different pitches are achieved. In this method, partitions are fabricated either before or after the photo color tuning. The polymer dispersing technique with large liquid crystal droplets can also be used with this method [40]. The major drawback of one layer mutiple color displays is that the reflection is low when only one color is on and the other colors are off.

In multiple color displays multiple layers are made by stacking three layers of cholesteric liquid crystals with pitches reflecting blue, green and red light [41–44]. One-layer displays with the three colors are fabricated first. Then they are glued togther. In order to decrease parallax, thin substrates, preferably substrates with conducting coating on both sides, should be used to decrease the distance between the liquid crystal layers. Because of the high reflection of the electrodes (usually ITO), the stacking order from bottom to top should be red, green and blue. The reflection spectra of a three-layer cholesteric display is shown in Figure 8.12 [42].

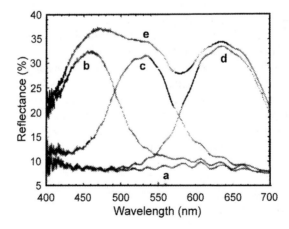

Figure 8.12 Reflection spectra of the stacked multiple color cholesteric display. Curve a: all off, curve b: blue on only, curve c: green on only, curve d: red on only, curve e: all on [42]

2.7 Polymer-dispersed cholesteric liquid crystal

Cholesteric liquid crystals can be dispersed in polymers to make reflective displays [45–47]. The dispersed cholesteric liquid crystal exists in isolated droplets. At zero field, the structure of the liquid crystal is determined by the anchoring condition and the droplet size. When the anchoring is tangential and the radius of the droplet is much larger than the pitch, the liquid crystal is in a spherulite texture where the helical axis is along the radial direction everywhere inside the droplet. The liquid crystal is weakly scattering in this texture. If the liquid crystal has a negative dielectric anisotropy, when a sufficiently high field is applied, the liquid crystal is switched to the planar texture with the helical axis in the cell normal direction, and becomes reflecting. Because the negative dielectric anisotropy is usually small, the drive voltage is high. Furthermore, the cholesteric liquid crystal is not bistable in relatively small droplets.

3 Transition Among Cholesteric Textures

Cholesteric liquid crystals exhibit three major textures as discussed above. The state of a cholesteric liquid crystal is mainly determined by the surface anchoring, the cell thickness and the applied field. There are many possible transitions among the textures, as shown in Figure 8.13. The transitions are very interesting and are of importance for both fundamental science and applications [48]. In order to design drive schemes for the bistable cholesteric reflective display, it is essential to understand the dynamics of the transitions. The cholesteric liquid crystals considered here have positive dielectric anisotropies unless otherwise specified.

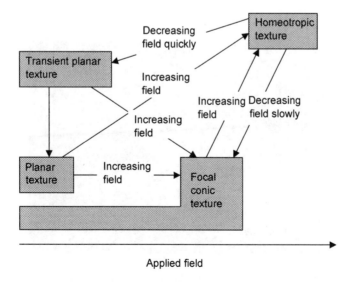

Figure 8.13 Schematic diagram showing the possible transitions among the cholesteric textures

3.1 Transition between planar texture and focal conic texture

Consider a cholesteric liquid crystal initially in the planar texture. When an external electric field is applied across the cell, the electric energy

$$-\frac{1}{2}\Delta\varepsilon\varepsilon_o(\boldsymbol{E}\cdot\boldsymbol{n})^2$$

is zero (high), because the liquid crystal director is perpendicular to the field everywhere. The electric energy can be reduced (becoming negative) if the liquid crystal is aligned parallel to the field. In the focal conic texture, the electric energy is reduced, because in some regions the liquid crystal is parallel to the applied field, while the helical structure is retained. Therefore the planar texture is unstable under the applied field. The liquid crystal will transform from the planar texture to the focal conic texture when the applied field is sufficiently high. There are two possible mechanisms for the transition from the planar texture to the focal conic texture: oily streak and Helfrich deformation.

A microphotograph of the oily streak in a cholesteric liquid crystal is shown in Figure 8.14. The dark rod in the bottom is a glass fiber spacer. The structure of the oily streak on a cross section is shown in Figure 8.15. By bending the cholesteric layers, the electric energy is reduced as a sacrifice of the elastic energy, surface energy at the cell surface and wall energy in the vertical middle plane [9]. The transition is a nucleation process. Sufficiently large oily streaks have to be created by irregularities, such as spacers, impurities and surface defects, in order to overcome the energy barrier. The applied field has to be higher than a threshold given by

$$V^2_{oily}=\frac{2(\varepsilon_{//}+\varepsilon_{\perp})}{\varepsilon_{\perp}(\varepsilon_{//}-\varepsilon_{\perp})\varepsilon_o}h\left[2(\sqrt{2}-1)w+\frac{3K}{4P_o}\ln\,2\right] \tag{8.29}$$

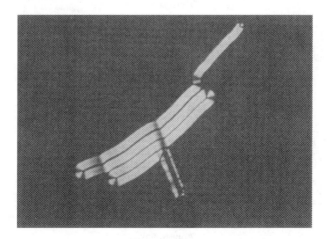

Figure 8.14 Microphotograph of the oily streak in the cholesteric liquid crystal. The bright finger is the oily streak. The dark background is the planar texture

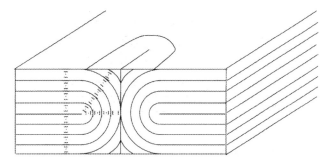

Figure 8.15 Schematic diagram showing the structure of the oily streak on a cross section

where h is the cell thickness, w is the surface energy and K is the elastic constant. In deriving the equation, the isotropic elastic constant approximation is used. Because it is a nucleation transition, the transition time is long. Once the applied voltage is above V_{oily}, the oily streaks grow until the liquid crystal is switched to the focal conic texture.

A microphotograph of the Helfrich deformation is shown in Figure 8.16. The Helfrich deformation is a two-dimensional undulation in the plane parallel to the cell surface [49, 50]. The structure of the liquid crystal in a vertical plane is shown in Figure 8.17. The helical pitch is dilated in some regions and compressed in other regions. The energies involved are elastic energy, which increases with the amplitude of the undulation, and electrical energy, which decreases with the amplitude of the undulation. The electrical energy decreases with increasing applied voltage. There is a threshold $V_{Helfrich}$ above which a decrease of the electrical energy is able to compensate for an increase of the elastic energy, and therefore Helfrich deformation takes place. $V_{Helfrich}$ is given by

$$V_{Helfrich} = \frac{4\pi^2 (2K_{22}K_{33})^{1/2}}{\Delta\varepsilon\varepsilon_o} \frac{h}{P_o}$$
(8.30)

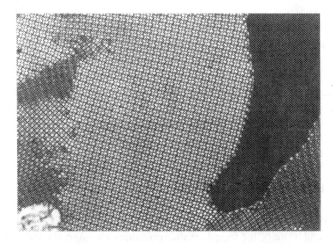

Figure 8.16 Microphotograph of the Helfich deformation. On the right-upper corner there is a dark area where the pitch is different from the rest of the cell, due to the quantisation of the pitch caused by the homogeneous anchoring of the cell, and the material is still in the undistorted planar texture

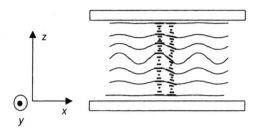

Figure 8.17 Schematic diagram showing the structure of Helfrich deformation in a plane perpendicular to the cell surface

The wavelength of the undulation is $\lambda = (2K_{33}/K_{22})^{1/4}(hP_o)^{1/2}$. The threshold is pitch-dependent. It is usually experimentally observed that $V_{Helfrich}$ is higher than V_{oily}. If the applied voltage is increased gradually, the oily streaks appear. If the voltage is increased abruptly above $V_{Helfrich}$, the Helfrich deformation dominates. Furthermore, Helfrich deformation is a homogeneous process and therefore it is much faster than the oily streak transition. Once the applied voltage is above $V_{Helfrich}$, the amplitude of the undulation increases with increasing voltage, and eventually the amplitude diverges and the liquid crystal transforms into the focal conic texture.

In cholesteric reflective display applications, it is desirable that the threshold of the transition from the planar texture to the focal conic texture is sufficiently high for the cholesteric liquid crystal to remain in the planar texture and not to exhibit flicker under low voltages.

Once the cholesteric liquid crystal is in the focal conic texture, depending on the surface anchoring condition, it may remain there after the applied field has been removed [13]. In bistable cholesteric reflective displays where either weak tangential or homeotropic alignment layers are used or polymers are dispersed in the liquid crystal, the liquid crystal remains in the focal conic texture when the applied voltage is turned off [23, 25]. There are two ways to switch it back to the planar texture. If the cholesteric liquid crystal has a positive dielectric anisotropy, a high voltage has to be applied to switch it to the homeotropic texture, then it relaxes back to the planar texture [51]. This will be discussed in more detail later. If the liquid crystal is a dual frequency material, exhibiting positive dielectric anisotropy at low-frequency voltages and negative dielectric anisotropy at high-frequency voltages, it can be switched back directly to the planar texture by applying a high-frequency voltage [52, 53]. If the cell has strong homogeneous alignment layers, the focal conic texture is not stable and the liquid crystal relaxes slowly back to the planar texture [13].

3.2 Transition between the focal conic texture and homeotropic texture

When the liquid crystal is in the focal conic texture and the externally applied electric field is increased, more and more of the liquid crystal molecules are aligned parallel to the field, and the pitch of the liquid crystal becomes longer, as shown in Figure 8.18. When the applied field is above a threshold E_C, the helical structure is unwound; the pitch becomes infinitely long and the liquid crystal is switched to the homeotropic texture [1, 10].

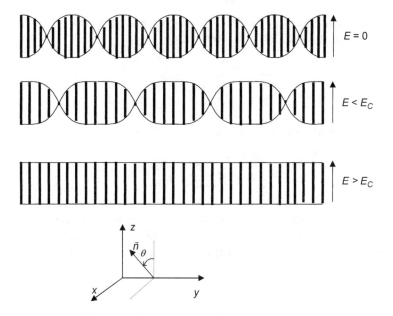

Figure 8.18 Schematic diagram showing the process of unwinding the helical structure in the focal conic-homeotropic transition

We first consider the unwinding of the helical structure ideally; as the applied field is increased, the pi-walls (the regions around the positions where the liquid crystal director is in the x direction) are propelled apart horizontally and annihilated at the far boundaries; there is no confinement restriction and the liquid crystal is in the state with the lowest free energy. In the focal conic texture, the liquid crystal director is given by

$$n_x = \sin \theta(y), \quad n_y = 0, \quad n_z = \cos \theta(y)$$

The free energy is given by

$$
\begin{aligned}
f &= \frac{1}{2} K_{22} [\mathbf{n} \cdot (\nabla \times \mathbf{n}) + q_o]^2 - \frac{1}{2} \Delta \varepsilon \varepsilon_o (\mathbf{n} \cdot \mathbf{E})^2 \\
&= \frac{1}{2} K_{22} (\theta' - q_o)^2 + \frac{1}{2} \Delta \varepsilon \varepsilon_o E^2 \sin^2 \theta + \text{const.}
\end{aligned}
\tag{8.31}
$$

where $\theta' = (\partial \theta / \partial y)$. The constant is not important and can be neglected. Introducing the dimensionless variables $\psi = f / K_{22} q_o^2$, $\xi = q_o y$, $e = E / E_o$, where

$$E_o = \frac{\pi}{2} q_o \left(\frac{K_{22}}{\varepsilon_o \Delta \varepsilon} \right)^{1/2} = \frac{\pi^2}{P_o} \left(\frac{K_{22}}{\varepsilon_o \Delta \varepsilon} \right)^{1/2}$$

Then we have

$$\zeta = \frac{1}{2} \left(\frac{d\theta}{d\xi} - 1 \right)^2 + \frac{1}{2} \left(\frac{\pi e}{2} \right)^2 \sin^2 \theta \tag{8.32}$$

Minimising the free energy using the Euler–Lagrange equation, assuming the chirality of the liquid crystal is right-handed, it can be obtained that

$$\frac{d\theta}{d\xi} = \left[\left(\frac{\pi e}{2} \sin \theta \right)^2 + A \right]^{1/2}$$ (8.33)

where A is the integration constant. The free energy density is

$$\psi = \frac{1}{2}(1 + A) - \frac{d\theta}{d\xi} + \left(\frac{\pi e}{2} \right)^2 \sin^2 \theta$$ (8.34)

A is field-dependent. When $e = 0$, $(d\theta/d\xi) = 1$, then $A = 1$. When $e \geq e_c = E_C/E_o$, $(d\theta/d\xi) = 0$ and $\theta = 0$, hence $A = 0$. Hence as the applied field is increased from 0 to e_C, A changes from 1 to 0. The normalised periodicity of the focal conic texture is $(P/2)q_o$ and is given by

$$(P/2)q_o = \int_0^\pi \left[A + \left(\frac{\pi e}{2} \sin \theta \right)^2 \right]^{-1/2} d\theta$$ (8.35)

The free energy density also has the same period and the averaged free energy density is given by

$$
\begin{aligned}
\bar{\psi} &= \frac{\displaystyle\int_0^{P/2} \left[\frac{1}{2}(1 + A) - \frac{d\theta}{d\xi} + \left(\frac{\pi e}{2} \sin \theta \right)^2 \right] d\xi}{(P/2)q_o} \\[3mm]
&= \frac{\displaystyle\int_0^{P/2} \left[\frac{1}{2}(1 + A) - \frac{d\theta}{d\varsigma} + \left(\frac{\pi e}{2} \sin \theta \right)^2 \right] d\xi}{\displaystyle\int_0^\pi \left[A + \left(\frac{\pi e}{2} \sin \theta \right)^2 \right]^{-1/2} d\theta} \\[3mm]
&= \frac{\displaystyle\int_0^\pi \left[\frac{1}{2}(1 + A) - \frac{d\theta}{d\varsigma} + \left(\frac{\pi e}{2} \sin \theta \right)^2 \right] \frac{d\xi}{d\theta} d\theta}{\displaystyle\int_0^\pi \left[A + \left(\frac{\pi e}{2} \sin \theta \right)^2 \right]^{-1/2} d\theta} \\[3mm]
&= \frac{-\pi + \displaystyle\int_0^\pi \left\{ \left[A + \left(\frac{\pi e}{2} \sin \theta \right)^2 \right]^{1/2} + \frac{1}{2}(1 - A) \left[A + \left(\frac{\pi e}{2} \sin \theta \right)^2 \right]^{-1/2} \right\} d\theta}{\displaystyle\int_0^\pi \left[A + \left(\frac{\pi e}{2} \sin \theta \right)^2 \right]^{-1/2} d\theta}
\end{aligned}
$$ (8.36)

Minimising $\bar{\psi}$ with respect to A, we obtain

$$\int_0^{\pi/2} \left[A + \left(\frac{\pi}{2} e \sin \theta \right)^2 \right]^{1/2} d\theta = \frac{\pi}{2} \tag{8.37}$$

At any applied field e, the value of A can be found by solving Equation 8.37. Once the value of A is known, the liquid crystal n and the helical pitch P can be calculated. At the threshold e_C, $A = 0$, and from Equation 8.36 it can obtained that $e_C = 1$; the pitch is given by

$$(P/2)q_o = \int_0^\pi \left[A + \left(\frac{\pi e_c}{2} \sin \theta \right)^2 \right]^{-1/2} d\theta = \int_0^\pi \left(\frac{\pi}{2} \sin \theta \right)^{-1} d\theta = \infty$$

Therefore the critical field unwinding the helical structure is

$$E_C = E_o = \frac{\pi}{2} q_o \left(\frac{K_{22}}{\varepsilon_o \Delta \varepsilon} \right)^{1/2} = \frac{\pi^2}{P_o} \left(\frac{K_{22}}{\varepsilon_o \Delta \varepsilon} \right)^{1/2}. \tag{8.38}$$

In the real world, the pi-walls in the focal conic-homeotropic transition cannot be propelled in the horizontal direction. Instead, the pi-walls form circles, as shown in Figure 8.19, which can persist even when the applied field is higher than E_c [54]. As the applied field is increased futher, the radii of the pi-wall circles decrease, and eventually disappear at a threshold which depends on the cell thickness and anchoring condition.

The focal conic-homeotropic transition is reversable. The liquid crystal can transform directly from the homeotropic texture back to the focal conic texture. There is, however, a hysteresis in the transition between them, which plays a crucial role in dynamic drive schemes. Here we consider semi-quantitatively the homeotropic-focal conic transition. If there is twist in the middle of the homeotropic texture, as shown in Figure 8.20, it may lead to the transition if the maximum twist angle θ_m grows to $\pi/2$. Let us consider the free energy change in this process. The free energy density of the homeotropic state is

$$f_h = \frac{1}{2} K_{22} q_o^2 \tag{8.39}$$

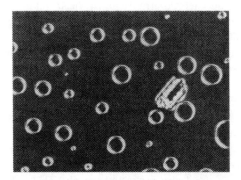

Figure 8.19 Microphotograph of the pi-wall circle in the focal conic-homeotropic transition [54]

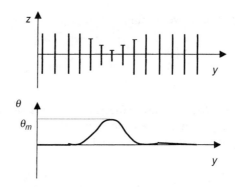

Figure 8.20 Liquid crystal director configuration and twist angle in space of a twist fluctuation in the homeotropic texture

The free energy density of the fluctuation state is

$$f_f = \frac{1}{2} K_{22} \left(q_o - \frac{d\theta}{dy} \right)^2 + \frac{1}{2} \Delta\varepsilon\varepsilon_o E^2 \sin^2\theta \tag{8.40}$$

The change of the total free energy is

$$\Delta F = \int_{-\infty}^{\infty} (f_f - f_h) dy$$

$$= \int_0^{\infty} \left[\frac{1}{2} K_{22} \left(\frac{d\theta}{dy} \right)^2 + \frac{1}{2} \Delta\varepsilon\varepsilon_o E^2 \sin^2\theta \right] dy - \int_0^{\infty} K_{22} \left(\frac{d\theta}{dy} \right) dy$$

$$= \int_0^{\infty} \left[\frac{1}{2} K_{22} \left(\frac{d\theta}{dy} \right)^2 + \frac{1}{2} \Delta\varepsilon\varepsilon_o E^2 \sin^2\theta \right] dy - K_{22}[\theta(\infty) - \theta(0)]$$

$$= \int_0^{\infty} \left[\frac{1}{2} K_{22} \left(\frac{d\theta}{dy} \right)^2 + \frac{1}{2} \Delta\varepsilon\varepsilon_o E^2 \sin^2\theta \right] dy > 0$$

This means that there is an energy barrier against the formation of the twisted state. The increase in the free energy in the region with negative twist is higher than the decrease of the free energy in the region with positive twist. Therefore the homeotropic-focal conic transition is a nucleation process and nucleation seeds are needed for the transition to take place.

Experiments have shown that the homeotropic-focal conic transition is indeed a nucleation process [54]. A microphotograph of a two-pi-wall in the homeotropic-focal conic transition is shown in Figure 8.21. Glass fiber spacers serve as the nucleation seeds. The critical field E_{HF} below which the nucleations start to appear is about $0.9E_C$. The structure of the two-pi-wall is shown in Figure 8.22. In the center of the two-pi-wall, the liquid crystal is aligned homeotropically, and therefore it looks dark under cross-polarisers. At a relatively high field, the two-pi-wall fingers grow only longitudinally. When the applied field is lowered further, they buckle and also grow transversely. That is how the hometropic-focal conic transition is realised.

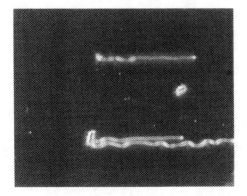

Figure 8.21 Microphotograph of the two-pi-walls in the homeotropic-focal conic transition. In the dark background, the liquid crystal is in the homeotropic texture [48]

Top view

Side view

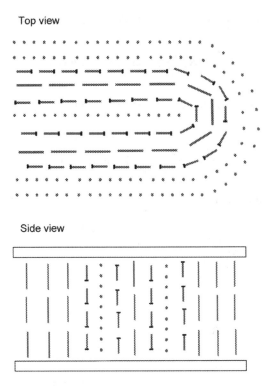

Figure 8.22 The liquid crystal director configuration of the two-pi-wall in the homeotropic-focal conic transition

The nucleation in the homeotropic-focal conic transition is not yet fully understood. Here we present only a simplified model to illustrate the physics involved. We use a free boundary condition created by a defect, where the liquid crystal molecule is free to rotate, and we consider the dynamics of the transition and calculate the threshold below which the focal conic structure can be formed. As schematically shown in Figure 8.23, the free boundary is at $y = 0$. We use a one-dimensional model where the twist angle θ is only a function of y and time t. The components of the liquid crystal director n are $n_x = \sin\theta$, $n_y = 0$ and $n_z = \cos\theta$. The dynamic equation is

$$\frac{\partial D}{\partial \dot{\theta}} = -\frac{\delta f}{\delta \theta} \tag{8.41}$$

where D is the dissipation function and $\dot{\theta} = (\partial\theta/\partial t)$. The left-hand side of the equation is the friction torque and the right-hand side is the elastic and electric torque. The dissipation function is given by

$$D = \frac{1}{2}\gamma\sum \dot{n}_i^2 = \frac{1}{2}\gamma\dot{\theta}^2 = \frac{1}{2}\gamma\left(\frac{\partial\theta}{\partial t}\right)^2 \tag{8.42}$$

where γ is the rotational viscosity coefficient. The dynamic equation is

$$\gamma\frac{\partial\theta}{\partial t} = K_{22}\frac{\partial^2\theta}{\partial y^2} - \Delta\varepsilon\varepsilon_o E^2\sin\theta\cos\theta \tag{8.43}$$

Using the dimensionless variables $\xi = q_o y$, $e = E/E_c$ and $\tau = t/(\gamma/K_{22}q_o^2)$, the dynamic equation becomes

$$\frac{\partial\theta}{\partial\tau} = \frac{\partial^2\theta}{\partial\xi^2} - \left(\frac{\pi}{2}e\right)^2\sin\theta\cos\theta \tag{8.44}$$

The free boundary condition is

$$\frac{\partial f}{\partial\left(\dfrac{\partial\theta}{\partial y}\right)}\Bigg|_{y=0} = -K_{22}\left(q_o - \frac{\partial\theta}{\partial y}\right)\Bigg|_{y=0} = 0$$

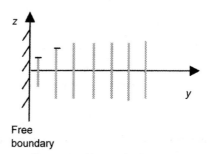

Free
boundary

Figure 8.23 The liquid crystal director configuration in the nucleation of a twisted structure from a free boundary

Hence

$$\frac{\partial \theta}{\partial y}\Big|_{y=0} = q_o,$$

that is

$$\frac{\partial \theta}{\partial \xi}\Big|_{\xi=0} = 1$$

We look for the threshold e_{hf} for the homeotropic-focal conic transition. When $e > e_{HF}$, the homeotropic texture is stable. In the equilibrium state, we have $(\partial\theta/\partial\tau) = 0$. From Equation 8.44, by integration, we obtain

$$\left(\frac{\partial \theta}{\partial \xi}\right)^2_{|\xi} - \left(\frac{\partial \theta}{\partial \xi}\right)^2_{|\xi=0} = \left(\frac{\pi}{2}e \sin \theta\right)^2_{|\theta} - \left(\frac{\pi}{2}e \sin \theta\right)^2_{|\theta_o} \tag{8.45}$$

For sufficiently large ξ, $\theta = 0$ and $(d\theta/d\xi) = 0$. Hence we obtain

$$\sin \theta_0 = \left(\frac{2}{\pi e}\right)^{1/2} \tag{8.46}$$

When e decreases, θ_o increases. Once θ_o becomes larger than $\pi/2$, helical twisting will be injected into the bulk. When $e \rightarrow e_{HF}$, we have $\theta_o \rightarrow \pi/2$. Hence we have the threshold

$$e_{HF} = \frac{2}{\pi} \tag{8.47}$$

When $e < e_{HF}$ the helical twisting is injected into the bulk from the surface. This value is lower than the one observed experimentally. Recently we used a three-dimensional model and a computer simulation technique to study the transition. The value of $e_{HF} \simeq 0.95$ is obtained for sufficiently thick cells [54].

We consider the homeotropic-focal conic transition time. At zero field, the dynamic Equation 8.44 is a diffusion equation. The normalised time for the twist structure to grow a distance L from the nucleation site into the bulk is approximately equal to $\tau_{HF} = (q_0 L)^2$. The real time is

$$T_{HF} = \frac{\gamma}{K_{22}q_0^2} \tau_{HF} = \frac{\gamma L^2}{K_{22}}$$

Hence if the averaged linear distance between nucleation seeds is L, the homeotropic-focal conic transition time is $T_{HF} = \gamma L^2 / K_{22}$.

For example, if $\gamma = 5 \times 10^{-2} NS/m^2$, $K_{22} = 10^{-11}N$ and $L = 5\,\mu m$, then $T_{HF} = 125$ ms. In summary, there is a hysteresis in the transition between the focal conic-homeotropic transition, and the homeotropic-focal conic transition is a slow nucleation process.

3.3 Transition between the homeotropic texture and the planar texture

For the liquid crystal in the homeotropic texture, when the applied field is reduced, there are two relaxation modes. One is the HF mode in which the liquid crystal relaxes into the focal conic texture as discussed in the previous section. The other is the HP mode in which the liquid crystal relaxes into the planar texture [48, 51, 55]. The movement of the liquid crystal in the HP mode is shown in Figure 8.24. The liquid crystal forms a conic helical structure with the helical axis in the cell normal direction. As the relaxation takes place, the polar angle θ increases. When the polar angle θ is zero, the liquid crystal is in the homeotropic texture. When the polar angle is $\pi/2$, the liquid crystal is in the planar texture.

We consider first the static conic helical structure. It is assumed that the polar angle θ is a constant independent of z and that the azimuthal angle ϕ varies along z with a constant rate q; that is, the twisting is uniform. The components of the director $\boldsymbol{n}$ are given by $n_x = \sin\theta\cos(qz)$, $n_y = \sin\theta\ \sin(qz)$ and $n_z = \cos\theta$. The free energy is given by

$$f = \frac{1}{2}K_{22}(q_o - q\sin^2\theta)^2 + \frac{1}{2}K_{33}q^2\sin^2\theta\cos^2\theta + \frac{1}{2}\Delta\varepsilon\varepsilon_o E^2\sin^2\theta \tag{8.48}$$

Using the dimensionless variables $K_3 = K_{33}/K_{22}$, $\gamma = q/q_o$, $\psi = f/K_{22}q_o^2$ and $e = E/E_c$, then the free energy becomes

$$\psi = \frac{1}{2}(1 - \lambda\sin^2\theta)^2 + \frac{1}{2}K_3\gamma^2\sin^2\theta\cos^2\theta + \frac{1}{2}\left(\frac{\pi}{2}e\right)^2\sin^2\theta \tag{8.49}$$

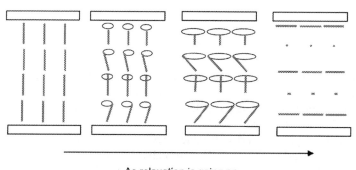

As relaxation is going on

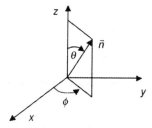

Figure 8.24 Schematic diagram showing the movement of the liquid crystal in the HP relaxation mode

By minimising ψ with respect to γ, we obtain

$$\lambda = \frac{1}{\sin^2\theta + K_3\cos^2\theta}$$

That is

$$q = \frac{q_0}{\sin^2\theta + K_3\cos^2\theta} \tag{8.50}$$

When the polar angle θ is very small, $q = q_0/K_3 = (K_{22}/K_{33})q_0$ and the pitch is $P = (K_{33}/K_{22})P_o$.
For most liquid crystals, $K_{33}/K_{22} \simeq 2$. Hence the pitch P of the conic helical structure with small polar angle is about twice as long as the intrinsic pitch P_o. When the polar angle θ is $\pi/2$, $q = q_0$; that is $P = P_o$. After being minimised with respect to λ, the free energy is

$$\psi = \frac{1}{2} + \frac{1}{2}\sin^2\theta\left[\left(\frac{\pi}{2}e\right)^2 - \frac{1}{K_3 + (1 - K_3)\sin^2\theta}\right] \tag{8.51}$$

We then examine whether there is any stable conic helical structure. We minimise the free energy with respect to the polar angle θ:

$$\frac{\partial\psi}{\partial(\sin^2\theta)} == \frac{1}{2}\left(\frac{\pi}{2}e\right)^2 - \frac{K_3}{2[K_3 + (1 - K_3)\sin^2\theta]^2}$$

$$\frac{\partial^2\psi}{\partial(\sin^2\theta)^2} = \frac{K_3(1 - K_3)}{[K_3 + (1 - K_3)\sin^2\theta]^2}$$

Because $K_3 > 1$, the second-order derivative is negative, and therefore there is no minimum free energy state, that is, no stable conic helical structure. The liquid crystal is either in the homeotropic texture with $\theta = 0$ or in the planar texture with $\theta = \pi/2$. In Figure 8.25 the free energy given by Equation 8.51 is plotted as a function of θ at three different fields. $e_{eq} = 2/\pi$

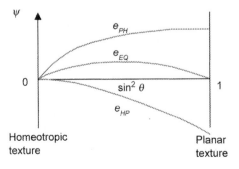

Figure 8.25 The free energy of the conic helical structure as a function of the polar angle θ at various applied fields

is the field at which the planar texture and the homeotropic texture have the same free energy. At this field there is, however, an energy barrier between the two states. As the field is decreased from e_{eq}, the free energy of the planar texture becomes lower than that of the homeotropic texture, but the energy barrier persists. When the field is sufficiently low, the energy barrier decreases to zero, and the homeotropic texture will become absolutely unstable. The critical field

$$e_{hp} = \frac{2}{\pi}(1/K_3)^{1/2} \tag{8.52}$$

can be obtained from the equation

$$\frac{\partial \psi}{\partial(\sin^2 \theta)}\Big|_{\theta=0} = 0$$

When the applied field is increased from e_{eq}, the free energy of the homeotropic texture becomes lower than that of the planar texture, but the energy barrier also persists. When the field is sufficiently high, the energy barrier decreases to zero, and the planar texture will become absolutely unstable. The critical field

$$e_{ph} = \frac{2}{\pi}\sqrt{K_3} \tag{8.53}$$

can be obtained from the equation

$$\frac{\partial \psi}{\partial(\sin^2 \theta)}\Big|_{\theta=\pi/s} = 0$$

For small K_3, e_{ph} may be lower than e_c, which means that it may be easier to switch the liquid crystal from the planar texture to the homeotropic texture than from the focal conic texture to the homeotropic texture. Experimental results support this conjecture, which will be discussed later.

We now consider the dynamics of the transition from the homeotropic texture to the planar texture. It is assumed that $\theta = \theta(z,t)$, $\phi = \phi(z,t)$. For simplicity, it is also assumed that $K_{11} = K_{33}$. The free energy density is

$$f = \frac{1}{2}K_{22}(q_o - \phi' \sin^2 \theta)^2 + \frac{1}{2}K_{33}(\phi'^2 \sin^2 \theta \cos^2 \theta + \theta'^2) + \frac{1}{2}\Delta\varepsilon\varepsilon_o E^2 \sin^2 \theta \tag{8.54}$$

The dissipation function is

$$D = \frac{1}{2}\gamma \sum \dot{n}_i^2 = \frac{1}{2}\gamma(\dot{\theta}^2 + \sin^2 \theta \dot{\phi}^2) \tag{8.55}$$

The dynamic equations are

$$\frac{\partial D}{\partial \dot{\theta}} = -\frac{\delta f}{\delta \theta} \quad \text{and} \quad \frac{\partial D}{\partial \dot{\phi}} = -\frac{\delta f}{\delta \phi}$$

Using the dimensionless variables

$$K_3 = K_{33}/K_{22}, \quad \eta = q_o z, \quad \tau = \frac{t}{\gamma/K_{22}q_o^2} \quad \text{and} \quad e = E/E_c$$

the dynamic equations become

$$\frac{\partial \theta}{\partial \tau} = K_3 \frac{\partial^2 \theta}{\partial \eta^2} + \left[2(K_3 - 1) \left(\frac{\partial \phi}{\partial \eta} \sin \theta \right)^2 - K_3 \left(\frac{\partial \phi}{\partial \eta} \right)^2 + 2\phi' - \left(\frac{\pi}{2} e \right) \right]^2 \sin \theta \cos \theta \quad (8.56)$$

$$\frac{\partial \phi}{\partial \tau} = [K_3 + (1 - K_3) \sin^2 \theta] \frac{\partial^2 \phi}{\partial \eta^2} + \frac{2 \cos \theta}{\sin \theta} \frac{\partial \theta}{\partial \eta} \left[2(1 - K_3) \frac{\partial \phi}{\partial \eta} \sin^2 \theta + K_3 \frac{\partial \phi}{\partial \eta} - 1 \right] \quad (8.57)$$

In the homeotropic-planar transition, initially, θ is very small. We first consider Equation 8.57. Because of the factor $(1/\sin \theta)$, ϕ changes rapidly with time until

$$K_3 \frac{\partial \phi}{\partial \eta} - 1 = 0$$

Hence in the beginning of the transition

$$\frac{\partial \phi}{\partial \eta} = \frac{1}{K_3}$$

that is

$$\frac{\partial \phi}{\partial z} = \frac{1}{K_3} q_o$$

This is the same as that obtained in the static state, and is also confirmed by computer simulation. Considering Equation 8.56, if the homeotropic-planar transition takes place, then it is required that $(\partial \theta/\partial \tau) > 0$. The first term in Equation 8.56 is a diffusion term and its role is averaging. Hence it is required that

$$\left[2(K_3 - 1) \left(\frac{\partial \phi}{\partial \eta} \sin \theta \right)^2 - K_3 \left(\frac{\partial \phi}{\partial \eta} \right)^2 + 2\phi' - \left(\frac{\pi}{2} e \right)^2 \right] \sin \theta \cos \theta > 0$$

Initially, θ is very small and

$$\frac{\partial \phi}{\partial \eta} = \frac{1}{K_3}$$

Therefore it is required that

$$-K_3 \left(\frac{1}{K_3} \right)^2 + 2 \frac{1}{K_3} - \left(\frac{\pi}{2} e \right)^2 = \frac{1}{K_3} - \left(\frac{\pi}{2} e \right)^2 > 0$$

From this we obtain that $e < e_{HP} = (2/\pi)(1/K_3)^{1/2}$ which is the same as that obtained in the static case and is confirmed experimentally [51]. Once the applied field has been reduced below e_{hp}, there is no longer an energy barrier against the transition. The transition is a homogeneous one and is quick. The transition time τ_{HP} can be estimated in the following way: for small θ

$$\frac{\partial \theta}{\partial \tau} \simeq \left[\frac{1}{K_3} - \left(\frac{\pi}{2} e \right)^2 \right] \sin \theta \cos \theta \simeq \left[\left(\frac{1}{K_3} \right) - \left(\frac{\pi}{2} e \right)^2 \right] \theta$$

Thus the transition time is

$$\tau_{hp} = 1 \left/ \left[\frac{1}{K_3} - \left(\frac{\pi}{2} e \right)^2 \right] \right.$$

At $e = 0$, $\tau_{hp} = K_3$. The un-normalised transition time is

$$T_{hp} = \frac{\gamma}{K_{22} q_o^2} \tau_{hp} = \gamma \frac{K_{33}}{(2\pi)^2 K_{22}} \frac{P_o^2}{K_{22}} \simeq \gamma \frac{P_o^2}{K_{22}}$$

For example, if $\gamma = 5 \times 10^{-2} NS/m^2$, $K_{22} = 10^{-11}$ N and $P_o = 0.5\,\mu$m, then $T_{HF} = 1.25$ ms. Therefore the homeotropic-planar transition is much faster than the homeotropic-focal conic transition. Simulations show that when θ changes from 0 to $(\pi/2)$, the helical twist rate $(\partial \phi / \partial z)$ only changes from 0 to $(K_{22}/K_{33})P_o$, corresponding to the pitch $(K_{33}/K_{22})P_o$ which is called the transient planar texture [56]. If the applied field is sufficiently low, then the liquid crystal will relax from the transient planar texture to the stable planar texture with the intrinsic pitch P_o.

The transient planar texture has a pitch $(K_{33}/K_{22})P_o$ and is less twisted than the (stable) planar texture with the pitch P_o. The free energy of the transient planar texture is higher than that of the (stable) planar texture and the liquid crystal will relax further to the (stable) planar texture if the applied field is turned off. In a cell with homogeneous alignment layers, the relaxation is a nucleation process and is slow [51, 57–59]. Helical twisting is added to the liquid crystal from nucleation seeds. In a cell with homeotropic alignment layers, helical twisting is injected into the liquid crystal from the surface and the transition time can be reduced [33].

If the liquid crystal is in the homeotropic texture and the applied field is reduced, it can relax either into the focal conic texture through the HF mode or into the planar texture through the HP mode. The HP mode is fast and occurs when the applied field is lower than e_{HP} while the HF mode is slow and occurs when the applied field is lower than e_{HF} which is higher than e_{HP}. The transition times are schematically shown in Figure 8.26. If the applied field is reduced quickly from a high field to a field below e_{HP}, the HP mode dominates and the liquid crystal ends in the planar texture. If the applied field is reduced quickly from the high field to a field in the region between e_{HP} and e_{HF}, or if it is reduced slowly and spends a sufficiently long time in the region, the HF mode dominates and the liquid crystal ends in the focal conic texture. In bistable cholesteric reflective displays, the way to switch the liquid crystal from the focal conic texture to the planar texture is by first applying a high field to

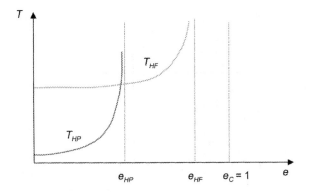

Figure 8.26 Field-dependence of the homeotropic-focal conic and homeotropic-planar transition times

switch it to the homeotropic texture, and then turning off the field quickly to allow it to relax to the planar texture.

4 Drive Schemes

As discussed in previous sections, cholesteric liquid crystals exhibit two bistable states at zero field: the reflecting planar texture and the non-reflective focal conic texture. They can be used to make multiplexed displays on a passive matrix. In this section, we discuss drive schemes which are also an important part in displays. It is necessary that the material does not flicker when being addressed. It is also desirable that inexpensive driver circuitry should be used. Drive schemes are designed based on the electro-optical response of the materials.

4.1 Electro-optical response of bistable cholesteric liquid crystals

Cholesteric liquid crystals are bistable at zero field, and their response to voltage pulses must be measured in order to design drive schemes for them. The measurement scheme is shown in Figure 8.27. First, a resetting voltage V_{reset} is applied to reset the cholesteric display. When the voltage of the pulse is appropriately high, independent of its initial state, the material is switched to the homeotropic texture during the pulse and reset to the planar texture afterward. When the voltage of the pulse is appropriately low, independent of its initial state, the material is switched to the focal conic texture during the pulse and remains there. Then the addressing voltage pulse is applied to study the response of the material. The measurement of the reflectance of the material is taken at a time when the reflectance no longer changes with time after the addressing pulse.

A typical response of a bistable cholesteric liquid crystal to voltage pulses is shown in Figure 8.28 [23]. The horizontal axis is the amplitude of the addressing voltage pulse and the vertical axis is the reflectance of the liquid crystal measured at 0 V about 1 s after the

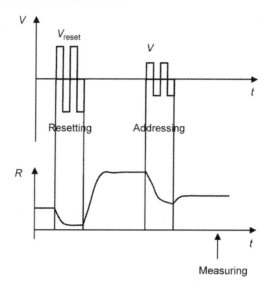

Figure 8.27 Schematic diagram showing the measurement of the response of the bistable cholesteric display

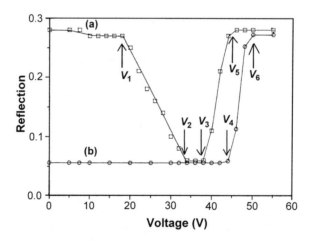

Figure 8.28 The response of the bistable cholesteric liquid crystal to voltage pulses

application of the addressing pulse. The width of the addressing voltage pulse is 40 ms. Curve (a) is the response when the material is initially in the planar texture (reset to the planar texture by a high resetting voltage before the application of each addressing voltage pulse). When the voltage of the addressing pulse is lower than $V_1 = 18$ V, the material remains in the planar texture and the reflectance retains the maximum value. When the voltage is above V_1, as the voltage of the pulse is increased, more and more domains are switched to the focal conic texture and therefore the reflectance decreases. In this region, gray scale reflections can easily be obtained. When the voltage is between $V_2 = 34$ V and $V_3 = 38$ V, the material is completely switched to the focal conic texture and the reflectance is decreased to a minimum. When the voltage is increased above V_3, as the voltage of the pulse is increased, more and more domains are switched to the homeotropic texture during the pulse, and afterwards the domains relax to the planar texture; the remaining domains are switched to the focal conic texture during the pulse and afterward remain there. Therefore the reflectance increases. When the voltage is increased above $V_5 = 46$ V, the material is switched completely to the homeotropic texture during the pulse and afterward relaxes back to the planar texture, and therefore the reflectance reaches the maximum again.

Curve (b) is the response when the material is initially in the focal conic texture (reset to the focal conic texture by a low resetting voltage before the application of each addressing voltage pulse). When the voltage of the addressing pulse is lower than $V_4 = 44$ V, the material remains in the focal texture and the reflectance retains the minimum value. When the voltage is above V_4, more and more domains are switched to the homeotropic texture during the pulse, and then the domains relax to the planar texture; the remaining domains retain the focal conic texture during the pulse and after. Therefore the reflectance increases. When the voltage is increased above $V_6 = 50$ V, the material is switched completely to the homeotropic texture during the pulse and then relaxes to the planar texture, and therefore the reflectance reaches a maximum. As discussed in the previous section, the liquid crystal can be switched from the planar texture directly to the homeotropic texture at a voltage which is lower than that needed to switch the material from the focal conic texture to the homeotropic texture. This may explain the difference between V_5 and V_6. In later discussion, we use the definitions $V_S = V_3$ and $V_R = V_6$. Independent of the initial state, a pulse with the voltage V_S drives the liquid crystal into the focal conic texture and a pulse with the voltage V_R drives the liquid crystal into the planar texture. It should be emphasised that reflection-pulse voltage curves depend on the time interval of the pulse. As the pulse width is reduced, the characteristic voltages are shifted to higher values. When the pulse width is shorter than a certain value, it becomes impossible to switch the liquid crystal completely from the planar texture to the focal conic texture.

The dynamic response of the liquid crystal is shown in Figure 8.29 where the width of the voltage pulses is 40 ms [23]. In Figure 8.29(a) the voltage of the applied pulse is 35 V. If the liquid crystal is initially in the planar texture, it is switched by the pulse to the focal conic texture. Although the fall time is only a few ms, if the pulse width is the same as the fall time, the liquid crystal goes back to the planar texture after the pulse. In the first few ms, the liquid crystal is tilted away from the planar texture and the reflection decreases, but it has not reached the focal conic texture. If the liquid crystal is initially in the focal conic texture, it remains there during the pulse. In Figure 8.29(b) the voltage of the applied pulse is 50 V. Independent of its initial state, the liquid crystal is switched to the homeotropic texture during the pulse, and afterwards relaxes into the planar texture. The relaxation time from the homeotropic texture to the planar texture is around 200 ms.

4.2 Conventional drive scheme

In the conventional drive scheme for the bistable cholesteric displays, the displays are addressed line by line [23]. The voltage for the row being addressed is

$$V_{ra} = \frac{1}{2}(V_R + V_S)$$

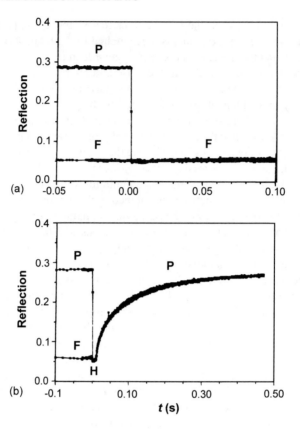

Figure 8.29 Dynamic responses of the bistable cholesteric liquid crystal: (a) to a low ($V_S = 35$ V) voltage pulse, (b) to a high ($V_R = 50$ V) pulse. P: planar texture, F: focal conic texture, H: homeotropic texture [23]

The voltage for the rows not being addressed is $V_{rna} = 0$. The column voltage to address a pixel to the reflecting planar texture is

$$V_{cp} = -\frac{1}{2}\Delta V$$

where $\Delta V = V_R - V_S$. The column voltage to address a pixel to the non-reflecting focal conic texture is

$$V_{cf} = \frac{1}{2}\Delta V$$

As an example, we consider a 2×2 bistable cholesteric display as shown in Figure 8.30. When row 1 is being addressed, the row voltage on row 1 is V_{ra}. The voltage across pixel 11 is

$$V_{11} = V_{ra} - V_{cp} = \frac{1}{2}(V_R + V_S) - \left[-\frac{1}{2}(V_R - V_S)\right] = V_R$$

and therefore the pixel is addressed to the planar texture. The voltage across pixel 12 is

$$V_{12} = V_{ra} - V_{cf} = \frac{1}{2}(V_R + V_S) - \left[-\frac{1}{2}(V_R - V_S) \right] = V_S$$

and therefore the pixel is addressed to the focal conic texture. When row 2 is being addressed, the row voltage on row 2 is V_{ra}. The voltage across pixel 21 is

$$V_{21} = V_{ra} - V_{cf} = \frac{1}{2}(V_R + V_S) - \left[\frac{1}{2}(V_R - V_S) \right] = V_S$$

and therefore the pixel is addressed to the focal conic texture. The voltage across pixel 22 is

$$V_{22} = V_{ra} - V_{cp} = \frac{1}{2}(V_R + V_S) - \left[-\frac{1}{2}(V_R - V_S) \right] = V_R$$

and therefore the pixel is addressed to the planar texture. The row voltage on row 1 is V_{rna}. The voltage across pixel 11 is

$$V_{rna} - V_{cf} = 0 - \left[\frac{1}{2}(V_R - V_S) \right] = -\frac{1}{2}\Delta V$$

and the voltage across pixel 12 is

$$V_{rna} - V_{cp} = 0 - \left[-\frac{1}{2}(V_R - V_S) \right] = -\frac{1}{2}\Delta V$$

As shown in Figure 8.26,

$$\frac{1}{2}\Delta V = \frac{1}{2}(V_R - V_S) = \frac{1}{2}(50 - 35) = 7.5 \text{ V}$$

which is lower than $V_1 = 18$ V, and therefore there is no cross-talk problem. It is important that

$$\frac{1}{2}\Delta V \leq V_1.$$

The frame time of the conventional drive scheme for an n-line bistable cholesteric display is $T_{frame} = n\Delta t$, where Δt is the time interval to address one line. For a typical bistable cholesteric liquid crystal, Δt is around 40 ms. For a 1000 line display, the frame time is $T_{frame} = 1000 \times 40 \text{ ms} = 40 \text{ s}$, which is very long. For some low viscosity cholesteric liquid crystals, Δt can be reduced to 10 ms, but the frame time is still long. One of the major applications of bistable cholesteric displays is the electronic book where the frame time should be around 1 s. For this application, new drive schemes must be designed.

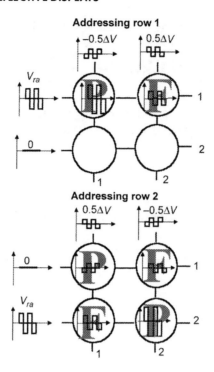

Figure 8.30 Schematic diagram showing the conventional drive scheme for the 2 × 2 bistable cholesteric display

Cholesteric (Ch) liquid crystal displays exhibit gray scale reflectance. In addressing a cholesteric display with gray scales, resetting is required, because the pulse voltage–reflectance curve depends on the initial state of the display, as shown in Figure 8.28. There are three schemes in addressing Ch displays with gray scale [30, 41]. As an example, we consider the cholesteric display whose characteristic is shown in Figure 8.28. In the first scheme, gray scale is achieved using pulses with voltage in the region between V_1 and V_2. First, all the pixels are addressed into the planar texture by applying a high-voltage pulse. Then the display is addressed line by line as in the conventional drive scheme. The row voltage for the row being addressed is $V_a = (V_1 + V_2)/2$ and the row voltage for the rows not being addressed is 0. The column voltage is in the range from $-(V_2-V_1)/2$ to $(V_2-V_1)/2$. When the column voltage is $(V_2-V_1)/2$, the voltage across the pixel is

$$(V_1 + V_2)/2 - (V_2 - V_1)/2 = V_1$$

and the pixel remains in the planar texture with the maximum reflectance. When the column voltage is $-(V_2-V_1)/2$, the voltage across the pixel is

$$(V_1 + V_2)/2 - [-(V_2 - V_1)/2] = V_2$$

and the pixel is addressed completely into the focal conic texture with the minimum reflectance. When the column voltage is between $(V_2-V_1)/2$ and $-(V_2-V_1)/2$, the voltage across the pixel is between V_1 and V_2, and the pixel is addressed into states with gray scale reflectance, in which planar domains and focal conic domains coexist. The advantage of this scheme is that the row voltage is low.

In the second scheme, gray scale is achieved using pulses with voltage in the region between V_3 and V_5. First, all the pixels are addressed into the planar texture by applying a high-voltage pulse. Then the display is addressed line by line as in the conventional drive scheme. The row voltage for the row being addressed is $V_a = (V_3 - V_5)/2$ and the row voltage for the rows not being addressed is 0. The column voltage is in the range from $-(V_5 - V_3)/2$ to $(V_5 - V_3)/2$. When the column voltage is $(V_5 - V_3)/2$, the voltage across the pixel is

$$(V_3 + V_5)/2 - (V_5 - V_3)/2 = V_3$$

and the pixel is addressed completely into the focal conic texture with the minimum reflectance. When the column voltage is $-(V_5 - V_3)/2$, the voltage across the pixel is

$$(V_3 + V_5)/2 - [-(V_5 - V_3)/2] = V_5$$

and the pixel is addressed completely into the homeotropic texture during the pulse and afterwards relaxes into the planar texture with the maximum reflectance. When the column voltage is between $(V_5 - V_3)/2$ and $-(V_5 - V_3)/2$, the voltage across the pixel is between V_3 and V_5, and the pixel is addressed into states with gray scale reflectance.

In the third scheme, a gray scale is achieved using pulses with voltage in the region between V_4 and V_6. First, all the pixels are addressed into the focal conic texture by applying an intermediate voltage pulse. Then the display is addressed line by line as in the conventional drive scheme. The row voltage for the row being addressed is $V_a = (V_4 + V_6)/2$ and the row voltage for the rows not being addressed is 0. The column voltage is in the range from $-(V_6 - V_4)/2$ to $(V_6 - V_4)/2$. When the column voltage is $-(V_6 - V_4)/2$, the voltage across the pixel is

$$(V_4 + V_6)/2 - (V_6 - V_4)/2 = V_4$$

and the pixel remains in the focal conic texture with the minimum reflectance. When the column voltage is $-(V_6 - V_4)/2$, the voltage across the pixel is

$$(V_6 + V_4)/2 - [-(V_6 - V_4)/2] = V_6$$

and the pixel is addressed completely into the homeotropic texture during the pulse and afterwards relaxes into the planar texture with the maximum reflectance. When the column voltage is between $(V_4 - V_6)/2$ and $-(V_6 - V_4)/2$, the voltage across the pixel is between V_4 and V_6, and the pixel is addressed into states with gray scale reflectance. The advantage of this scheme is that the reflectance of the focal conic texture can be minimised by applying a suitable refreshing pulse and the addressing pulse can be shortened by using higher voltages.

The gray scale drive schemes can be implemented using either amplitude modulation, as discussed above, or time modulation. In the time modulation, the column voltage pulse consists of two parts: one with high voltage and the other with low voltage. The relative time widths of the two parts determine the gray scale [30].

4.3 Dynamic drive schemes

Dynamic drive schemes are designed based on the fast homeotropic-transient planar transition and the hysteresis in the transition between the focal conic texture and the homeotropic texture [60]. The three-phase dynamic drive scheme is shown schematically in Figure 8.31 [61, 62]. The state of the liquid crystal in the drive sequence is shown in Table 8.1.

In the preparation phase, the applied voltage V_p is sufficiently high to switch the cholesteric liquid crystal to the homeotropic texture. In the selection phase, the voltage is either V_{on}, which is higher than V_{HP}, or V_{off}, which is lower than V_{HP}. V_{HP} is the voltage below which the liquid crystal relaxes from the homeotropic texture to the planar texture. The time interval of the selection phase is short, typically around 1 ms. If the selection voltage V_s is V_{on}, the liquid crystal remains in the homeotropic texture. If the selection voltage is V_{off}, the liquid crystal relaxes quickly into the transient planar texture. In the evolution phase, the voltage V_e is chosen to be between the voltage V_{HF}, below which the liquid crystal relaxes from the homeotropic texture to the focal conic texture, and V_{FH}, above which the liquid crystal transforms from the focal conic texture to the homeotropic texture. If the liquid crystal remains in the homeotropic texture in the selection phase, it remains in the homeotropic texture in the evolution phase, because V_e is higher than V_{HF}. If the liquid crystal relaxes into the transient planar texture in the selection phase, it is switched to the focal conic texture, but not to the homeotropic texture, in the evolution phase, because the transient planar texture is unstable in the field and V_e is lower than V_{FH}. The transition from the transient planar texture to the focal conic texture is similar to the transition from the planar texture to the focal conic texture as discussed in Section 3.1. After the evolution phase, the applied voltage is 0 or low. If the liquid crystal remains in the homeotropic texture

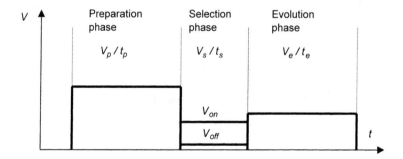

Figure 8.31 Schematic diagram of the three-phase dynamic drive scheme

Table 8.1 The state of the cholesteric liquid crystal in the three-phase dynamic drive scheme

	Preparation phase	Selection phase	Evolution phase	Afterward
Obtain On-state	Homeotropic texture	High voltage V_{on} Homeotropic texture	Homeotropic texture	Planar texture
Obtain Off-state	Homeotropic texture	Low voltage V_{off} Transient planar texture	Focal conic texture	Focal conic texture

in the evolution phase, it relaxes into the planar texture. If the liquid crystal is switched to the focal conic texture in the evolution phase, it remains in the focal conic texture.

In the preparation phase, the voltage and time interval are chosen to switch the cholesteric liquid crystal into the homeotropic texture. Typically V_p is higher than V_{FH} which is the threshold to switch the liquid crystal from the focal conic texture to the homeotropic texture, as discussed in Section 3.2. This voltage is also higher than V_{PH} which is the threshold to switch the liquid crystal from the planar texture to the homeotropic texture. For a given cholesteric material, the higher the preparation voltage, the shorter the time interval can be. For a 5 µm cholesteric display reflecting green light, the typical preparation voltage V_p is about 45 V and the time interval t_p is about 40 ms.

In the selection phase, the voltage and time interval are chosen in such a way that the liquid crystal remains in the homeotropic texture under the high selection voltage V_{on} and transforms into the transient planar texture under the low selection voltage V_{off}. For a cholesteric display whose V_{FH} is about 40 V, the obtained reflectance versus the selection voltage in the dynamic drive scheme is shown in Figure 8.32 [63], where the preparation phase is 50 V/60 ms and the evolution phase is 31 V/40 ms. When the selection phase is 20 ms wide, the voltage V_{on} to select the on-state is 23 V and the voltage V_{off} to select the off-state is 15 V. As we will discuss later, the column voltage to select the on-state is $-(V_{on}-V_{off})/2=-4$ V, and the column voltage to select the off-state is $(V_{on}-V_{off})/2=4$ V. These voltages do not cause cross-talking problems. When the selection phase is 1.0 ms wide, V_{on} and V_{off} are shifted to 16 V and 11 V, respectively. When the selection phase is too short, the liquid crystal is no longer able to relax to the transient planar texture in the selection phase, and the reflectance of the obtained off-state is high. The figure shows that gray scales are possible in the dynamic drive scheme.

The selection phase can be divided into sub-phases such as pre-selection and post-selection. Sometimes the pre-selection phase is referred to as the post-preparation phase. The physics behind these drive schemes is that the cholesteric liquid crystal in the

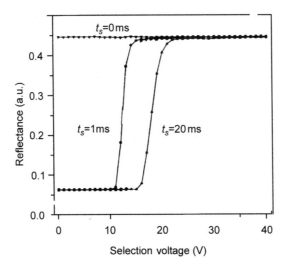

Figure 8.32 The reflectance of the state obtained in the dynamic drive scheme versus the selection voltage under various selection times [63]

relaxation from the homeotropic texture to the transient planar texture through the conic helical structure can be switched back to the homeotropic texture by applying a voltage. This required voltage is higher when the liquid crystal relaxes further toward the transient planar texture. In the four-phase dynamic drive scheme, the selection time is reduced to half of the time in the three-phase dynamic drive scheme [64]. In the five-phase dynamic drive scheme, the selection time is reduced by one order of magnitude [65].

In the evolution phase, the applied voltage must be able to hold the liquid crystal in the homeotropic texture and to switch the liquid crystal from the transient planar texture to the focal conic texture. Hence the evolution voltage must be carefully chosen to be in the hysteresis loop of the transition between the focal conic texture and the homeotropic texture. The evolution time must be sufficiently long to be able to switch the liquid crystal from the transient planar texture to the focal conic texture. For the same display discussed above, the contrast ratio of the selected on- and off-states and the evolution voltage is shown in Figure 8.33. The preparation phase is 50 V/60 ms. The selection phase is 1 ms wide and the voltage V_{on} to select the on-state is 16 V and the voltage V_{off} to select the off-state is 11 V. The evolution time is 60 ms. When the evolution voltage is region I and IV, the liquid crystal always relaxes into the planar texture after the addressing regardless of the selection voltage. In region II, for a low selection voltage, the liquid crystal relaxes from the transient planar texture to the stable planar texture in the evolution phase and therefore the high reflectance planar texture is selected. For a high selection voltage, the liquid crystal transforms from the homeotropic texture to the focal conic texture in the evolution phase, and therefore the low reflectance focal conic texture is selected. In region III, for a low selection voltage, the liquid crystal transforms from the transient planar texture to the focal conic texture in the evolution phase, and therefore the low reflectance focal conic texture is selected. For a high selection voltage, the liquid crystal remains in the homeotropic texture in the evolution phase and therefore the high reflectance planar texture is selected. In this region, high contrast can be

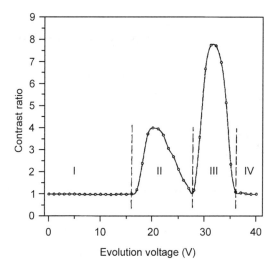

Figure 8.33 The contrast ratio between the selected on- and off-states versus the evolution voltage in the dynamic drive scheme [63]

achieved. The time interval of the evolution phase can be chosen to be very long. The drive scheme even works well when the evolution time is as long as the frame time [66].

In the dynamic drive scheme, the time intervals of the preparation and evolution phases are long, typically more than 40 ms. This is, however, not a problem. The preparation and evolution times can be shared by putting multiple lines in these phases because the voltages are the same for all pixels, independent of the state to which the pixels will be addressed. This time-sharing pipeline algorithm is schematically shown in Figure 8.34. The column voltages are $V_4 = (V_{on} - V_{off})/2$ to select the off-state and $-V_4 = -(V_{on} - V_{off})/2$ to select the on-state, respectively, and have the frequency f_1. The row voltage for the preparation phase is V_1 which satisfies the relation $V_p = (V_1^2 + V_4^2)^{1/2}$, and has the frequency f_2 which is chosen to be very different from f_1. The number of lines put into the preparation phase is equal to or larger than $n_p = t_p/t_s$. The row voltage for the selection phase is V_2 which is equal to $(V_{on} + V_{off})/2$ and has the frequency f_1. Only one line is put into the selection phase. The voltage for the evolution phase is V_3 which satisfies the relation $V_e = (V_3^2 + V_4^2)^{1/2}$ and has the frequency f_2. The number of lines put into the selection phase is equal to or larger than $n_p = t_e/t_s$. The row voltage for the rows after addressing is 0, and therefore the voltage across the pixel in the rows is V_4 which does not cause cross-talk problems. Thus the frame time for an n-line display is given by

$$t_{frame} = t_p + t_e + n \times t_s \tag{8.58}$$

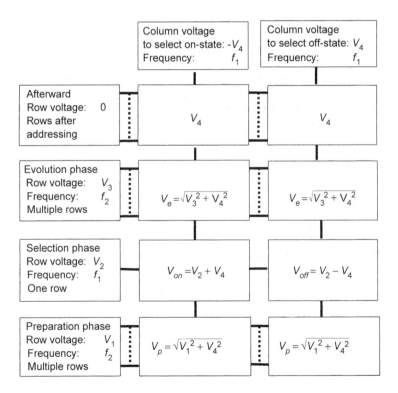

Figure 8.34 Schematic diagram showing how the cholesteric display is addressed in the dynamic drive scheme

As an example: if $t_p = 40$ ms, $t_s = 1$ ms, $t_e = 40$ ms and $n = 1000$, then the frame time is $t_{frame} = 40 + 40 + 1000 \times 1 = 1080$ ms, which is sufficiently fast for electronic book applications.

There are several more drive schemes for bistable cholesteric reflective displays, which are designed based on the dynamic behavior of cholesteric liquid crystals [67–70]. Cumulative drive schemes have also been designed for the cholesteric displays [71, 72], where the images are changed gradually by multiple pulses.

References

1 P. G. de Gennes and J. Prost, "The physics of liquid crystals" (Oxford University Press, New York, 1993).

2 F. Reinitzer, *Monatsch Chem.* **9**, 421 (1898).

3 A. W. Hall, J. Hollingshurst, and J. W. Goodby, Chiral and achiral calamitic liquid crystals for display applications, "Handbook of liquid crystal research," ed. by P. J. Collings and J. S. Patel (Oxford University Press, New York, 1997).

4 S. Chandrasekhar "Liquid crystals," 2nd ed. (Cambridge University Press, New York, 1997).

5 O. Lehmann, *Z. Phys. Chem. Stoechiom, Verwandtschafsl.* **4**, 62 (1889).

6 G. Friedel, *Ann. Phys. (Paris)* **18**, 273 (1922).

7 M. Kleman, "Points, lines and walls" (Wiley, New York, 1983).

8 L. M. Blinov and V. G. Chigrinov, "Electrooptical effects in liquid crystal materials" (Springer-Verlag, New York, 1994).

9 O. D. Lavrentovich and D.-K. Yang, Cholesteric cellular patterns with electric-field-controlled line tension, *Phys. Rev. E* **57**, Rapid Communications, R6269 (1998).

10 R. B. Meyer, Distortion of a cholesteric structure by a magnetic field, *Appl. Phys. Lett.* **14**, 208 (1969).

11 W. Greubel, U. Wolf, and H. Kruger, Electric field induced texture changes in certain nematic/ cholesteric liquid crystal mixtures, *Mol. Cryst. Liq. Cryst.* **24**, 103 (1973).

12 G. A. Dir, J. J. Wysock, J. E. Adams, J. H. Becker, W. E. Haas, B. Mechlowitz, L. B. Leader, E. D. Saeva, and J. L. Daily, Cholesteric liquid crystal texture change displays, *Proc. of SID* **13**(2), 106 (1972).

13 D.-K. Yang, J. L. West, L. C. Chien, and J. W. Doane, Control of the reflectivity and bistability in displays based on cholesteric liquid crystals, *J. Appl. Phys.* **76**, 1331 (1994).

14 E. B. Priestley, Introduction to the optical properties of cholesteric and chiral nematic liquid crystals, "Introduction to liquid crystals," ed. by E. B. Priestley, P. J. Wojtoicz and P. Sheng (Plenum, New York, 1979).

15 D.-K. Yang and X.-D. Mi, Modeling of the reflection of cholesteric liquid crystals using Jones matrix, *J. Phys. D.* **33**, 672 (2000).

16 D. W. Berreman, Optics in stratified and anisotropic media: 4×4-matrix formulation, *J. Opt. Soc. Am.* **62**, 502 (1972).

17 D. W. Berreman and T. J. Scheffer, Bragg reflection of light from single-domain cholesteric liquid crystal films, *Phys. Rev. Lett.* **25**, 577 (1970).

18 D. W. Berreman and T. J. Scheffer, Reflection and transmission by single-domain cholesteric liquid crystal films: theory and verification, *Mol. Cryst. Liq. Cryst.* **11**, 395 (1970).

19 W. D. St. John, W. J. Fritz, Z. J. Lu, and D.-K. Yang, Optical properties of bistable cholesteric reflective displays, *Phys. Rev. E* **51**, 1191 (1995).

20 M. Xu, F. D. Xu, and D.-K. Yang, Effects of cell structure on the reflection of cholesteric liquid crystal display, *J. Appl. Phys.* **83**, 1938 (1998).

21 D.-K. Yang, J. W. Doane, Z. Yaniv, and J. Glasser, Cholesteric reflective display: drive scheme and contrast, *Appl. Phys. Lett.* **65**, 1905 (1994).

22 D.-K. Yang, L.-C. Chien, and J. W. Doane, Cholesteric liquid crystal/polymer gel dispersion bistable at zero field, *Proc. Intl. Display Res. Conf.*, 49 (1991).

23 D.-K. Yang and J. W. Doane, Cholesteric liquid crystal/polymer gel dispersions: reflective displays, *SID Intl. Symp. Digest Tech. Papers* **23**, 759 (1992).

24 J. W. Doane, D.-K. Yang, and Z. Yaniv, Front-lit flat panel display from polymer stabilized cholesteric textures, *Proc. Japan Display'92*, 73 (1992).

25 Z.-J. Lu, W. D. St. John, X.-Y. Huang, D.-K. Yang, and J. W. Doane, Surface modified reflective cholesteric displays, *SID Intl. Symp. Digest Tech. Papers*, **26**, 172 (1995).

26 R. Q. Ma and D.-K. Yang, Polymer stabilized bistable black-white cholesteric reflective display, *SID Intl. Symp. Digest Tech. Papers* **28**, 101 (1997).

27 R. Q. Ma and D.-K. Yang, Optimization of polymer stabilized bistable black-white cholesteric reflective display, *J. SID* **7**, 61 (1999).

28 X.-Y. Huang, N. Miller, A. Khan, D. Davis, and J. W. Doane, Gray scale of bistable reflective cholesteric displays, *SID Intl. Symp. Digest Tech. Papers* **29**, 810 (1998).

29 M. Xu and D.-K. Yang, Optical properties of the gray-scale states of cholesteric reflective displays, *SID Intl. Symp. Digest Tech. Papers* **30**, 950 (1999).

30 J. Gandhi, D.-K. Yang, X.-Y. Huang, and N. Miller, Gray scale drive schemes for bistable cholesteric reflective displays, *Proc. Asia Display'98*, 127 (1998).

31 M. H. Lu, H. J. Yuan, and Z. Yaniv, Color reflective liquid crystal display, U. S. Patent 5,493,430 (1996).

32 M. Pfeiffer, D.-K. Yang, J. W. Doane, R. Bunz, and E. Lueder, A high information content reflective cholesteric display, *SID Intl. Symp. Digest Tech. Papers* **26**, 706 (1995).

33 X.-D. Mi and D.-K. Yang, Cell designs for fast reflective cholesteric LCDs, *SID Intl. Symp. Digest Tech. Papers* **29**, 909–912 (1998).

34 I. Sage, Thermochromic liquid crystal devices, "Liquid crystals - applications and uses," Vol. 3, ed. by B. Bahadur (World Scientific, New Jersey, 1990).

35 Y. Kim, J. Francl, B. Taheri, and J. L. West, A method for the formation of polymer walls in liquid crystal/polymer mixtures, *Appl. Phys. Lett.* **72**, 2253 (1998).

36 Y. Ji, J. Francl, W. J. Fritz, P. J. Bos, and J. L. West, Polymer walls in higher-polymer-content bistable reflective cholesteric displays, *SID Intl. Symp. Digest Tech. Papers* **27**, 611 (1996).

37 L.-C. Chien, U. Muller, M.-F. Nabor, and J. W. Doane, Multicolor reflective cholesteric displays, *SID Intl. Symp. Digest Tech. Papers* **26**, 169 (1995).

38 F. Vicentini and L.-C. Chien, Tunable chiral materials for multicolor reflective cholesteric displays, *Liq. Cryst.* **24**, 483 (1998).

39 L.-C. Chien, F. Vicentini, Y. Lin, and U. Muller, Color pixelization of cholesteric materials, *Proc. 3rd Color Imaging Conference*, 58 (1995).

40 D.-K. Yang, Z.-J. Lu, and J. W. Doane, Bistable polymer dispersed cholesteric liquid crystal displays, U. S. Patent 8,852,589 (1999).

41 K. Hashimoto, M. Okada, K. Nishguchi, N. Masazumi, E. Yamakawa, and T. Taniguchi, Reflective color display using cholesteric liquid crystals, *SID Intl. Symp. Digest Tech. Papers* **29**, 897 (1998).

42 D. Davis, K. Kahn, X.-Y. Huang, and J. W. Doane, Eight-color high-resolution reflective cholesteric LCDs, *SID Intl. Symp. Digest Tech. Papers* **29**, 901 (1998).

43 J. L. West and V. Bodnar, Optimization of stacks of reflective cholesteric films for full color displays, *Proc. 5th Asian Symp. on Information Display* 29 (1999).

44 D. Davis, K. Hoke, A. Khan, C. Jones, X. Y. Huang, and J. W. Doane, Multiple color high resolution reflective cholesteric liquid crystal displays, *Proc. Intnl. Display Research Conf.*, 242 (1997).

45 P. P. Crooker and D. K. Yang, Polymer-dispersed chiral liquid-crystal color display, *Appl. Phys. Lett.* **57**, 2529 (1990).

46   D. K. Yang and P. P. Crooker, Field-induced texture of polymer-dispersed chiral liquid crystal microdroplets, *Liq. Cryst.* **9**, 245 (1991).

47   H.-S. Kitzerow, Polymer-dispersed liquid crystals, from the nematic curvilinear aligned phase to ferroelectric films, *Liq. Cryst.* **16**, 1 (1994).

48   D.-K. Yang, X. Y. Huang, and Y.-M. Zhu, Bistable cholesteric reflective displays: material and drive schemes, *Annual Review of Materials Science* **27**, 117 (1996).

49   W. Helfrich, Deformation of cholesteric liquid crystals with low threshold voltage, *Appl. Phys. Lett.* **17**, 531 (1970).

50   J. P. Hurault, Static distortions of a cholesteric planar structure induced by magnetic or ac electric fields, *J. Chem. Phys.* **59**, 2068 (1973).

51   D.-K. Yang and Z.-J. Lu, Switching mechanism of bistable cholesteric reflective displays, *SID Intl. Symp. Digest Tech. Papers* **26**, 351 (1995).

52   M. Xu and D.-K. Yang, Electro-optical properties of dual-frequency cholesteric liquid crystal reflective display and drive scheme, *Jpn. J. Appl. Phys.* **38**, 6827 (2000).

53   M. Xu and D.-K. Yang, Dual frequency cholesteric light shutters, *Appl. Phys. Lett.* **70**, 720 (1997).

54   X.-D. Mi, Dynamics of the transitions among cholesteric liquid crystal textures, Dissertation, Kent State University, 2000.

55   M. Kawachi and O. Kogure, Hysteresis behavior of texture in field-induced nematic-cholesteric relaxation, *Jpn. J. Appl. Phys.* **16**, 1673 (1977).

56   M. Kawachi, O. Kogure, S. Yosji, and Y. Kato, Field-induced nematic-cholesteric relaxation in a small angle wedge, *Jpn. J. Appl. Phys.* **14**, 1063 (1975).

57   P. Watson, V. Sergan, J. E. Anderson, J. Ruth, and P. J. Bos, A study of the dynamics of the reflection color, helical axis oreintation, and domain size in cholesteric LCDs, *SID Intl. Symp. Digest Tech. Papers* **29**, 905 (1998).

58   P. Watson, J. E. Anderson, V. Sergan, and P. J. Bos, The transition mechanism of the transient planar to planar director configuration change in cholesteric liquid crystal displays, *Liq. Cryst.* **26**, 1307 (1999).

59   M.-H. Lu, Bistable reflective cholesteric liquid crystal display, *J. Appl. Phys.* **81**, 1063 (1997).

60   X.-Y. Huang, D.-K. Yang, and J. W. Doane, Transient dielectric study of bistable reflective cholesteric displays and design of rapid drive scheme, *Appl. Phys. Lett.* **69**, 1211 (1995).

61   X.-Y. Huang, D.-K. Yang, P. Bos, and J. W. Doane, Dynamic drive for bistable reflective cholesteric displays: a rapid addressing scheme, *SID Intl. Symp. Digest Tech. Papers* **26**, 347 (1995).

62   X.-Y. Huang, D.-K. Yang, P. J. Bos, and J. W. Doane, Dynamic drive for bistable cholesteric displays: a rapid addressing scheme, *J. SID* **3**, 165 (1995).

63   X.-Y. Huang, Field-induced transitions in cholesteric liquid crystals: dynamics and application to displays, Dissertation, Kent State University, 1996.

64   X.-Y. Huang, D.-K. Yang, M. Stefanov, and J. W. Doane, High-performance dynamic drive scheme for bistable reflective cholesteric displays, *SID Intl. Symp. Digest Tech. Papers* **27**, 359 (1996).

65   Y.-M. Zhu and D.-K. Yang, High speed dynamic drive scheme for bistable reflective cholesteric displays, *SID Intl. Symp. Digest Tech. Papers* **28**, 97 (1997).

66   J. Ruth, R. Hewitt, and P. J. Bos, Low cost dynamic drive scheme for reflective bistable cholesteric liquid crystal displays, *Proc. Flat Panel Display*, 89 (1997).

67   F. H. Yu and H. S. Kwok, New drive scheme for reflective bistable cholesteric LCDs, *SID Intl. Symp. Digest Tech. Papers* **28**, 659 (1997).

68   W. C. Yip and H. S. Kwok, Modulation techniques for gray-scale bistable cholesteric displays, *SID Intl. Symp. Digest Tech. Papers* **31**, 133 (2000).

69   A. Kozachenko, V. Sorokin, and P. Oleksenko, Control of reflectivity utilizing hysteresis property in reflective cholesteric displays, *Proc. Intnl. Display Research Conf.* 35 (1997).

70 A. Rybalochka, V. Sorokin, S. Valyukh, A. Sorokin, and V. Nazarenko, Dynamic drive scheme for fast addressing of cholesteric displays, *SID Intl. Symp. Digest Tech. Papers* **31**, 818 (2000).

71 X.-Y. Huang, A. Khan, N. Miller, C. Jones, and J. W. Doane, Cumulative drive scheme for bistable reflective cholesteric displays, *Proc. Intnl. Display Research Conf.*, 30 (2000).

72 Y.-M. Zhu and D.-K. Yang, Cumulative drive scheme for bistable reflective cholesteric LCDs, *SID Intl. Symp. Digest Tech. Papers* **29**, 798 (1998).

9

Bistable Nematic Displays

1 Introduction

Multiplexibility of liquid crystal materials is necessary for their use in high information content displays with passive matrices. Conventional twisted nematic (TN) materials lack this property. They neither exhibit bistable states nor have steep voltage–transmittance curves. They cannot be used to make multiplexed displays with a passive matrix. Sophisticated thin film transistors (TFT) must be used in order to multiplex TNs. Super twisted nematic (STN) materials have steep voltage–transmittance curves and can be multiplexed. They suffer, however, the drawbacks of low contrast ratio, narrow viewing angle and limited information content.

In recent years there has been a lot of research and development activity in developing bistable nematic liquid crystal displays (LCD). Several types of bistable nematic LCD have been demonstrated, which have good performance characteristics. They are the bistable twisted-untwisted nematic LCD [1–5], the zenithal bistable nematic LCD [6, 7], the surface induced bistable nematic LCD [8–11], the mechanically bistable nematic LCD and the bistable STN-LCD [12–16]. A bistable nematic material has two stable states with different optical properties. Once it is driven into a stable state, it remains there. It can be used to make highly multiplexed displays on a passive matrix. There is no limitation on the information content. Bistable N-LCDs exhibit high contrast ratios and large viewing angles. The drawback is that most bistable N-LCDs do not have gray scale capability.

2 Bistable Twisted-untwisted Nematic Liquid Crystal Display

In the bistable twisted-untwisted nematic (BTN) liquid crystal display, the two stable states (precisely speaking, they are meta-stable) are selected by making use of the hydrodynamic motion of the liquid crystal [1, 3, 17]. Under one hydrodynamic condition, the liquid crystal is switched to one twisted state; under another hydrodynamic condition, the liquid crystal is switched to the other twisted state. An example is shown in Figure 9.1. One stable state is the 0° twist state shown by Figure 9.1(a) and the other stable state is the 360° state shown by Figure 9.1(c). The bistable twisted-untwisted TN was first developed by Berreman and Heffner [1]. Besides this particular design, there are other possible designs. Generally speaking, the twist angles of the two bistable states are $(\phi, 2\pi + \phi)$ [18, 19]. ϕ is the angle between the alignment directions of the alignment layers on the bottom and top substrates of the cell. The angle is usually in the region between $-\pi/2$ and $\pi/2$. The twist angle difference between the two bistable states is 360°.

2.1 Bistability and switching mechanism

We first consider the (0°, 360°) BTN where the alignment directions on the two cell surfaces are parallel, as shown in Figure 9.1. Chiral dopants are added to the nematic liquid crystal to obtain the intrinsic pitch P such that $h/P = 1/2$, where h is the cell thickness [1, 18, 20]. The real stable state is the 180° twist state, as shown in Figure 9.1(d); it has a free energy lower than both the 0° and 360° twist states.

Hydrodynamics plays an important role in the switching between the bistable states in the bistable TN. Liquid crystals have a unique and complicated hydrodynamic behavior in which the rotational motion of the liquid crystal director and the translational motions of the liquid crystal are coupled [21–23]. On one hand, a rotation of the liquid crystal produces a viscous stress that results in a translational motion. On the other hand, a translational velocity gradient produces a viscous torque and affects the rotation of the director. In the BTN, independent of the initial state of the liquid crystal ($\Delta\varepsilon > 0$), once an electric field slightly higher than the threshold V_{th} of the Freedericksz transition is applied, the liquid crystal is switched to the homeotropic state as shown in Figure 9.1(b). In this state, the liquid crystal is aligned homeotropically only in the middle of the cell where there is no twist. The liquid crystal near the surface of the cell has some twisting. Once the field is turned off, the

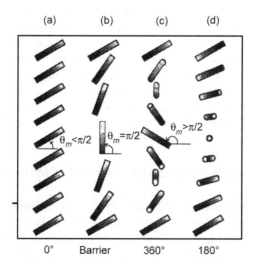

Figure 9.1 The liquid crystal director configurations of the states in the bistable TN

liquid crystal relaxes into the 0° twist state, because the 0° twist state and the homeotropic state are topologically the same while the 180° twist state is topologically different. If a very high field, higher than a saturation voltage V_{sa}, is applied to the liquid crystal, then in the homeotropic state, the liquid crystal in most regions except very near the cell surfaces is aligned homeotropically and has low elastic energy. The liquid crystal director changes orientation rapidly in space near the surface and has a very high elastic energy. When the applied field is removed suddenly, the liquid crystal director near the cell surface rotates very quickly because of the high elastic torque while the liquid crystal director in the middle rotates slowly because of the low elastic torque [3]. Thus a translational motion is induced, as shown in Figure 9.2, and will affect the rotation of the liquid crystal in the middle in such a way that the angle is increasing instead of decreasing. If the liquid crystal is nematic without chiral agents, the rotation of the director near the surface in one direction and the rotation of the director at the middle in the opposite direction produces a distortion of the director, which is not energetically favored. The angle of the director in the middle eventually decreases again, resulting in the kickback phenomenon [23–25]. If the liquid crystal has an intrinsic twist, the angle of the director in the middle can increase further and the liquid crystal is switched into the 360° twist state, as shown in Figure 9.1(c).

It is difficult to calculate the involved hydrodynamics analytically. We present a qualitative analysis here. In the Ericksen–Leslie theory, the viscous stress tensor is given by

$$\sigma'_{\alpha\beta} = \alpha_1 n_\alpha n_\beta n_\mu n_\rho A_{\mu\rho} + \alpha_2 n_\alpha N_\beta + \alpha_3 n_\beta N_\alpha + \alpha_4 A_{\alpha\beta} + \alpha_5 n_\alpha n_\mu A_{\mu\beta} + \alpha_6 n_\beta n_\mu A_{\mu\alpha} \quad (9.1)$$

α_i ($i = 1, 2, 3, 4, 5, 6$) are Leslie viscosity coefficients. n_α ($\alpha = 1, 2, 3$) are the components of the liquid crystal director. $A_{\alpha\beta}$ are the components of the symmetric velocity gradient tensor and are given by

$$A_{\alpha\beta} = \frac{1}{2}\left(\frac{\partial v_\beta}{\partial x_\alpha} + \frac{\partial v_\alpha}{\partial x_\beta}\right) \quad (9.2)$$

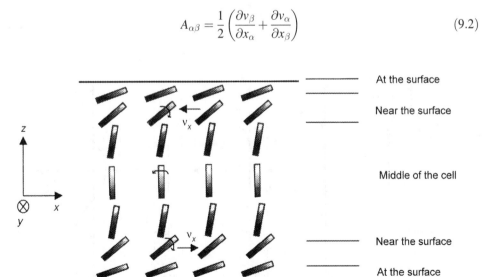

Figure 9.2 Schematic diagram showing the translational and rotational motions in the bistable TN

N_α are the components of the net angular velocity of the director and are given by

$$N_\alpha = \frac{dn_\alpha}{dt} - W_{\alpha\beta}n_\beta \qquad (9.3)$$

$W_{\alpha\beta}$ are the components of the anti-symmetric velocity gradient and are given by

$$W_{\alpha\beta} = \frac{1}{2}\left(\frac{\partial v_\beta}{\partial x_\alpha} - \frac{\partial v_\alpha}{\partial x_\beta}\right) \qquad (9.4)$$

The angular velocity ω of the flow is related to $W_{\alpha\beta}$ by

$$\omega = \frac{1}{2}\nabla \times v = (W_{yz,} W_{zx,} W_{xy}) \qquad (9.5)$$

Right after the applied high voltage has been turned off, the components of the liquid crystal director as a function of z are shown in Figure 9.3(a) and (b), respectively. Except for very close to the cell surface, n_x is small and there is no flow at the beginning of the relaxation. The net angular velocity of the liquid crystal director is given by

$$N = \frac{dn}{dt} = \frac{\partial n_x}{\partial t}\hat{x} \qquad (9.6)$$

Near the cell surface, $\partial n_x/\partial t > 0$. The significant components of the stress tensor are

$$\sigma_{xz} = \sigma'_{xz} = \alpha_3 n_z \frac{\partial n_x}{\partial t} \qquad (9.7)$$

$$\sigma_{zx} = \sigma'_{zx} = \alpha_2 n_z \frac{\partial n_x}{\partial t} \qquad (9.8)$$

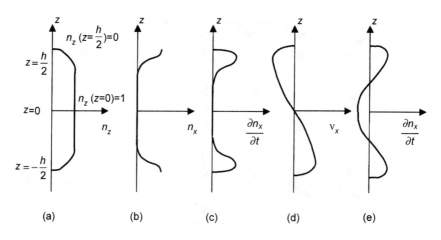

(a) (b) (c) (d) (e)

Figure 9.3 The profiles of the liquid crystal director and velocity in the relaxation of the bistable TN

The equation of motion is

$$\rho\left(\frac{\partial v_\alpha}{\partial t} + v_\beta \frac{\partial v_\alpha}{\partial x_\beta}\right) = \rho\frac{\partial v_\alpha}{\partial t} = \frac{\partial \sigma_{\alpha\beta}}{\partial x_\beta} \tag{9.9}$$

The significant translational motion is induced in the x direction and the acceleration is given by

$$\rho\frac{\partial v_x}{\partial t} = \frac{\partial \sigma_{x\beta}}{\partial x_\beta} = \frac{\partial}{\partial z}\left(\alpha_3 n_z \frac{\partial n_x}{\partial t}\right) \tag{9.10}$$

Consider the case $\alpha_3 > 0$ (the final result is the same for negative α_3). In the top half of the cell, because n_z decreases with increasing z

$$\frac{\partial}{\partial z}\left(\alpha_3 n_z \frac{\partial n_x}{\partial t}\right) < 0$$

and therefore $v_x < 0$, i.e. the flow is in the $-x$ direction. In the bottom half of the cell, because n_z increases with increasing z

$$\frac{\partial}{\partial z}\left(\alpha_3 n_z \frac{\partial n_x}{\partial t}\right) > 0$$

and therefore $v_x > 0$, i.e. the flow is in the x direction. Because of the variation of the liquid crystal director along the z axis, A gradient of v_x along z is developed as shown in Figure 9.3(d). In the middle of the cell, the significant components of A are

$$A_{xz} = A_{zx} = \frac{1}{2}\frac{\partial v_x}{\partial z} \tag{9.11}$$

This gradient produces a viscous torque on the liquid crystal director in the middle of the cell. The dynamic equation of the rotation is

$$I\frac{d}{dt}\left(n \times \frac{dn}{dt}\right) = n \times h - \gamma_1 n \times N - \gamma_2 n \times (A \cdot n) \tag{9.12}$$

where h is the molecular field from the elastic distortion and externally applied fields. On the right-hand side of Equation 9.12, the first term is the torque produced by the molecular field, the second term is the torque produced by the viscosity of the rotation of the director and the third term is the torque produced by the viscosity of the translational motion. In over-damped rotation, the inertia term on the left-hand side of Equation 9.12 is small and can be neglected. In the middle of the cell, at zero applied voltage, the elastic distortion is small at the beginning of the relaxation, and therefore the molecular field is negligible. Equation 9.12 becomes

$$\gamma_1 n \times N = -\gamma_2 n \times (A \cdot n) \tag{9.13}$$

In the middle of the cell, the significant component of $A \cdot n$ is

$$(A \cdot n)_x = \frac{1}{2} n_z \frac{\partial v_x}{\partial z} \qquad (9.14)$$

From Equation 9.13 we obtain

$$\frac{\partial n_x}{\partial t} = -\frac{\gamma_2}{2\gamma_1} n_z \frac{\partial v_x}{\partial z} \qquad (9.15)$$

γ_1 is positive and γ_2 is negative. In the middle of the cell $\partial v_x/\partial z > 0$, and therefore $\partial n_x/\partial t < 0$. Near the cell surface, $\partial n_x/\partial t > 0$. This explains why at the cell surface the polar angle of the director decreases while in the middle of the cell the polar angle increases. Figure 9.3(c) shows $\partial n_x/\partial t$ versus z right after the applied high voltage is turned off and before the translational motion is developed. Figure 9.3(e) shows $\partial n_x/\partial t$ versus z midway in the relaxation and the translational motion is fully developed.

In order for the angle of the liquid crystal in the middle to reach the value close to π as shown in Figure 9.1(c), the liquid crystal in the middle must gain sufficient momentum at the beginning. If the applied voltage is not sufficiently high or a bias voltage is applied when the high voltage is turned off, the liquid crystal in the middle cannot obtain a sufficiently high angular velocity to transform into the 360° twist state, and therefore the liquid crystal ends in the 0° twist state [2, 18, 19]. Computer simulations have successfully modeled the hydrodynamic behavior observed in the bistable TN [3, 17].

Regarding the bistability, the important parameters are the angle ϕ between the aligning directions of the alignment layers, the pretilt angle θ_p of the alignment layer and the intrinsic pitch P of the liquid crystal [17, 18]. So far the bistability has been observed for $-\pi/2 \leq \phi \leq \pi/2$.

When the first (0°, 360°) bistable TN was developed, a high pretilt angle was used. Now both experimental and theoretical results have shown that the high pretilt angle is not necessary. A pretilt angle of a few degrees is sufficient [17].

The intrinsic pitch of the liquid crystal should be chosen heuristically in such a way that the $(\phi + \pi)$ twist state has the minimum free energy, i.e. $(\phi + \pi)$ is the intrinsic twist. Hence

$$\frac{2\pi}{P} h = \phi + \pi \qquad (9.16)$$

which gives $h/P = 1/2 + \phi/(2\pi)$. Bistability can be achieved for P in the regions shown in Table 9.1 [18]. The discrepancy between the experimental value and the simulation value may be attributed to the fact that the pretilt angle is not considered in the heuristic choice and that the viscosity coefficients are not precisely known. Another important reason is that the d/h value must be significantly higher in order to switch the liquid crystal into the high twisted state [47].

Table 9.1   The regions of the ratio between the cell thickness h and the pitch P where bistability can be achieved [18]

ϕ	$-\pi/2$	0	$\pi/2$
Heuristic value	0.25	0.5	0.75
Experimental value	0.285–0.305	0.55–0.7	0.9
Simulation value	0.425	0.85	unstable

As mentioned earlier, the $(\phi, \phi + 360°)$ states are only metastable. The $\phi + 180°$ state is the real stable state. Once the liquid crystal has been addressed into the metastable states, it can remain in those states for a few seconds [1]. There are energy barriers between the metastable states and the stable state. The transition from the metastable states to the $\phi + 180°$ stable state is a slow nucleation process. In display applications, this limited lifetime of the metastable states is not a problem if the displays are addressed constantly. Nevertheless, it is desirable that the two metastable states have long lifetimes. The lifetime can be extended in a few ways [20, 26]. When alignment layers with a high pretilt angle are used, the free energy of the $\phi + 180°$ state is increased due to the splay distortion. The free energy difference between the metastable states and the $\phi + 180°$ state is decreased, and therefore the lifetime of the metastable states is increased. The growth of the $\phi + 180°$ state depends on the cell thickness and is slower with larger cell thickness; thus the lifetime of the metastable states is increased. The trade-off of this method is that the response time of the material becomes longer. The lifetime of the metastable states can be improved using the bulk alignment of the polymer stabilisation technique and the multi-dimensional alignment of polymer wall technique. In future discussion, the $(\phi, \phi + 180°)$ states are called bistable states.

It has been also reported that a $(0°, 180°)$ bistable TN has been developed [27, 28]. In this bistable TN, the two stable states are the $0°$ twisted and $180°$ twisted states. The chiral dopant concentration is chosen such that the two states have the same energy. The switching between the two states also makes use of the hydrodynamic effect. When a sufficiently high voltage is applied, the liquid crystal is switched to the homeotropic state. If the applied voltage is turned off slowly, the liquid crystal relaxes into the $0°$ twisted state. If the applied voltage is turned off abruptly, the liquid crystal relaxes into the $180°$ twisted state. With the employment of one tilted strong anchoring alignment layer and one weak planar anchoring alignment layer, the time interval of the addressing pulse can be reduced to μs.

2.2   Optical properties

Before we consider the optical properties of the twisted-untwisted bistable TNs, we discuss the 2×2 Jones matrix formulation [29, 30]. In a coordinate with the z axis parallel to the propagation direction of the light, the electric field of the light is represented by

$$E = \begin{pmatrix} E_x \\ E_y \end{pmatrix} \tag{9.17}$$

where the phase factor $\exp[i(\omega t - kz)]$ is omitted and E_x and E_y are the electric field components in the x and y directions, respectively. When the light passes through a slab of nematic liquid crystal whose director is uniform and makes an angle β with the x axis, the Jones vector of the outgoing light is related to that of the incident light by [31]

$$
E_{out} = \begin{pmatrix} E_{xo} \\ E_{yo} \end{pmatrix} = \begin{pmatrix} \cos\beta & -\sin\beta \\ \sin\beta & \cos\beta \end{pmatrix} \begin{pmatrix} \exp-i\gamma/2 & 0 \\ 0 & \exp i\gamma/2 \end{pmatrix}
$$
$$
\times \begin{pmatrix} \cos\beta & \sin\beta \\ -\sin\beta & \cos\beta \end{pmatrix} \begin{pmatrix} E_{xi} \\ E_{yi} \end{pmatrix} \tag{9.18}
$$
$$
= S(\beta) \cdot G(\gamma) \cdot S^{-1}(\beta) E_i
$$

γ is the phase retardation and is given by

$$
\gamma = \frac{2\pi}{\lambda}(n_e - n_o)\Delta h \tag{9.19}
$$

where Δh is the thickness of the slab, λ is the wavelength of the light in a vacuum, and n_o and n_e are the ordinary and extraordinary refractive indices, respectively.

A uniform twisted nematic liquid crystal with thickness h and twist rate α can be divided into N thin slabs with thickness $\Delta h = h/N$, as shown in Figure 9.4. Within each slab, the liquid crystal director can be considered approximately uniform. The Jones vector of the outgoing light $\vec{E}_o$ is related to that of the incident light $\vec{E}_i$ by

$$
\begin{aligned}
E_o &= \left[S_N \cdot G_N \cdot S_N^{-1} \right] \cdot \left[S_{N-1} \cdot G_{N-1} \cdot S_{N-1}^{-1} \right] \cdot \ldots \cdot \left[S_1 \cdot G_1 \cdot S_1^{-1} \right] E_i \\
&= \left[S(N\beta) \cdot G(\gamma) \cdot S^{-1}(N\beta) \right] \cdot \left[S[(N-1)\beta] \cdot G(\gamma) \cdot S^{-1}([(N-1)\beta] \right] \\
&\quad \cdot \ldots \cdot \left[S(2\beta) \cdot G(\gamma) \cdot S^{-1}(2\beta) \right] \cdot \left[S(\beta) \cdot G(\gamma) \cdot S^{-1}(\beta) \right] \cdot E_i \\
&= S(N\beta) \cdot \left[G(\gamma) \cdot S^{-1}(\beta) \right]^N \cdot E_i
\end{aligned} \tag{9.20}
$$

where $\beta = \alpha\Delta h$. Define a new matrix A by

$$
A = G(\gamma) \cdot S^{-1}(\beta) = \begin{pmatrix} \exp(-i\gamma/2)\cos\beta & -\exp(-i\gamma/2)\sin\beta \\ \exp(i\gamma/2)\sin\beta & \exp(i\gamma/2)\cos\beta \end{pmatrix} \tag{9.21}
$$

Using the Cayley–Hamilton theory [32, 33], A^N can be expanded as

$$
A^N = \lambda_1 I + \lambda_2 A \tag{9.22}
$$

where λ_1 and λ_2 are found from the equations

$$
q_1^N = \lambda_1 + \lambda_2 A q_1 \tag{9.23}
$$

$$
q_2^N = \lambda_1 + \lambda_2 A q_2 \tag{9.24}
$$

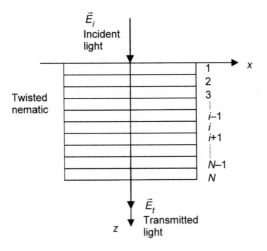

Figure 9.4 The coordinate system used in Jones matrix method

where q_1 and q_2 are the eigenvalues of A. We obtain

$$
A^N = -\frac{\sin(N-1)\theta}{\sin\theta}I + \frac{\sin N\theta}{\sin\theta}A
$$

$$
= \begin{pmatrix}
\dfrac{\sin N\theta}{\sin\theta}\cos\beta\exp(-i\gamma/2) - \dfrac{\sin(N-1)\theta}{\sin\theta} & \dfrac{\sin N\theta}{\sin\theta}\sin\beta\exp(-i\gamma/2) \\
-\dfrac{\sin N\theta}{\sin\theta}\sin\beta\exp(i\gamma/2) & \dfrac{\sin N\theta}{\sin\theta}\cos\beta\exp(i\gamma/2) - \dfrac{\sin(N-1)\theta}{\sin\theta}
\end{pmatrix}
$$

$$(9.25)$$

where θ is given by

$$
\cos\theta = \cos\beta\cos\left(\frac{\gamma}{2}\right) \tag{9.26}
$$

We also have

$$
S^N(\beta) = S(N\beta) = \begin{pmatrix} \cos N\beta & -\sin N\beta \\ \sin N\beta & \cos N\beta \end{pmatrix} \tag{9.27}
$$

When $N\to\infty$, $\beta\to 0$ and $\gamma\to 0$. We have $\sin\beta=\beta$, $\cos\beta=1$ and $\sin\gamma=\gamma$, $\cos\gamma=1$. Also from Equation 9.26 we have

$$
\theta = \left[\beta^2 + \left(\frac{\gamma}{2}\right)^2\right]^{1/2} \tag{9.28}
$$

The total twist angle is

$$\Phi = N\beta = \alpha h \tag{9.29}$$

The total phase retardation angle is

$$\Gamma = N\gamma = \frac{2\pi}{\lambda}(n_e - n_o)h \tag{9.30}$$

Therefore

$$\Theta = N\theta = \left[\Phi^2 + \left(\frac{\Gamma}{2}\right)^2\right]^{1/2} \tag{9.31}$$

$$
\begin{aligned}
\left(A^N\right)_{11} &= \frac{\sin N\theta}{\sin \theta} \cos \beta \left[\cos\left(\frac{\gamma}{2}\right) - i \sin\left(\frac{\gamma}{2}\right)\right] \\
&\quad - \frac{\sin N\theta \cos \theta - \sin \theta \cos N\theta}{\sin \theta} \\
&= \frac{\sin N\theta}{\sin \theta} \cos \theta - i \frac{\sin N\theta}{\sin \theta} \cos \beta \sin\left(\frac{\gamma}{2}\right) \\
&\quad + \cos N\theta - \frac{\sin N\theta \cos \theta}{\sin \theta} \\
&= \cos \Theta - i \frac{\gamma}{2\theta} \sin \Theta = \cos \Theta - i \frac{\Gamma}{2\Theta} \sin \Theta
\end{aligned}
\tag{9.32}
$$

$$\left(A^N\right)_{12} = \frac{\sin N\theta}{\sin \theta} \sin \beta \exp\left(-i\gamma/2\right) = \frac{\beta}{\theta} \sin \Theta = \frac{\Phi}{\Theta} \sin \Theta \tag{9.33}$$

$$\left(A^N\right)_{21} = -\frac{\Phi}{\Theta} \sin \Theta \tag{9.34}$$

$$\left(A^N\right)_{22} = \cos \Theta + i \frac{\Gamma}{2\Theta} \sin \Theta \tag{9.35}$$

Equation 9.20 becomes

$$
E_o = \begin{pmatrix} \cos \Phi & -\sin \Phi \\ \sin \Phi & \cos \Phi \end{pmatrix} \begin{pmatrix} \cos \Theta - i \dfrac{\Gamma}{2\Theta} \sin \Theta & \dfrac{\Phi}{\Theta} \sin \Theta \\ -\dfrac{\Phi}{\Theta} \sin \Theta & \cos \Theta + i \dfrac{\Gamma}{2\Theta} \sin \Theta \end{pmatrix} E_i \tag{9.36}
$$

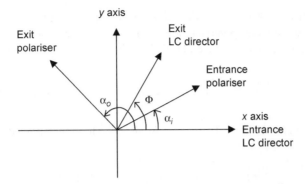

Figure 9.5 The arrangement of the polarisers and alignment directions of the TN display

For a display geometry as shown in Figure 9.5, the incident light is

$$\boldsymbol{E}_i = \begin{pmatrix} \cos \alpha_i \\ \sin \alpha_i \end{pmatrix} \tag{9.37}$$

The amplitude of the exiting light after the exit polariser is given by

$$
E_o = (\cos \alpha_o \quad \sin \alpha_o) \begin{pmatrix} \cos \Phi & -\sin \Phi \\ \sin \Phi & \cos \Phi \end{pmatrix}
$$

$$
\times \begin{pmatrix} \cos \Theta - i\dfrac{\Gamma}{2\Theta}\sin \Theta & \dfrac{\Phi}{\Theta}\sin \Theta \\ -\dfrac{\Phi}{\Theta}\sin \Theta & \cos \Theta + i\dfrac{\Gamma}{2\Theta}\sin \Theta \end{pmatrix} \begin{pmatrix} \cos \alpha_i \\ \sin \alpha_i \end{pmatrix}
$$

$$
= [\cos(\alpha_o - \Phi) \quad \sin(\alpha_o - \Phi)] \begin{bmatrix} \cos \Theta \cos \alpha_i + \dfrac{\Phi}{\Theta}\sin \Theta \sin \alpha_i - i\dfrac{\Gamma}{2\Theta}\sin \Theta \cos \alpha_i \\ \cos \Theta \sin \alpha_i - \dfrac{\Phi}{\Theta}\sin \Theta \cos \alpha_i + i\dfrac{\Gamma}{2\Theta}\sin \Theta \sin \alpha_i \end{bmatrix}
$$

$$\tag{9.38}$$

The intensity of the light is [34, 35]

$$
\begin{aligned}
I_o = |E_o|^2 =\ & \cos^2(\alpha_o - \alpha_i - \Phi) \\
& - \sin^2\left[\left(\Phi^2 + \left(\frac{\Gamma}{2}\right)^2\right)^{1/2}\right] \sin[2(\alpha_o - \Phi)] \sin(2\alpha_i) \\
& - \frac{\Phi^2}{\left[\Phi^2 + (\Gamma/2)^2\right]} \sin^2\left[\left(\Phi^2 + \left(\frac{\Gamma}{2}\right)^2\right)^{1/2}\right] \cos[2(\alpha_o - \Phi)] \cos(2\alpha_i) \\
& - \frac{\Phi}{2\left[\Phi^2 + (\Gamma/2)^2\right]^{1/2}} \sin\left[2\left(\Phi^2 + \left(\frac{\Gamma}{2}\right)^2\right)^{1/2}\right] \sin[2(\alpha_o - \alpha_i - \Phi)]
\end{aligned} \tag{9.39}
$$

Now we use this result for the optical properties of the bistable TNs. In the $(0°, 360°)$ bistable TN, $\phi = 0°$; crossed polarisers are used and they make an angle of $45°$ with the liquid crystal director in the entrance plane [2]. $-\alpha_i = \alpha_o = \pi/4$. Because

$$\cos (2\alpha_i) = \cos \left(\frac{-\pi}{2}\right) = 0 \quad \text{and} \quad \sin [2(\alpha_o - \alpha_i - \Phi)] = \sin (\pi - 2\Phi) = \sin (2\Phi)$$

When the liquid crystal is in the $0°$ twist state, $\Phi = 0$. The transmittance is given by

$$T(0°) = \cos^2 \left(\frac{\pi}{4} + \frac{\pi}{4} - 0\right) - \sin^2 \left(\frac{\Gamma}{2}\right) \sin \left[2\left(\frac{\pi}{4} - 0\right)\right] \sin \left[2\left(\frac{-\pi}{4}\right)\right] = \sin^2 \left(\frac{\Gamma}{2}\right)$$

$$(9.40)$$

Hence if one chooses

$$\frac{\Gamma}{2} = \frac{\pi h(n_e - n_o)}{\lambda} = \frac{\pi}{2}$$

the maximum transmittance of 1 is obtained. When the liquid crystal is in the $360°$ twist state, $\Phi = 2\pi$. The transmittance is given by

$$T(360°) = \cos^2 \left(\frac{\pi}{4} + \frac{\pi}{4} - 2\pi\right) - \sin^2 \left[\pi \left(2^2 + \left(\frac{1}{2}\right)^2\right)^{1/2}\right]$$

$$\times \sin \left(-\frac{7\pi}{2}\right) \sin \left(-\frac{\pi}{2}\right) = \sin^2 [2.06\pi] = 0.037$$

$$(9.41)$$

There are many choices for the geometry of the bistable TN. As another example [36], one can choose the entrance polariser parallel to the liquid crystal director in the entrance plane and the exit polariser perpendicular to it, i.e. $\alpha_i = 0$ and $\alpha_o = \pi/2$, and $\Gamma/2 = \sqrt{2}\pi$. When the liquid crystal is in the $0°$ twist state, $\Phi = 0$, and its transmittance is given by

$$T(0°) = 0 \tag{9.42}$$

When the liquid crystal is in the $360°$ twist state, $\Phi = 2\pi$, and its transmittance is given by

$$T(360°) = -\frac{(2\pi)^2}{\left[(2\pi)^2 + (\sqrt{2}\pi)^2\right]} \sin^2 \left[\left((2\pi)^2 + (\sqrt{2}\pi)^2\right)^{1/2}\right] \cos (-3\pi) \cos (0) = 0.65$$

$$(9.43)$$

Because the liquid crystal has twisted planar orientation in both states, the viewing angle of the bistable TN is fairly good. The contrast ratio versus viewing angle in the horizontal viewing direction is shown in Figure 9.6 [36].

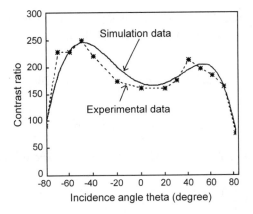

Figure 9.6 The contrast ratio versus viewing angle for the $(0°, 360°)$ bistable TN [36]

For the $(-90°, 270°)$ bistable TN, $\phi = -90°$; crossed polarisers are used and they make an angle of $45°$ with the liquid crystal director in the entrance plane [18]. $-\alpha_i = \alpha_o = \pi/4$. Because

$$\cos\,(2\alpha_i) = \cos\,\left(\frac{-\pi}{2}\right) = 0$$

$$\sin\,[2(\alpha_o - \alpha_i - \Phi)] = \sin\,(\pi - 2\Phi) = \sin\,(2\Phi)$$

When the liquid crystal is in the $-90°$ twist state, $\Phi = -\pi/2$. The transmittance is given by

$$T(-90°) = \cos^2\left(\frac{\pi}{4} + \frac{\pi}{4} + \frac{\pi}{2}\right) - \sin^2\left(\frac{\Gamma}{2}\right)\sin\left[2\left(\frac{\pi}{4} + \frac{\pi}{2}\right)\right]$$

$$\times \sin\left[2\left(\frac{-\pi}{4}\right)\right] = \cos^2\left(\frac{\Gamma}{2}\right) \tag{9.44}$$

Hence if one chooses

$$\frac{\Gamma}{2} = \frac{\pi h(n_e - n_o)}{\lambda} = \frac{\pi}{2}$$

the minimum transmittance of 0 is obtained. When the liquid crystal is in the $270°$ twist state, $\Phi = 3\pi/2$. The transmittance is given by

$$T(270°) = \cos^2\left(\frac{\pi}{4} + \frac{\pi}{4} - \frac{3\pi}{2}\right) - \sin^2\left[\pi\left(\left(\frac{3}{2}\right)^2 + \left(\frac{1}{2}\right)^2\right)^{1/2}\right]$$

$$\times \sin\left(-\frac{5\pi}{2}\right)\sin\left(-\frac{\pi}{2}\right) = \cos^2\,[2.06\pi] = 0.965 \tag{9.45}$$

For the $(90°, 450°)$ bistable TN, $\phi = 90°$; parallel polarisers are used and they are parallel to the liquid crystal director in the entrance plane [18]. $\alpha_i = \alpha_o = 0$. Because $\sin(2\alpha_i) = \cos(0) = 0$

$$\sin\,[2(\alpha_o - \alpha_i - \Phi)] = -\sin\,(2\Phi)$$

$$\cos\,(\alpha_o - \alpha_i - \Phi) = \cos\,(\Phi)$$

When the liquid crystal is in the $90°$ twist state, $\Phi = \pi/2$. The transmittance is given by

$$
\begin{aligned}
T(90°) &= \cos^2\left(\frac{\pi}{2}\right) - \frac{(\pi/2)^2}{[(\pi/2)^2 + (\Gamma/2)^2]}\,\sin^2\,[((\pi/2)^2 + (\Gamma/2)^2)^{1/2}] \\
&\quad \times \cos\left[2\left(0 - \frac{\pi}{2}\right)\right]\,\cos\,(0) \qquad\qquad (9.46) \\
&= \frac{(\pi/2)^2}{[(\pi/2)^2 + (\Gamma/2)^2]}\,\sin^2\,[((\pi/2)^2 + (\Gamma/2)^2)^{1/2}]
\end{aligned}
$$

Hence if one chooses

$$\frac{\Gamma}{2} = \frac{\pi h(n_e - n_o)}{\lambda} = \frac{\sqrt{3}\pi}{2}$$

the minimum transmittance of 0 is obtained. When the liquid crystal is in the $450°$ twist state, $\Phi = 5\pi/2$. The transmittance is given by

$$
\begin{aligned}
T(450°) &= \cos^2\left(\frac{5\pi}{2}\right) - \frac{(5\pi/2)^2}{[(5\pi/2)^2 + (\Gamma/2)^2]}\,\sin^2\,[((\pi/2)^2 + (\sqrt{3}\pi/2)^2)^{1/2}] \\
&\quad \times \cos\left[2\left(0 - \frac{5\pi}{2}\right)\right]\,\cos\,(0) \qquad\qquad (9.47) \\
&= \frac{25}{28}\,\sin^2\left[(28)^{1/2}\frac{\pi}{2}\right] = 0.718
\end{aligned}
$$

The transmittance of bistable TNs depends strongly on the wavelength of the incident light. The dispersion of the transmittance as a function of the wavelength must be considered in designing bistable TNs under white incident light. Tang, Chui and Kwok have performed a study to optimise the performance of the bistable TN under white incident light [37–40]. In designing transmittive bistable TNs, the following condition should be satisfied: for linearly polarised incident light, in both states the polarisation of the exiting light must be linear and must be perpendicular to each other. The Mueller matrix method was used in their theoretical calculation. One of the best performances was obtained with the following parameters: the angle of the entrance polariser is $45°$, the angle of the exit polariser is $-33.75°$, the retardation $h\Delta n$ is $0.273\,\mu m$ and the twist angle ϕ is $-11.25°$ [37]. The transmission spectra of the two bistable $(-11.25°, 348.75°)$ states are shown in Figure 9.7.

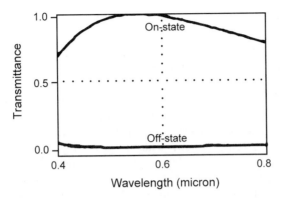

Figure 9.7 The calculated transmission spectra of the bistable states of the bistable TN [37]

The bistable TN can be used to make reflective displays, if the reflective display is designed so that there is only one polariser and the rear reflector is a mirror. In one stable state, the light is linearly polarised after passing through the liquid crystal for the first time for the linearly polarised incident light produced by the polariser. After the light is reflected by the mirror, it is linearly polarised in the same direction. When light passes through the liquid crystal again, it is still linearly polarised parallel to the polariser, and therefore it passes the polariser and the material has a high transmittance. In the other stable state, the light is circularly polarised after passing through the liquid crystal for the first time for the linearly polarised incident light produced by the polariser. After the light is reflected by the mirror, it becomes circularly polarised with the opposite handedness. When the light passes through the liquid crystal again, it becomes linearly polarised perpendicular to the polariser, and therefore it is absorbed by the polariser and the material has a low transmittance. One of the best performances was obtained with the following parameters: the angle of the entrance polariser is $-24.24°$, the retardation $h\Delta n$ is $0.3108\,\mu m$ and the twist angle ϕ is $-67.2°$ [37]. The reflection spectra of the two bistable $(-67.2°, 292.8°)$ states are shown in Figure 9.8.

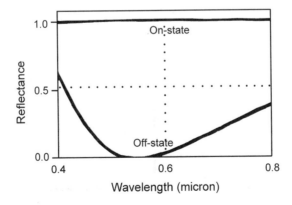

Figure 9.8 The calculated reflection spectra of the bistable states of the bistable TN [37]

2.3 Drive schemes

As discussed in Chapter 8, the bistability of a display material does not guarantee multi-plexibility. Proper drive schemes must be designed in order to make multiplexed displays on a passive matrix. A good drive scheme should possess the properties of fast addressing speed, low drive voltage, no cross-talk and a simple waveform. According to the number of phases in the addressing, there are three major types of drive scheme for the bistable TN: one-phase, two-phase and three-phase drive schemes.

The one-phase drive scheme is shown schematically in Figure 9.9. The state of the liquid crystal is changed by one voltage pulse [2, 18, 41]. A low voltage V_L addressing pulse switches the material to the low twisted state, while a high voltage V_H addressing pulse switches the material into the high twisted state. The low voltage is slightly higher than the threshold V_{th} of the Freedericksz transition. When the low voltage is applied, independent of the initial state of the liquid crystal, the liquid crystal is switched to a homeotropic state where the liquid crystal is aligned homeotropically only in a small region in the middle of the cell. When the low voltage is turned off, the liquid crystal relaxes into the low twisted state because it does not have sufficiently high potential. The high voltage is higher than the saturation voltage V_{sa}, which is much higher than the threshold of the Freedericksz transition. When the high voltage is applied, independent of the initial state, the liquid crystal is switched to a homeotropic state where the liquid crystal is aligned homeotropically in most regions of the cell except for those very close to the cell surfaces, and gains a high potential. When the high voltage is turned off, the liquid crystal relaxes into the high twisted state because of the backflow created by the hydrodynamic effect. As an example, consider a $(-90°, 270°)$ bistable TN with h/P value near 0.25 and cell thickness 5 μm. The $-90°$ twist state corresponds to the dark state, as described by Equation 9.44. The $270°$ twisted state corresponds to the bright state, as described by Equation 9.45. When the time interval of the addressing pulse is 20 ms, the $V_L = 4$ V pulse switches the material to the low twisted state and $V_H = 10$ V switches the material to the high twisted state. The addressing waveform and the dynamic response of the bistable TN are shown in Figure 9.10 [18]. In addressing a multiplexed display, the row voltage for the row being selected is $V_{rs} = (V_L + V_H)/2 = 7$ V; the voltage for the row not being selected is $V_{rns} = 0$ V. The column voltage is $V_{con} = -(V_H - V_L)/2 = -3$ V to select the bright state and $V_{coff} = (V_H - V_L)/2 = 3$ V to select the dark state. The material should have no cross-talk problem under the 3 V column voltage, i.e. the threshold of the Freedericksz transition should be higher than 3 V. The problem of this drive scheme is that the addressing pulse is long and the addressing speed is slow.

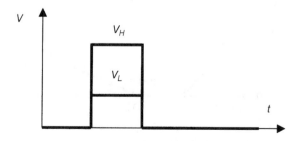

Figure 9.9 The schematic diagram of the one-phase drive scheme for the bistable TN

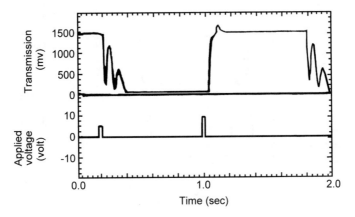

Figure 9.10 The dynamic response of the bistable TN to the addressing pulses in the one-phase drive scheme [18]

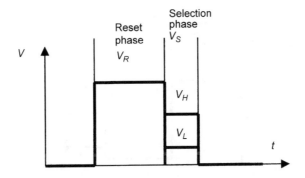

Figure 9.11 The schematic diagram of the two-phase drive scheme for the bistable TN

The two-phase drive scheme is shown in Figure 9.11 [18, 20]. In the reset phase, a high voltage V_R (greater than V_{sa}) is applied to switch the liquid crystal to the homeotropic state. When the reset voltage is turned off, the liquid crystal begins to relax. In the selection phase, if the selection voltage V_S is the low voltage V_L, there is no hindrance to the backflow and the liquid crystal relaxes to the high twisted state. If the selection voltage V_S is the high voltage V_H, the applied voltage hinders the backflow and the liquid crystal relaxes to the low twisted state. The dynamic response of the $(-90°, 270°)$ bistable TN to the addressing pulses in this drive scheme is shown in Figure 9.12 [18]. The reset pulse is 10 V high and 20 ms wide. The time interval of the selection phase is 4 ms. The low voltage V_L is 0 V and the high voltage V_H is 3 V. Although the reset phase is long, multiple lines can be put into the reset phase such that the time is shared. This is the pipeline algorithm as discussed in the dynamic drive scheme in Chapter 8. The addressing speed is 4 ms per line, which is faster than the addressing speed of 20 ms per line in the one-phase drive scheme. An alternative to this two-phase drive scheme is to decrease the high voltage linearly with time. If the high voltage is decreased to 0 sufficiently fast, the backflow is not hindered and the liquid crystal relaxes to the high twisted state. If the high voltage is decreased to 0 sufficiently slowly, the back-flow is low and the liquid crystal relaxes to the low twisted state.

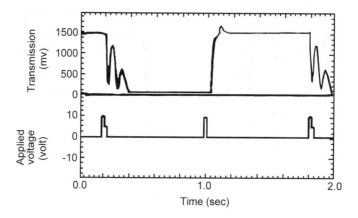

Figure 9.12   The dynamic response of the bistable TN to the addressing pulses in the two-phase drive scheme [18]

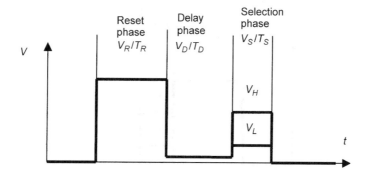

Figure 9.13   The schematic diagram of the three-phase drive scheme for the bistable TN

The three-phase drive scheme is shown in Figure 9.13 [42]. It consists of three phases: reset, delay and selection. The physics behind this drive scheme is that at the beginning of the relaxation after the reset phase, the liquid crystal is allowed to relax freely, and the backflow can be controlled by a voltage in the late stage of the relaxation. Therefore the time interval of the selection phase is reduced. In the reset phase, the high voltage V_R (greater than V_{sa}) switches the liquid crystal into the homeotropic texture and the angle θ_m of the liquid crystal director at the middle plane becomes $\pi/2$, as shown in Figure 9.14. In the delay phase, the applied voltage V_D is 0 and the liquid crystal starts to relax. In this phase, because of the backflow, θ_m first increases then decreases and then increases again. In the selection phase, if the applied voltage V_S is the high voltage V_H, the backflow is hindered, and θ_m decreases, as shown by curve (a). After the selection phase, θ_m decreases to 0 and the liquid crystal is addressed to the low twisted state. If the applied voltage V_S is the low voltage V_L, the backflow is not hindered and θ_m continues increasing, as shown by curve (b). After the selection phase, θ_m becomes close to π and the liquid crystal is addressed to the high twisted state. As an example, for a $(0°, 360°)$ bistable TN, the reset phase is 25 V/2.0 ms, the delay phase is 1.6 V/420 μm and the selection phase is 140 μs wide. If

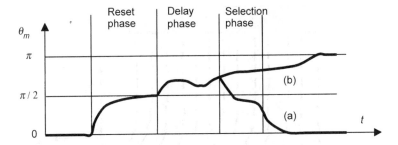

Figure 9.14 The angle θ_m of the liquid crystal director at the middle plane in the bistable TN as a function of time in the three-phase drive scheme

the selection voltage V_S is the high voltage $V_H = 5.0$ V, the liquid crystal is addressed to the $0°$ twisted state. If the selection voltage V_S is the low voltage $V_L = 4.3$ V, the liquid crystal is addressed to the $360°$ twisted state. The time of the reset and delay phases can be shared using the pipeline algorithm. The addressing speed of the three-phase drive scheme is 140 µs per line, which is much faster than the speed of the one- and two-phase drive schemes. A video rate display becomes possible with this drive scheme.

3 Surface Stabilised Nematic Liquid Crystals

In liquid crystal displays, liquid crystals are sandwiched between two substrates. A certain alignment of the liquid crystal at the surface of the substrates is usually necessary for a display to operate properly. Bistable nematic liquid crystals can be created by using surface alignment layers. They are divided into two categories: zenithal bistable nematic display and azimuthal bistable nematic display.

3.1 Zenithal bistable nematic display

The zenithal (Z) bistable nematic liquid crystal was developed by G. P. Bryan-Brown *et al.*, using surface stabilisation [7]. One substrate of the cell has an alignment layer with homeotropic anchoring and the other substrate is a one-dimensional grating, as shown in Figure 9.15. The groove of the grating is along the y direction. The grooves are made from a photoresist. The non-symmetric profile of the grooves is obtained by using UV light incident obliquely at $60°$ in the photolithography. A surfactant is coated on top of the grooves to obtain homeotropic anchoring.

The liquid crystal in the Z-bistable cell has two stable states at zero field. One is the high tilt state shown in Figure 9.15(a). The other is the low tilt state shown in Figure 9.15(b). This state possesses a polarisation P produced by the flexoelectric effect, given by $P = e_1 n(\nabla \cdot n) + e_2(\nabla \times n) \times n$. The electric energy is lower when the angle between the applied electric field and the polarisation is small. Thus the state of the liquid crystal is sensitive to the polarity of the applied voltage pulse. The bulk liquid crystal responds to applied electric fields dielectrically; it is not sensitive to the polarity of the applied voltage (the bulk flexoelectric polarisation is small). It is switched to the homeotropic state when a sufficiently high voltage is

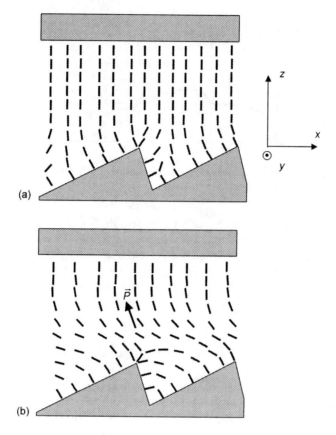

Figure 9.15 Schematic diagram of the liquid crystal director configurations of the two bistable states of the Z-bistable nematic liquid crystal

applied. It is the liquid crystal near the grating substrate which is sensitive to the polarity of the applied voltage. When a sufficiently high voltage with one polarity is applied, the liquid crystal is switched to a homeotropic state, with the liquid crystal near the grating substrate having a configuration similar to that in the high tilt state; it relaxes into the high tilt state after the applied voltage has been removed. When a sufficiently high voltage with the opposite polarity is applied, the liquid crystal is switched to a homeotropic state, with the liquid crystal near the grating substrate having a configuration similar to that in the low tilt state; it relaxes into the low tilt state after the applied voltage is removed.

In building a transmissive Z-bistable display, crossed polarisers are used. The polarisers make an angle of 45° with the grating groove direction. Hence the plane containing the liquid crystal director in the low tilt state also makes an angle of 45° with the polarisers. The cell thickness and birefringence of the liquid crystal are chosen in such a way that the retardation of the low tilt state is π, and therefore the transmittance of the low tilt state is high. The retardation of the high tilt state is small and therefore its transmittance is low. A response to the Z-bistable display to voltage pulses is schematically shown in Figure 9.16. The width of the addressing voltage pulse is about 100 μs for a field about 10 V/μm, promising a video rate. The turn-on time (from the homeotropic state to the low tilt state)

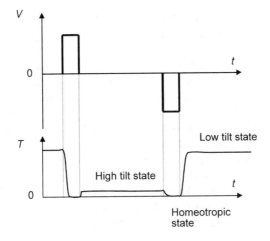

Figure 9.16 Schematic dynamic response of the Z-bistable TN display

is about 20 ms and the turn-off state (from the homeotropic state to the low tilt state) is about 1 ms. The material can also be used to make reflective displays with the retardation adjusted properly.

The Z-bistable nematic liquid crystal can also be used to make displays by using a different geometry; the cell is made of the grating substrate and another substrate with a homogeneous anchoring [6, 43]. The aligning direction of the homogeneous anchoring makes an angle of 90° with the liquid crystal director near the grating surface in the low tilt state. Thus a hybrid TN is formed. In making a transmissive display, two crossed polarisers are used. The groove of the grating is arranged parallel to one of the polarisers. When the liquid crystal is in the low tilt state, the material acts as a polarisation guide and the transmittance of the display is high. When the liquid crystal is in the high tilt state, the polarisation of the incident light is rotated only slightly and therefore the transmittance is low. The selection of the states is made by using DC voltage pulses in the same way as described before. In this design, higher contrast is achieved. One advantage of this display is that written images at zero field are not erased by flow when the display is squeezed.

3.2 Azimuthal bistable nematic display

The alignment of a liquid crystal at the cell surface is achieved by the intermolecular interaction between the molecules of the alignment layer and the liquid crystal molecules and the geometrical shape of the surface of the alignment layer through the elastic energy of the liquid crystal. For an alignment layer having uni-directional grooves (grating) on the alignment layer, the liquid crystal is aligned along the groove direction and has a low elastic energy. For an alignment layer with grooves in two perpendicular directions (bigrating), two alignment directions can be created with properly controlled groove amplitude and pitch. The liquid crystal can be anchored along either direction. Thus two stable orientation states can be achieved [44]. A problem with this bistable nematic liquid crystal is that if the pre-tilt angle of the alignment layer is $0°$, it cannot be addressed by using electric voltage applied in the normal cell direction.

Alignment layers with bi-directional anchoring and non-zero pre-tilt can be made by using asymmetric (blazed) grooves. Consider such an alignment layer as shown by the bottom plate in Figure 9.17(1a); the blaze directions are along x (with the azimuthal angle $\phi = 0°$) and y (with the azimuthal angle $\phi = 90°$), respectively. One alignment direction has the azimuthal angle $\phi = 45°$ and a non-zero pre-tilt angle θ_0. The other alignment direction has the azimuthal angle $\phi = -45°$ and zero pre-tilt angle θ_0. In a cell with such an alignment layer on the two substrates, the liquid crystal is bistable. Because of the non-zero pre-tilt angle, there are splay deformations in the two stable states, which produce flexo-electric polarizations. In the state shown in Figure 9.17(1a) the flexo-electric polarisation is upward, while in the state shown in Figure 9.17(1b) the flexo-electric polarisation is downward. In principle, these states can be selected by using a DC voltage pulse. This effect, however, has not yet been observed.

Multi-stable surface anchorings can also be produced by obliquely evaporating SiO on glass substrates [8, 9, 11]. By controlling the zenithal angle α of the evaporation and the thickness d of SiO, Barberi *et al.* have achieved various anchorings [46, 47]. When α is small and d is large, the anchoring direction is perpendicular to the evaporation plane (with the azimuthal angle ϕ of $90°$) and parallel to the substrate with zero pre-tilt angle. When α is large, the anchoring direction is oblique in the evaporation plane (with the azimuthal angle ϕ of $0°$ and non-zero pre-tilt angle). In the transition region with intermediate α and small d, bistable or multi-stable anchorings can be obtained.

Under one SiO evaporation condition, three anchorings are obtained: one stable and two metastable as shown by the bottom plate in Figure 9.17(2a). The stable anchoring is the planar anchoring with $\phi = 0$ and $\theta = 0$, and is denoted by P. One of the two metastable anchorings has a non-zero pre-tilt angle θ_0 and an azimuthal angle $\phi = -45°$, and is denoted by O. The other metastable anchoring has the same non-zero pre-tilt angle θ_0 and the azimuthal angle $\phi = 45°$, and is denoted by O'. The display cell is constructed of two parallel substrates with the alignment layer. The top substrate is rotated $45°$ around the normal with respect to the bottom substrate. The liquid crystal is the cell is bistable. In one stable state, the liquid crystal is anchored in the P state on the bottom surface and in the O' state on the top surface as shown in Figure 9.17(2a). In the other stable state, the liquid crystal is anchored in the O state on the bottom surface and in the P state on the top surface as shown in Figure 9.17(2b). There are splay distortions in these states. The induced flexo-electric polarisation is pointing upward in one state and downward in the other state. Therefore these states can be selected by using DC voltage pulses through the flexo-electric effect.

Under another SiO evaporation condition, two stable anchorings are obtained shown by the bottom plate in Figure 9.17(3a). One anchoring is in the direction with a non-zero pre-tilt angle θ_0 and an azimuthal angle $-\phi_0$, and is denoted by O'. The other anchoring is in the direction with the same pre-tilt angle θ_0 and the azimuthal angle ϕ_0, and is denoted by O. In a cell constructed with two anti-parallel substrates with the alignment layer, the liquid crystal is bistable. In one of the bistable states (called the O' state), the liquid crystal is uniformly aligned with the liquid crystal anchored in the O' state at both the bottom and top surfaces as shown in Figure 9.17(3a). In the other stable state (called the O state), the liquid

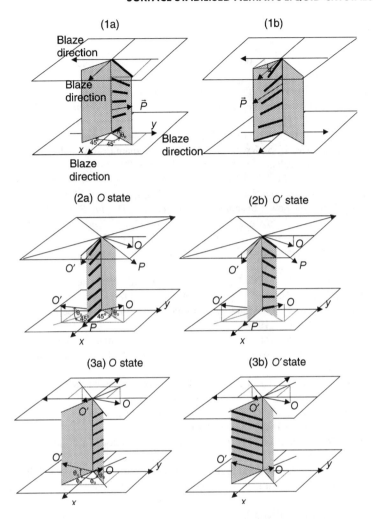

Figure 9.17 Schematic diagrams of the azimuthal bistable nematic displays

crystal is uniformly aligned with the liquid crystal anchored in the O state at both the bottom and top surfaces. In order to select these two bistable states, two types of ions of opposite charge and chirality are doped to the nematic liquid crystal. The positive ion has a positive chirality and the negative ion has a negative chirality. Initially the liquid crystal is in the O' state. When a negative field is applied (the top plate has a low potential), the positive ions are driven to the top surface; the liquid crystal near the top surface experiences a positive torque and is switched to the O anchoring state. The negative ions are driven to the bottom surface; the liquid crystal near the bottom surface experiences a negative torque and is also switched to the O anchoring state. Thus the liquid crystal is switched to the O state. When a positive field is applied, the ions are driven to the opposite directions and the liquid crystal is switched back to the O' state.

References

1 D. W. Berreman and W. R. Heffner, New bistable cholesteric liquid-crystal display, *Appl. Phys. Lett.* **37**, 109 (1980).

2 T. Tanaka, Y. Sato, A. Inoue, Y. Momose, H. Nomura, and S. Iino, A bistable twisted nematic (BTN) LCD driven by a passive-matrix addressing, *Proc. Asia Display 95*, 259 (1995).

3 T. Tanaka, Y. Sato, T. Obikawa, H. Nomura, and S. Iino, Physical and electro-optical properties of bistable twisted nematic (BTN) LCD, *Proc. Intnl. Display Research Conf.*, M-64 (1997).

4 T.-Z. Qian, Z.-L. Xie, H.-S. Kwok, and P. Sheng, Dynamic flow and switching bistability in nematic liquid crystal cells, *Appl. Phys. Lett.* **71**, 596 (1997).

5 Y. J. Kim, S. M. Park, I. Lee, S. W. Suh, and S. D. Lee, Numerical modeling and optical switching characteristics of a bistable TN-LCD, *Proc. Euro Display 96*, 337 (1996).

6 G. P. Brown, Ultra low power bistable LCDs, *Proc. Intnl. Display Research Conf. 00*, 76 (2000).

7 G. P. Bryan-Brown, C. V. Brown, J. C. Jones, E. L. Wood, I. C. Sage, and P. Brett, Grating aligned bistable nematic device, *SID Intl. Symp. Digest Tech. Papers* **28**, 37 (1997).

8 R. Barberi and G. Durand, Electrically controlled bistable surface switching in nematic liquid crystals, *Appl. Phys. Lett.* **58**, 2907 (1991).

9 R. Barberi, M. Giocondo, and G. Durand, Flexoelectrically controlled bistable surface switching in nematic liquid crystals, *Appl. Phys. Lett.* **60**, 1085 (1992).

10 R. Barberi, M. Giocondo, J. Li, and R. Bartolino, Fast bistable nematic display with gray scale, *Appl. Phys. Lett.* **71**, 3495 (1997).

11 R. Barberi and G. Durand, Controlled textural bistability in nematic liquid crystals, "Handbook of liquid crystal research," ed. by P. J. Collings and J. S. Patel (Oxford University Press, New York, 1997).

12 G. D. Boyd, J. Cheng, and P. D. T. Ngo, Liquid-crystal orientational bistability and nematic storage effects, *Appl. Phys. Lett.* **36**, 556 (1980).

13 R. N. Thurston, J. Cheng, and G. D. Boyd, Mechanically bistable liquid crystal display structures, *IEEE Trans. Elec. Dev.* **ED-27**, 2069 (1980).

14 J. Cheng and R. N. Thurston, The propagation of disclinations in bistable switching, *J. Appl. Phys.* **52**, 2766 (1981).

15 P. A. Breddels and H. A. van Sprang, An analytical expression for the optical threshold in highly twisted nematic systems with nonzero tilt angles at the boundaries, *J. Appl. Phys.* **58**, 2162 (1985).

16 H. A. van Sprang and P. Breddels, Numerical calculations of director patterns in highly twisted nematic configurations with nonzero pretilt angles, *J. Appl. Phys.* **60**, 968 (1986).

17 J. C. Kim, G.-J. Choi, Y.-S. Kim, K. H. Kang, T.-H. Yoon, and K. G. Nam, Numerical modeling and optical switching characteristics of a bistable TN-LCD, *SID Intl. Symp. Digest Tech. Papers* **28**, 33 (1997).

18 Z. L. Xie and H. S. Kwok, New bistable twisted nematic liquid crystal displays, *J. Appl. Phys. Lett.* **84**, 77 (1998).

19 Z. L. Xie, Y. M. Dong, S. Y. Xu, H. J. Gao, and H. S. Kwok, $\pi/2$ and $5\pi/2$ twisted bistable nematic liquid crystal display, *J. Appl. Phys.* **87**, 2673 (2000).

20 C. D. Hoke, J. R. Kelly, J. Li, and P. J. Bos, Matrix addressing of a 0°–360° bistable twist cell, *SID Intl. Symp. Digest Tech. Papers* **28**, 29 (1997).

21 W. H. de Jeu, "Physical properties of liquid crystalline materials" (Gordon and Breach, New York, 1980).

22 P. G. de Gennes and J. Prost, "The physics of liquid crystals" (Oxford University Press, New York, 1993).

23 J. Kelly, S. Jamal, and M. Cui, Simulation of the dynamics of twisted nematic devices including flow, *J. Appl. Phys.* **86**, 4091 (1999).

24 D. W. Berreman, Liquid-crystal twist cell dynamics with backflow, *J. Appl. Phys.* **46**, 3746 (1975).

25 C. Z. van Doorn, Dynamic behavior of twisted nematic liquid crystal layers in switched fields, *J. Appl. Phys.* **46**, 3738 (1975).

26 C. D. Hoke and P. Bos, The effect of material constant on the long term stability of the bistable twist cell, *Proc. Intnl. Display Research Conf.*, 85 (1997).

27 I. Dozov, M. Nobili, and G. Durand, Fast bistable nematic display using monostable surface switching, *Appl. Phys. Lett.* **70**, 1179 (1997).

28 Ph. Martinot-Lagrade, I. Dozov, E. Polossat, M. Giocondo, I. Lelidis, and G. Durand, Fast bistable nematic display using monostable surface anchoring switching, *SID Intl. Symp. Digest Tech. Papers* **28**, 41 (1997).

29 R. C. Jones, *J. Opt. Soc. Am.* **32**, 486 (1942).

30 A. Yariv and P. Yeh, "Optical waves in crystals" (John Wiley & Sons, 1984).

31 S. Chandrasekhar, "Liquid crystals," 2nd ed. (Cambridge University Press, New York, 1997).

32 W. L. Brogan, "Modern control theory," 2nd ed. (Prentice-Hall, Englewood Cliffs, NJ, 1985).

33 Private communication with X.-D. Mi.

34 H. L. Ong, Optical properties of general twisted nematic liquid-crystal displays, *Appl. Phys. Lett.* **51**, 1398 (1987).

35 H. L. Ong, Origin and characteristics of the optical properties of general twisted nematic liquid crystals, *J. Appl. Phys.* **64**, 614 (1988).

36 Z. L. Xie, C. Y. Zheng, S. Y. Xu, H. J. Gao, and H. S. Kwok, $0°–360°$ bistable nematic liquid crystal display with large $d\Delta n$ and high contrast, *J. Appl. Phys.* **88**, 1722 (2000).

37 S. T. Tang, H. W. Chiu, and H. S. Kwok, Optically optimized transmittive and reflective bistable twisted nematic liquid crystal display, *J. Appl. Phys.* **87**, 632 (2000).

38 Z. L. Xie, H. J. Gao, S. Y. Xu, and S.H. Kwok, Optimization of reflective bistable nematic liquid crystal displays, *J. Appl. Phys.* **86**, 2373 (1999).

39 H. Cheng and H. Gao, Optical properties of reflective bistable twisted nematic liquid crystal display, *J. Appl. Phys.* **87**, 7476 (2000).

40 Z. L. Xie, H. J. Gao, B. Z. Chang, and S. Y. Xu, A new BTN LCD with high contrast ratio and large cell gap, *Proc. Asia Display 98*, 303 (1998).

41 G.-D. Lee, K.-H. Park, K.-C. Chang, T.-H. Yoon, J. C. Kim, and E.-S. Lee, Optimization of drive scheme for matrix addressing of a bistable twisted nematic LCD, *Proc. Asia Display 98*, 299 (1998).

42 T. Tanaka, T. Obikawa, Y. Sato, H. Nomura, and S. Iino, An advanced driving method for bistable twisted nematic (BTN) LCD, *Proc. Asia Display 98*, 295 (1998).

43 E. L. Wood, G. P. Bryan-Brown, P. Brett, A. Graham, J. C. Jones, and J. R. Hughes, Zenithal bistable device (ZBD) suitable for portable applications, *SID Intl. Symp. Digest Tech. Papers* **31**, 124 (2000).

44 G. P. Bryan-Brown, M. J. Towler, M. S. Bancroft, and D. G. McDonnell, Bistable nematic alignment using bigratings, *Proc. Intnl. Display Research Conf. 94*, 209 (1994).

45 M. Monkade, M. Boix, and G. Durand, Order electricity and oblique nematic orientation on rough solid surfaces, *Europhys. Lett.* **5**, 697 (1988).

46 A. Gharbi, F. R. Fekih, and G. Durand, Dynamics of surface anchoring breaking in a nematic liquid crystal, *Liq. Cryst.* **12**, 515 (1992).

47 Private communication with Prof. P. Bos.

10

Fast Response Liquid Crystals

1 Introduction

The response time of a liquid crystal (LC) device affects human perception of the displayed images. For a desktop computer, the displayed images are usually static; the pace of information update is relatively slow. However, for an LCD TV, the motion pictures could change rapidly. If the LC response speed is too slow, the images will be blurred. It is highly desirable to have the LC device operate at the true video rate, i.e. less than 16 ms regardless of ambient temperature.

With the continuous improvement of low viscosity liquid crystal materials and thinner cell gap fabrication techniques, the response time (rise + decay) of a 4 μm thick transmissive 90° twisted nematic cell is approaching 30 ms and a homeotropic cell is about 20 ms. For a large screen projection display, the response time could be faster than 16 ms, benefiting from the thermal effect of the high power lamp. However, for a color sequential microdisplay employing a single LCD panel, the required frame time is less than 4 ms in order to achieve a 85 Hz operation rate. This presents a technical challenge to nematic LCDs, particularly in the low-temperature regime.

The ferroelectric liquid crystal display is a bistable device and has a microsecond response time [1]. Thus, it is an ideal candidate for a color-sequential display [2, 3]. Both Displaytech [4] and Sony [5] have demonstrated several high-resolution color-sequential projection displays using a ferroelectric liquid crystal on silicon (LCOS). Table 10.1 summarises the performance parameters of a Displaytech SXGA panel. However, two problems remain to be overcome before the FLC can become the mainstream approach. The first problem is the DC voltage effect [6, 7]. In a FLC device, DC voltage pulses with opposite polarities are applied in every other frame in order to balance the DC level. A problem arises from the migration and accumulation of ionic impurities. After prolonged operation these

Table 10.1   Performance parameters of a Displaytech FLCD panel

Resolution	1280 × 1024
Active area	16.90 × 13.52 mm
Array diagonal	0.85 inch
Pixel pitch	13.2 μm
Pixel size	12.4 μm square
Inter-pixel gap	0.8 μm
Fill factor	84.4%
Color depth	8 bits/color
Contrast ratio	>100 : 1
Frame rate	60 Hz
Input signal voltage	3.3–5.0 V
Operating temp.	20 to 60°C
Storage temp.	−20 to 70°C

accumulated ions change the switching characteristics of the FLC layer and cause serious image sticking [8]. The second problem is on the thin cell gap requirement. The operation voltage of a typical FLC material is 5 V/μm. Thus, in order to drive a FLC using the 5 V CMOS silicon backplane, a LC cell gap as thin as 1 μm is required [9]. The Displaytech cell is actually 0.8 μm. Such a thin cell gap is challenging to fabricate and may limit the manufacturing yield.

Up-to-date major liquid crystal displays as discussed in Chapters 3–9 use nematics. Nematic liquid crystals offer several attractive features for display applications, including a low operation voltage, analog gray scales, simple molecular alignment, abundant material selection, and a large cell gap tolerance. Two major drawbacks that demand continuous improvements are the response time and the viewing angle. Extensive efforts have been put into viewing angle improvement. Using the transverse field effect, a viewing angle comparable to that of CRTs has been achieved. Various approaches are addressed in Chapter 12.

In this chapter we focus on improving the response time of nematic liquid crystals. To optimise the LC parameters for a fast response time, a figure of merit (FoM) is defined. This FoM takes the temperature-dependent birefringence, viscosity and elastic constant into account. Material effects (viscosity, elastic constant, and birefringence), molecular alignment, temperature, voltage and cell gap all play significant roles in determining the LC response time. Each factor is analysed in detail.

2 Figure of Merit

In a nematic LCD, switching between different gray levels determines the real response time. From the small angle approximation [10, 11], the rise and decay times of a LC device are dependent on the cell gap d, rotational viscosity γ_1, elastic constant K, threshold voltage V_{th}, bias voltage V_b and applied voltage V as follows:

$$\tau_{rise} = \frac{\gamma_1 d^2 / K \pi^2}{(V/V_{th})^2 - 1} \qquad (10.1)$$

$$\tau_{decay} = \frac{\gamma_1 d^2/K\pi^2}{|(V_b/V_{th})^2 - 1|} \tag{10.2}$$

In a displayed gray scale image, each pixel may have a different response time, depending on the driving and bias voltages. Normally, the LC rise time is faster than the decay time owing to the driving voltage factor. However, if the voltage swing is small such that V/V_{th} is less than $\sqrt{2}$, then τ_{rise} could be slower than τ_{decay} according to Equations 10.1 and 10.2.

Figure 10.1 shows eight gray levels of a transmissive TN cell partitioned by equal transmission. In the case of switching from gray level 1 to 2 and back to 1, that corresponds to $V = V_2$ and $V_b = V_1$ in Equations 10.1 and 10.2. No doubt, this is the slowest among all the gray levels switching because V_b is close to V_{th}. Recall that Equations 10.1 and 10.2 are derived by small angle approximation. This shows divergence on decay time when V_b approaches V_{th}. In reality, this implies that the decay time is slow, but not infinity as described by Equation 10.2 [12]. From Figure 10.1, switching between gray levels 7 and 8 has the fastest rise and decay times. Although switching from gray level 1 to 2 is slow, the associated brightness change is from the brightest to the next brightest. Thus, the observers are less sensitive in detecting the difference.

After the voltage effect, the response time of a LC cell is primarily determined by the cell gap and the visco-elastic coefficient (γ_1/K). As discussed in Chapter 2, the elastic constants of a LC basically follow the order $K_{33}>K_{11}>K_{22}$. Thus, for a given LC mixture, its alignment also makes an important contribution to response time.

For a transmissive 90° TN cell, the product of the cell gap d and the birefringence Δn needs to satisfy the Gooch–Tarry first minimum condition [13]: $d\Delta n = (\sqrt{3}/2)\lambda \simeq 0.5\,\mu m$. Let us consider the free relaxation process, which means that the bias voltage $V_b = 0$. Under such circumstances, the decay time has the following form:

$$\tau_{decay} = \frac{1}{4\pi^2}\frac{\gamma_1}{K\Delta n^2} \simeq \frac{0.025}{\text{FoM}} \tag{10.3}$$

In Equation 10.3, the figure of merit is defined as $\text{FoM} = K\Delta n^2/\gamma_1$.

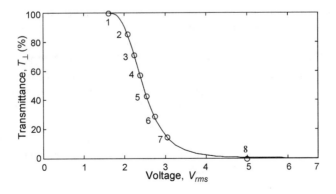

Figure 10.1 The voltage-dependent transmittance of a 90° TN cell. $d\Delta n = 480\,nm$, $\lambda = 550\,nm$, 1–8 represents 8 gray levels

3 Temperature Effect

Temperature has a dramatic effect on the figure of merit defined above. Since the nematic is a thermotropic liquid crystal, all the physical properties depend on the temperature. As the temperature increases, the birefringence, elastic constant and viscosity all decrease, but at different rates. From Chapter 2, the temperature dependencies of Δn, K and γ_1 are as follows:

$$\Delta n = (\Delta n)_o S \tag{10.4}$$

$$K = aS^2 \tag{10.5}$$

$$\gamma_1 = bS \exp\left(E/k_o T\right) \tag{10.6}$$

Here $(\Delta n)_o$ and a are respectively the birefringence and elastic constant at $T=0$ (or $S=1$), b is the proportionality constant for rotational viscosity, S is the order parameter, E is the activation energy of rotational viscosity, and k_o is the Boltzmann constant. When the operation temperature T is not too close to the clearing temperature T_c, the order parameter S can be approximated as [14]:

$$S = (1 - T/T_c)^\beta \tag{10.7}$$

In Equation 10.7, β is a material constant. In many LCs studied, $\beta \simeq 0.25$ is insensitive to materials. Substituting Δn, K, γ_1 and S into Equation 10.3, the following temperature-dependent figure of merit and free relaxation time are derived:

$$\text{FoM} = (a/b)(\Delta n)_o^2 (1 - T/T_c)^{3\beta} \exp\left(-E/k_o T\right) \tag{10.8}$$

$$\tau_{decay} = (b/40a)(\Delta n)_o^{-2} (1 - T/T_c)^{-3\beta} \exp\left(E/k_o T\right) \tag{10.9}$$

From Equation 10.9, four parameters (the birefringence, operation temperature, clearing point and activation energy of viscosity) are found to have a great impact on the LC response time. The birefringence effect basically originates from the d^2 dependence. The temperature and activation energy dependencies are exponential. Thus, their effect is paramount.

From Equation 10.9, a higher operation temperature would undoubtedly lead to a shorter response time. However, in a notebook or desktop computer, the LC operation temperature is limited to about 35°C due to the low-power backlight employed. For a large screen projector, the operation temperature could reach 50–60°C and a faster response is achieved due to the elevated temperature effect.

Equation 10.8 has a maximum FoM at an optimum operation temperature, called T_{op}. To derive T_{op}, let us set $\partial(\text{FoM})/\partial T = 0$ and find

$$T_{op} = \frac{E}{6\beta k_o}[(1 + 12\beta k_o T_c/E)^{1/2} - 1] \tag{10.10}$$

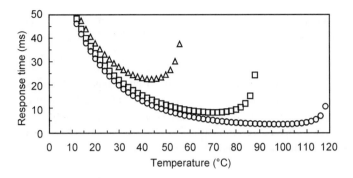

Figure 10.2 The temperature-dependent decay time of a 4 µm transmissive TN LC cell with $T_c = 60°C$ (triangles), 90°C (squares) and 120°C (circles). $\beta = 0.25$ and $E = 0.40\,\text{eV}$

The quantity $12\beta k_o T_c / E$ in Equation 10.10 is small, so the square root term can be expanded into a power series, and keeping the lowest order terms

$$T_{op} \simeq T_c[1 - 3\beta k_o T_c / E + \ldots] \tag{10.11}$$

From Equation 10.11, if $T_c = 100°C$, $\beta = 0.25$ and $E = 300\,\text{meV}$, then T_{op} is estimated to be about 80°, i.e. 20° below T_c. A high T_c mixture exhibits a high T_{op}. Operating at high T_{op} improves the merit factor significantly. The trade-off is in the need to heat the display.

To demonstrate the high T_c advantage, let us compare the temperature-dependent decay time of a 4 µm 90° TN LC cell with $T_c = 60$, 90 and 120°C, assuming $\beta = 0.25$ and $E = 0.40\,\text{eV}$ for these LC mixtures. The results are depicted in Figure 10.2. The decay time decreases as the temperature increases, reaches a minimum, and then climbs as T approaches T_c. In the low-temperature regime, γ_1 / K drops more quickly than $(\Delta n)^2$ as T increases, resulting in a decreasing response time. As T approaches T_c, the reverse situation occurs and the response time increases. For the examples shown in Figure 10.2, the mixture with $T_c = 60°C$ has a minimum response time of 22 ms at $T_{op} = 45°C$. By contrast, the $T_c = 90$ and 120°C mixtures have response times $\tau \simeq 9$ and 3.5 ms at $T_{op} = 70$ and 100°C, respectively. A high operation temperature is favorable from the response time viewpoint owing to decreased γ_1 / K, but is less desirable for practical applications. For a desktop monitor, the backlight heats the LC cell to $T = 35°C$. At this modest temperature, the LC response time is nearly two times faster than that at $T = 20°C$.

4 Material Effects

Based on Equation 10.3, the viscosity, elastic constant and birefringence all make significant contributions to FoM. Let us discuss the viscosity first.

4.1 Viscosity effect

For most display applications, low rotational viscosity is always favorable. From Equation 10.6, the rotational viscosity depends heavily on the activation energy. A small change in activation energy will affect the viscosity greatly. To make a fair comparison, let us assume that all the three LC cells compared have $d = 4\,\mu m$, $T_c = 90°C$ and $\beta = 0.25$, but different activation energies: $E = 0.38, 0.40$ and $0.42\,eV$. Computer simulation results are depicted in Figure 10.3. Circles, squares and triangles represent results for $E = 0.42, 0.40$ and $0.38\,eV$, respectively. The LC with the lowest activation energy has a clear advantage on the response time.

However, the macroscopic properties of a LC compound are all closely related. Changing a molecular constituent could affect several physical properties simultaneously. Table 10.2 illustrates such a complicated relationship. The bi-cyclohexane-ethyl-phenyl homologues, with structures shown below, are used as examples [15]:

$$C_3H_7 -\!\!\left\langle\;\right\rangle\!\!-\!\!\left\langle\;\right\rangle\!\!-C_2H_4-\!\!\left\langle\;\right\rangle\!\!-X$$

When X is equal to CH_3, OCH_3, F, CF_3, OCF_3 and CN, their birefringence, viscosity, dielectric anisotropy and clearing temperature all change. The CN group is an effective electronic acceptor and has a large dipole moment. It enhances dielectric anisotropy and birefringence. However, the cyano compounds tend to form dimers. As a result, their

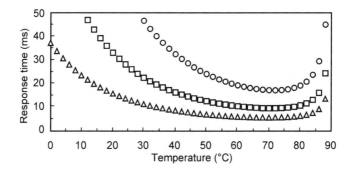

Figure 10.3 The temperature-dependent response time of three LCs with activation energy $E = 0.38\,eV$ (triangles), $0.40\,eV$ (squares) and $0.42\,eV$ (circles). $T_c = 90°C$ and $\beta = 0.25$

Table 10.2 Physical properties of some alkyl, fluoro and cyano LC homologues. Here ν stands for flow viscosity at $T = 20°C$

$$C_3H_7 -\!\!\left\langle\;\right\rangle\!\!-\!\!\left\langle\;\right\rangle\!\!-C_2H_4-\!\!\left\langle\;\right\rangle\!\!-X$$

X	Δn	ν (mm^{2}s^{-1})	$\Delta\varepsilon$	T_c (°C)	$(\Delta n)^2/\nu$ (10^{-4})
CH_3	0.105	14	0	153	7.88
OCH_3	0.125	23	0	176	6.79
F	0.10	18	6	137	5.55
CF_3	0.10	23	12	117	4.35
OCF_3	0.10	14	8	137	7.14
CN	0.17	75	12	196	3.85

viscosity is substantially higher than that of the fluoro group. On the other hand, the fluoro group is an electron donor. It withdraws electrons from the core. Therefore, the fluoro compound has a lower birefringence than its alkyl counterpart, as shown in Table 10.2. The OCF_3 group is found to have a low viscosity, high resistivity and large dielectric anisotropy. Thus, it is a good candidate for TFT-LCD applications. Some OCF_3 compounds do not have a nematic phase, but are still useful as dopants.

4.2   Elastic constant effect

For a given LC compound, the elastic constants follow the order $K_{33}>K_{11}>K_{22}$. For some linear LC compounds, $K_{33}>2K_{22}$ [16]. Thus, LC alignment also plays an important role in affecting the response time and contrast ratio. To take advantage of a large K_{33}, the homeotropic alignment needs to be implemented. Under the crossed polariser condition, the homeotropic alignment possesses an unprecedented contrast ratio (greater than 1000 : 1). In a conventional homeotropic alignment utilising a longitudinal electric field [17], only LC mixtures with negative dielectric anisotropy are allowed. However, several operation modes employing the fringing field effect permit positive $\Delta\varepsilon$ LCs to be used in homeotropic alignment [18]. The material selection is therefore greatly widened.

The positive $\Delta\varepsilon$ liquid crystals have a resultant dipole along the molecular axis. On the other hand, the negative $\Delta\varepsilon$ LC compounds often have two fluoro groups in the lateral positions, as discussed in Chapter 2. As a result, the positive $\Delta\varepsilon$ LC compounds are likely to have a lower viscosity than their negative counterparts. Using a positive $\Delta\varepsilon$ LC in a homeotropic alignment shortens the response time noticeably. A four-domain device with a wide viewing angle, high contrast ratio and 19 ms frame time has been achieved using a 5 μm cell gap [19].

4.3   Birefringence effect

The effect of birefringence on the response time is through the cell gap dependence. A larger birefringence enables a thinner cell gap to be used. From Equation 10.1, the cell gap dependence is quadratic. Therefore, if the cell gap constraint can be removed, the role of birefringence will be very significant. However, if manufacturing yield restrains the cell gap to be $d \geq 4$ μm, then the birefringence effect is not essential.

For a transmission-type 90° TN cell, the required $d\Delta n$ value is 0.866λ. For displays in the visible region, $d\Delta n \simeq 0.5$ μm. If the cell gap is fixed at $d = 4$ μm, then $\Delta n = 0.125$. Many LC mixtures meet this modest requirement on birefringence. However, for telecommunication applications at $\lambda = 1.55$ μm, the required $d\Delta n$ value is increased to 1.34 μm. For a 4 μm cell, the minimum Δn is 0.336. In addition, the birefringence in the IR region is usually about 15% lower than that in the visible region. Therefore, a high birefringence ($\Delta n > 0.39$ at $\lambda = 550$ nm), while preserving low viscosity materials, becomes critical.

5   Voltage Effect

As shown in Equations.10.1 and 10.2, both the applied voltage and the bias voltage have important effects on the rise and decay times of a LC cell. Two methods have been found to be powerful in improving LC response times: the bias voltage effect and the transient nematic effect [9]. In the case that $V_b > V_{th}$, the LC molecules are already tilted to a certain degree by the bias voltage. Thus, they quickly follow the incoming activation pulses. As a result, the rise time is shortened significantly. The higher the bias voltage, the faster the response time. However, the remaining phase retardation is also reduced.

From Equation 10.2, when the bias voltage is close to the threshold, $V_b \simeq V_{th}$ (whether under or above), the decay time becomes very slow. One way to overcome this bottleneck is to use the undershoot method, also known as the transient nematic effect [20].

Let us take the TN cell shown in Figure 10.1 as an example. The switching between V_1 and V_2 is the slowest among all the gray scales. For the first gray level, V_1 is assumed to be slightly under V_{th}. The slow rise time results from the small V_2/V_{th} ratio and the slow decay time is due to $V_b/V_{th} \simeq 1$. To speed up the rise time, an overdrive voltage with $V > V_2$ but a narrower pulse width is applied. The rise time is shortened to be approximately the same as the driving pulse width. During decay, the bias voltage is removed temporarily for a period of time until the final state is reached and a holding voltage is applied afterwards to stabilise the LC molecules at V_1 level. Using this overdrive and undershoot driving scheme, the response time could be improved by three to five times. Of course, the burden is loaded on the driving electronics.

6 Surface Stabilisation

Adding dopants or polymer networks to stabilise the boundary layers and enhance the restoring torque of nematic LC molecules has been explored by the IBM Japan [21] and Brown University [22] groups. This surface stabilisation effect has been applied to the in-plane switching (IPS) mode and it produced a big improvement in the response time. When the polymer effects are strong compared to surface anchoring, and when the applied electric field is weaker than the effective polymer field ($E < E_p$), the relaxation times for the center of the cell are derived as [20]:

$$t_{on} \simeq \frac{\gamma_1}{\varepsilon_0 |\Delta\varepsilon| (E^4 + E_p^4)^{1/2}} \tag{10.12}$$

$$t_{off} \simeq \frac{\gamma_1}{\varepsilon_0 |\Delta\varepsilon| E_p^2} \tag{10.13}$$

Owing to the existence of the polymer field, both rise and decay times are improved. The polymer field helps to pull back the LC molecules and therefore accelerates the relaxation process.

In the in-plane switching cell demonstrated by the IBM group [19], 0.16 wt% of chiral agent S811 is added to the LC mixture ZLI-4850. The cell gap is 12 μm, the pitch length is 48 μm, and the electrode gap is 30 μm. Results indicate that the relaxation time is shortened by eight times, while the rise time remains basically unaffected. However, the addition of the chiral agent causes light leakage at the dark state and subsequently degrades the contrast ratio. This is because of a small twist angle induced by the chiral agent.

On the other hand, the Brown group mixes 0.5–2 wt% reactive agent, RM-257 (EM Industries) to MLC-6233-000. In the IPS cell studied, the interdigitated ITO electrode is

32 µm wide, the electrode separation is 68 µm, and LC cell gap is $d = 5$ µm. The results show that the decay times are significantly reduced through polymer stabilisation. This reduction is more pronounced with an increasing polymer load. Without polymer stabilisation, the decay time (90–10%) is 75 ms. As the polymer concentration is increased to 0.5, 1.0 and 2.0%, the decay time is decreased to 70, 55 and 35 ms, respectively.

It is generally known that the rise time decreases as the applied electric field increases. However, with the polymer networks the rise time is shortened as the polymer load is increased. For example, the rise time (10–90%) of the bulk sample at low fields (less than 0.8 V/µm) is $\simeq 90$ ms and the 2% mixture is 40 ms. At high fields (greater than 0.8 V/µm), the rise time is $\simeq 10$ ms, insensitive to polymer concentration.

Two problems associated with the polymer stabilisation are the increased operation voltage and reduced transmittance. As the polymer concentration increases, the operation voltage is gradually increased and the transmittance is reduced. For example, without polymer networks, the transmittance reaches 28%. With 0.5, 1.0 and 2.0% polymer loads, the transmittance is reduced to 26, 21 and 9%, respectively. In most liquid crystal displays, light efficiency is probably the most critical requirement. In order to keep high transmittance while improving the response time, the polymer concentration should be kept below 1%. Under such circumstances, the gain in response time is not too significant.

7 Cell Gap Effect

For some virtual displays using a liquid crystal on silicon (LCOS) [23, 24], the field sequential approach enables a high-resolution light valve to be used for displaying color images. This approach leads to simple packaging and low cost, and eliminates the pixel registration problem as encountered in the three light valve configuration. However, the burden is loaded on the LC response time. To avoid color break up and image flickering, the frame rate should be at least 85 Hz, i.e. the LC response time needs to be shorter than 4 ms [25].

Based on Equations 10.1 and 10.2, four factors can be considered to achieve a fast response time. They are the cell gap, viscosity, elastic constant and driving voltage. Let us discuss each factor separately.

7.1 Thin cell approach

The thin cell is a natural way to obtain a short response time. Let us ignore the constraint of manufacturing technology and concentrate on the theoretical performance limit. As technology advances, reliable thin cell manufacturing will be developed with a high yield.

Three LC operation modes are often considered for virtual microdisplays: the homeotropic cell, the film-compensated homogeneous cell [26, 27], and the mixed-mode twisted nematic (MTN) cell [28]. Table 10.3 compares their minimum $d\Delta n$ requirement for achieving a 1π phase change, light modulation efficiency and a high contrast ratio. In principle, for a film-compensated LC device using the pure birefringence effect, its phase retardation can be calculated from following equation:

$$\delta = 4\pi(d_1\Delta n_1 - d_2\Delta n_2)/\lambda \tag{10.14}$$

Table 10.3 Comparison of five LC modes for virtual displays: homeotropic ($\perp$) cell, homogeneous ($\parallel$) cell, film-compensated homogeneous cell, 90° MTN cell with $\beta = 0°$ and 90° MTN cell with $\beta = 20°$. η = light modulation efficiency under crossed polarisers. NB = normally black, NW = normally white

	$\perp$-cell	$\parallel$-cell	$\parallel$-cell + film	90° MTN $\beta = 0°$	90° MTN $\beta = 20°$
$d\Delta n$ (nm)	140	140	180	170	240
$\eta(\%)$	100	100	100	68	88
Mode	NB	NW	NW	NW	NW
CR	>300:1	6:1	50:1	250:1	300:1

The $d_{1,2}\Delta n_{1,2}$ in Equation 10.14 represent the LC and the film. As analysed below, each cell has its own merits and demerits.

Homeotropic cell

For a reflective homeotropic cell under crossed polarisers, the dark state occurs at $V < V_{th}$. Since a good dark state is achieved automatically, the theoretically required $d\Delta n$ is $\lambda/4$, or about 140 nm at $\lambda = 550$ nm. However, this minimum $d\Delta n$ value implies that $R_\perp = 100\%$ is achieved at $V = \infty$. Therefore, in order to lower the operation voltage to less than 5 V, a thicker LC layer is required. Figure 10.4 plots the voltage-dependent reflectance of a homeotropic cell using a Merck high resistivity TFT-LCD mixture MLC-6608 ($\Delta\varepsilon = -4.2$, $\Delta n = 0.083$) as an example. The on-state voltage decreases as the $d\Delta n$ value increases. A disadvantage of the larger $d\Delta n$ is a slower response time, unless a higher Δn LC is used.

Although the homeotropic cell shows a favorable performance in both the high contrast ratio and potentially fast response time through the small $d\Delta n$ requirement and the K_{33} effect, the LC materials employed need to have a negative $\Delta\varepsilon$. From the structure-property relationships discussed in Chapter 2, the negative LCs exhibit a slightly larger viscosity than their positive $\Delta\varepsilon$ counterparts, and the molecular alignment is more complicated.

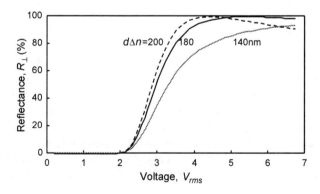

Figure 10.4 Voltage-dependent reflectance of homeotropic cells with $d\Delta n = 200$, 180 and 140 nm. LC = MLC-6608, $\Delta\varepsilon = -4.2$ and $\Delta n = 0.083$

Homogeneous cell

For a homogeneous cell, its theoretical $d\Delta n$ value is also 140 nm. However, unlike the homeotropic cell, the dark state of the thin homogeneous cell takes place in the high-voltage regime, as discussed in Chapter 3. The boundary layers cannot easily be oriented by the applied voltage. Thus, the uncompensated homogeneous cell would have a relatively low contrast ratio (CR $\simeq 6:1$) at $V = 5\,V_{rms}$, as shown in Figure 10.5. For a fair comparison with the homeotropic cell, an LC with $\Delta\varepsilon = 4.2$ is used for simulations. To boost the contrast ratio to $50:1$ and maintain the operation voltage lower than $5\,V_{rms}$, a phase compensation film with $d_2\Delta n_2 \simeq 40$ nm is needed. The addition of this compensation film boosts the required $d\Delta n$ of the LC to about 180 nm [29]. As a result, the response time is about 65% slower than that of the uncompensated homogeneous cell. To further lower the dark state voltage, the larger $d\Delta n$ value of the LC and film needs to be employed. Although the operation voltage is decreased, the response time is sacrificed, unless a higher Δn LC is used.

TN cells

For 90° MTN cells with β angles 0 and 20°, their $d\Delta n$ values are 170 nm and 240 nm, respectively. Here β represents the angle between the polarisation axis and the front LC axis. Owing to the natural self-phase compensation of the 90° TN cell, no phase compensation film is needed.

Color sequential display

Assuming that the birefringence of the LC mixtures developed for a color sequential display is around 0.1, the cell gap requirement is 1.4 µm for the homeotropic and homogeneous cells, 1.8 µm for the film-compensated homogeneous, and 1.7 and 2.4 µm for the 90° MTN cells with $\beta = 0$ and 20°, respectively. For such thin cells, the response time should be less

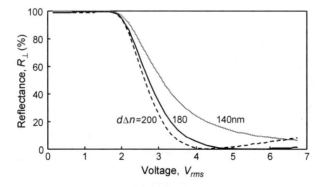

Figure 10.5 Voltage-dependent reflectance of homogeneous cells with $d\Delta n = 200$, 180 and 140 nm. For the LC with $d\Delta n = 200$, 180 and 140 nm, the compensation film has $d\Delta n = -60$, -40, and 0 nm, respectively. Assume LC $\Delta\varepsilon$ is 4.2

than 2 ms. Further reducing the cell gap would lead to an even faster response time. An obvious trade-off is the reduced light modulation efficiency. In a virtual display, LEDs are normally used as light sources because of their good color chromaticity, low power consumption, high brightness and fast response time. If the display brightness is not an issue, a very thin cell gap can be employed using a high Δn LC mixture. A nematic LCD with a sub-millisecond response time has been demonstrated, as shown in Figure 10.6 [30]. Sequential color operation over the -10 to $70°C$ temperature range is possible.

For a TN cell with twist angle deviating from $90°$, a compensation film is needed in order to achieve high contrast ratio. For example, the film-compensated $45°$ MTN cell shown in Figure 4.21 exhibits a 100% light modulation efficiency, a reasonably low operation voltage and a fast response time. Thus, it should be attractive for both virtual and large screen projection displays.

7.2 Cell gap measurements

How to accurately measure the $d\Delta n$ and twist angle of a thin cell is a crucial issue. To control the cell gap, during the manufacturing process spacer balls are sprayed uniformly either around the display area (for large display panels) or along the peripheral glue lines of the substrate (for micro-displays). After assembling the cell, the final cell gap could deviate from the specified diameter of the spacer balls. It is highly desirable to characterise the final $d\Delta n$ value and twist angle of the display devices.

The interferometer method has commonly been used to measure the gap of empty cells [31]. By counting the interference fringes, an accurate cell gap can be obtained. This method can be extended to measure a filled twisted nematic (TN) cell by applying a sufficiently high voltage so that the LC directors are virtually aligned perpendicular to the substrates. However, in order to calculate the cell gap precisely, the director profiles in the high-voltage state need to be known. Another drawback of this method is that it does not provide twist angle information. The Jones matrix method has been developed to simultaneously measure the twist angle and $d\Delta n$ of filled transmissive TN cells [32, 33]. Experimental techniques using a single wavelength or wavelength scan have been demonstrated [34, 35]. For reflective displays, a reflector is deposited on the inner side of the back substrate and the light does

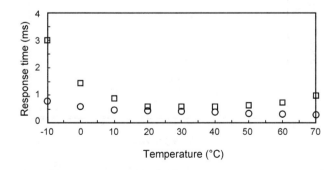

Figure 10.6 Response time of a normally white reflective TN cell for a color sequential virtual display. Circles: voltage-on state; squares: voltage-off state

not transmit through the cell. Thus, the methods developed to characterise transmissive displays are not directly applicable to reflective cells.

Principles of measurement

In a reflective cell, the normalised reflectance $R_\perp$ under the crossed-polariser configuration has been derived by the Jones matrix method as follows [36]:

$$R_\perp = \left(\Gamma \frac{\sin X}{X}\right)^2 \left(\sin 2\beta \cos X - \frac{\phi}{X}\cos 2\beta \sin X\right)^2 \qquad (10.15)$$

In Equation 10.15, $\Gamma = 2\pi d\Delta n/\lambda$, $X = (\phi^2 + (\Gamma/2)^2)^{1/2}$, ϕ is the twist angle and β is the angle between the input polarisation and the front LC director.

Two special cases ($\beta = 0$ and $\pm 45°$) are found especially useful to determine the cell gap. In these two special orientations, Equation 10.15 is reduced to

$$R_{\perp,0} = \left(\Gamma\phi\frac{\sin^2 X}{X^2}\right)^2 \qquad (10.16)$$

$$R_{\perp,45} = \left(\Gamma\frac{\sin 2X}{2X}\right)^2 \qquad (10.17)$$

In theory, the two unknowns (Γ and ϕ) could be determined simultaneously by measuring $R_{\perp,0}$ and $R_{\perp,45}$ at a given wavelength. However, in an experiment it is difficult to obtain accurate measurements on the normalised reflectance due to the following factors: multiple reflections from substrates, the interference effect between the transparent electrode and the back reflector, and the finite TFT aperture ratio. To avoid the normalisation problem, one could measure the relative reflectance of $R_{\perp,0}$ or $R_{\perp,45}$, and search for its minimum (zero), or simply take their ratio. Taking the ratio of Equations 10.16 and 10.17 would eliminate the normalisation factor and lead to the following simple form:

$$R_{\perp,0}/R_{\perp,45} = \left(\frac{\phi}{X}\tan X\right)^2 \qquad (10.18)$$

A preferred approach depends on the detailed $d\Delta n$ value for the given TN cell, as illustrated in the following examples.

First, let us consider the case $R_{\perp,0} = 0$. This corresponds to $X = p\pi$, where $p = 1, 2, 3$, etc. For simplicity, let us consider the case of $p = 1$, the following equation is obtained:

$$\left(\frac{\phi}{\pi}\right)^2 + \left(\frac{d\Delta n}{\lambda}\right)^2 = 1 \qquad (10.19)$$

To elucidate the significance of Equation 10.19, let us use the 45° reflective TN cell [37] as an example. The 45° TN liquid crystal light valve has been developed for normally black projection displays. It can also be extended to a normally white direct-view display employing only one polariser, although the contrast ratio is reduced substantially. For a 45° TN cell, we find that $R_{\perp,0} = 0$ occurs at $d\Delta n = 0.968\lambda$. If a 45° TN cell has $d\Delta n \simeq 500\,\mathrm{nm}$, then its $R_{\perp,0} = 0$ would occur at $\lambda \simeq 517\,\mathrm{nm}$. This is within the range of a commercial computer-controlled visible spectro-photometer, such as the Perkin-Elmer Lambda-9.

Second, let us consider the case of $R_{\perp,45} = 0$. From Equation 10.17, this corresponds to $X = (q/2)\pi$; $q = 1, 2, 3$, etc. For $q = 1$, $X = \pi/2$ and the following equation holds:

$$\left(\frac{\phi}{\pi}\right)^2 + \left(\frac{d\Delta n}{\lambda}\right)^2 = \frac{1}{4} \tag{10.20}$$

This special case is particularly useful for the displays with small $d\Delta n$ value, such as the mixed-mode twisted-nematic (MTN) cells [38] developed for normally white projection and direct-view displays. Let us use the 63.6° MTN reflective cell [39] as an example. For a 63.6° MTN cell, its first $R_{\perp,45} = 0$ occurs at $d\Delta n = 0.354\lambda$. If the 63.6° MTN cell has $d\Delta n \simeq 180\,\mathrm{nm}$, then the lowest order reflectance minimum should occur at $\lambda \simeq 508\,\mathrm{nm}$.

In an experiment one can scan the cell to obtain $R_{\perp,0}(\lambda)$ and $R_{\perp,45}(\lambda)$, and then fit data using Equations 10.16 and 10.17 with ϕ and $d\Delta n$ as parameters. Owing to the nature of the multi-valued solutions of these equations, the methods described here are difficult to apply to an arbitrary cell.

Equations 10.16 and 10.17 have two variables: Γ and ϕ. This means that if the twist angle is known, then we can determine $d\Delta n$ by making just one spectral scan. If Γ and ϕ are both unknown, then we need to make two scans in order to solve the miscellaneous equations. It is highly desirable to determine Γ and ϕ independently.

To do so, let us examine Equation 10.18 further. When $\Gamma/2 \ll \phi$, X is reduced to ϕ and Equation 10.18 has a simple expression:

$$R_{\perp,0}/R_{\perp,45} \simeq \tan^2\phi \tag{10.21}$$

In such circumstances, the twist angle can be determined simply by the ratio $R_{\perp,0}/R_{\perp,45}$. In Equation 10.21, $\tan^2(\phi)$ is a sensitive function of ϕ. As ϕ changes from 45 to 50°, $\tan^2\phi$ is increased by 42%. This is a sensitive method for determining the twist angle. However, to satisfy the condition $\Gamma/2 \ll \phi$ is not easy, as it implies $d\Delta n \ll (\phi/\pi)\lambda$. For most reflective MTN cells, the twist angle is 60–90° and $d\Delta n$ is in the 150–250 nm range [40,41]. For a 60° MTN cell with $d\Delta n = 180\,\mathrm{nm}$, $\lambda \gg 540\,\mathrm{nm}$. Thus, to satisfy the condition $\Gamma/2 \ll \phi$, we have to measure the $R_{\perp,0}/R_{\perp,45}$ ratio using a long wavelength ($\lambda > 2\,\mu\mathrm{m}$) light source. An even longer wavelength will undoubtedly lead to a better accuracy. However, most glass substrates are opaque beyond $\lambda = 3\,\mu\mathrm{m}$, which imposes a practical limitation to this long-wavelength measurement technique.

Experiment

The experimental set up for measuring the cell gap and twist angle of a reflective cell is depicted in Figure 10.7. The light from an incandescent bulb passes through a polariser and then is reflected by a polarising beam splitter (PBS) or a prism to the LC cell. The

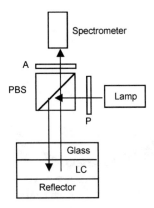

Figure 10.7 Experimental set up for measuring the cell gap and twist angle of a reflective cell. PBS = polarising beam splitter, P = polariser and A = analyser

linearly polarised light impinges on the cell at an angle β with respect to the front LC director. This is the β angle described in Equation 10.15. The light is then reflected back by the reflector embedded on the inner side of the LC cell. Owing to the accumulated phase retardation, the outgoing light has a different polarisation from the incident beam and is transmitted by the PBS. Since the employed PBS has a relatively low extinction ratio (about 50 : 1), an extra sheet analyser is used to improve the signal-to-noise ratio. This optical signal is sent to a scanning spectrometer to analyse its spectral contents. A reflectance spectrum between 330 and 1000 nm was measured. However, data fitting was made only in the 350 to 750 nm range; beyond this range the signals were too noisy to yield meaningful results.

The oscillations in the detected spectrum originate from the interference between beams reflected from two interfaces above and below the LC layer. To take the interference effect into account, Equations 10.16 and 10.17 should be revised to:

$$R_{\perp,0} = \left(\Gamma\phi\frac{\sin^2 X}{X^2}\right)^2 B(1 + 4A^2 + 4A\cos\Delta) \tag{10.22}$$

$$R_{\perp,45} = \left(\Gamma\frac{\sin 2X}{2X}\right)^2 B(1 + 4A^2 + 4A\cos\Delta) \tag{10.23}$$

Here $B = R_{bot}^2(1-R_{top})^4$

$$A = R_{bot}R_{top}\left(\cos^2 X + (\phi^2 - (\Gamma/2)^2)\sin^2 X/X^2\right)$$

and $\Delta = 2\pi dn/\lambda$; here $R_{top}\ll 1$ and $R_{bot}\simeq 1$ are the reflectivity of the top ITO and the bottom aluminum reflectors, and n is the average refractive index of the LC. It is interesting to note that $R_{\perp,0}/R_{\perp,45}$ remains the same as in Equation 10.18 with the interference effect included, since the interference factors shown in Equations 10.22 and 10.23 cancel each other out.

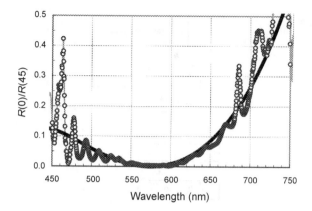

Figure 10.8 The measured wavelength-dependent normalised reflectance $(R_{\perp,0}/R_{\perp,45})$ of a 45° TN liquid crystal light valve. Dots are experimental data and solid lines are fittings using Equation 10.18 with $d = 4.12\,\mu\text{m}$. The LC mixture employed is ZLI-3807

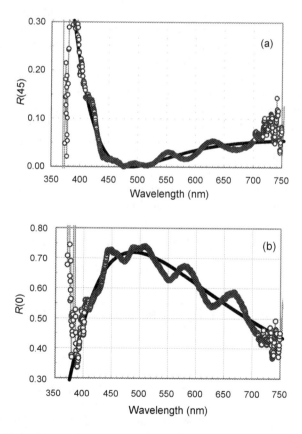

Figure 10.9 The wavelength-dependent reflectance (a) $R_{\perp,45}$ and (b) $R_{\perp,0}$ of a 63.6° MTN liquid crystal light valve. Dots are experimental data and solid lines are fittings using Equation 10.18 with $\phi = 63.7°$ and $d\Delta n = 172.5\,\text{nm}$. The LC mixture employed is ZLI-2359. The interference fringes are from interfacial reflections of the LC cell

To fit experimental data with the interference effect, one should use Equations 10.22 and 10.23. However, too many unknowns are involved. We found that in many cases it is equally accurate just to treat the interference as experimental noise and fit the data using the simple formula of Equations 10.16 and 10.17.

To demonstrate the experimental procedures, 45° TN and 63.6° MTN reflective light valves are used as examples. For the 45° TN cell, one side of the substrate was deposited with 4.0 μm SiO$_2$ spacers surrounding the substrates and the cell was filled with a Merck ZLI-3807 ($\Delta n = 0.133$) LC mixture. For the 63.6° MTN cell, the cell gap was reduced to 3.3 μm and a low birefringence ($\Delta n = 0.0512$) LC mixture ZLI-2359 was used.

For the 45° TN cell, the relative reflectances at $\beta = 0°$ and 45° were measured and the ratio $R_{\perp,0}/R_{\perp,45}$ was calculated. The results are shown in Figure 10.8. In this example, we assume that the twist angle ($\phi = 45°$) is known and treat $d\Delta n$ as a fitting parameter. From fitting, $d\Delta n = 0.548$ μm is found. Knowing that $\Delta n = 0.133$, the cell gap $d = 4.12$ μm is obtained. This result agrees with the deposited 4 μm SiO$_2$ spacers within 3%. One advantage of using the ratio method is that the interference factors in $R_{\perp,0}(\lambda)$ and $R_{\perp,45}(\lambda)$ cancel each other and the result is less sensitive to the interference.

For the 63.6° MTN cell, the measured and calculated $R_{\perp,0}(\lambda)$ and $R_{\perp,45}(\lambda)$ are plotted in Figure 10.9. As expected, the interference patterns are more obviously manifested. Solid lines in Figure 10.9 represent the fitting of experimental data with Equations 10.16 and 10.17 using ϕ and $d\Delta n$ as adjustable parameters. Through fitting, $\phi = 63.7°$ and $d\Delta n = 172.7$ nm (or $d = 3.4$ μm) are obtained. These results agree quite well with the controlled $\phi = 63.6°$ and $d = 3.3$ μm.

8 π-cell

In the π-cell approach [42], the pre-tilt angles in the top and bottom substrates are in the opposite directions, as shown in Figure 3.1. The opposite pre-tilt angle leads to two special features for the π-cell. First, its viewing angle is wide and symmetric. Second, its bend director profile in the voltage-on state eliminates the back-flow effect and therefore results in a fast response time. The back-flow effect in a homogeneous cell and a TN cell gives rise to optical bounce during the relaxation process and increases the decay times [43, 44].

To overcome the domain instability, a constant bias voltage of about 2 V is needed for a $d = 7$ μm optically compensated bend cell [45]. The LC used is TD-6004XX (Chisso) and the pre-tilt angle is 5°. By adjusting the $d\Delta n$ value of the compensation film, both normally white and normally black modes can be achieved [46]. Due to the bias voltage effect, the OCB cell exhibits a fast response time between gray level switching. The switching time between gray levels 1 and 2 is still less than 3.5 ms. The high level switching has a 1–1.5 ms response time at room temperature.

For a single domain device, the π-cell has been used for a high-speed light shutter to convert a black and white display into color [47]. However, for flat panel displays involving millions of pixels, the π-cell has serious technical difficulties in obtaining a uniform splay-to-bend transition among pixels [48]. How to initiate a uniform splay-to-bend transition and reduce the defects are challenging tasks before the OCB cell can fully deliver its promise for high-speed liquid crystal displays.

References

1 N. A. Clark and S. T. Lagerwall, Submicrosecond bistable electro-optic switching in liquid crystals, *Appl. Phys. Lett.* **36**, 899 (1980).

2 D. Banas, H. Chase, J. Cunningham, M. A. Handschy, R. Malzbender, M. R. Meadows, and D. Ward, Miniature FLC/CMOS color-sequential display systems, *J. SID* **5**, 27 (1997).

3 T. Toshihara, T. Makino, and H. Inoue, A 254-ppi full-color video rate TFT-LCD based on field sequential color and FLC display, *SID Tech. Digest* **31**, 1176 (2000).

4 M. Wand, W. N. Thurmes, R. T. Vohra, and K. M. More, Advances in ferroelectric liquid crystals for microdisplay applications, *SID Tech. Digest* **27**, 157 (1996).

5 O. Akimoto and S. Hashimoto, A 0.9" UXGA/HDTV FLC microdisplay, *SID Tech. Digest* **31**, 194 (2000).

6 J. Xie, M. Handschy, and L. Ji, Reduction of image sticking in FLC devices by kicking, *SID Tech. Digest* **31**, 989 (2000).

7 H. Zhang, K. D'have, H. Pauwels, D. D. Parghi, and G. Heppke, Influence of ion transport in AFLCDs, *SID Tech. Digest* **31**, 1000 (2000).

8 J. Xie, S. H. Perimutter, and M. R. Meadows, Flow of ionic impurities in a ferroelectric liquid crystal device, *SID Tech. Digest* **31**, 997 (2000).

9 H. Takanashi, K. Takahashi, T. Isozaki, S. Kurogoshi, and S. Hashimoto, *SID Tech. Digest* **31**, 992 (2000).

10 E. Jakeman and E. P. Raynes, Electro-optic response times of liquid crystals, *Phys. Lett.* **A39**, 69 (1972).

11 I. C. Khoo and S. T. Wu, "Optics and Nonlinear Optics of Liquid Crystals" (World Scientific, Singapore, 1993).

12 S. T. Wu, Design of a liquid crystal based electro-optic filter, *Appl. Opt.* **28**, 48 (1989).

13 C. H. Gooch and H. A. Tarry, The optical properties of twisted nematic liquid crystal structures with twisted angles $\leq 90°$, *J. Phys. D.* **8**, 1575 (1975).

14 I. Haller, Thermodynamic and static properties of liquid crystals, *Prog. Solid State Chem.* **10**, 103 (1975).

15 V. Reiffenrath, U. Finkenzeller, E. Poetsch, B. Rieger, and D. Coates, Synthesis and properties of liquid crystalline materials with high optical anisotropy, *Proceedings SPIE* **1257**, 84 (1990).

16 R. Tarao, H. Saito, S. Sawada, and Y. Goto, Advances in liquid crystals for TFT displays, *SID Tech. Digest* **25**, 233 (1994).

17 M. F. Schiekel and K. Fahrenschon, Deformation of nematic liquid crystals with vertical orientation in electric fields, *Appl. Phys. Lett.* **19**, 391 (1971).

18 S. H. Lee, H. Y. Kim, I. C. Park, B. G. Rho, J. S. Park, and C. H. Lee, Rubbing-free, vertically aligned nematic liquid crystal display controlled by in-plane field, *Appl. Phys. Lett.* **71**, 2851 (1997).

19 S. H. Hong, Y. H. Jeong, H. Y. Kim, H. M. Cho, W. G. Lee, and S. H. Lee, Electro-optic characteristics of 4-domain vertical alignment nematic liquid crystal display with interdigital electrode, *Appl. Phys. Lett.* **87**, 8259 (2000).

20 S. T. Wu and C. S. Wu, High speed liquid crystal modulators using transient nematic effect, *J. Appl. Phys.* (1989).

21 M. Hasegawa, Response-time improvement of the In-Plane-Switching mode, *SID Tech. Digest* **28**, 699 (1997).

22 M. J. Escuti, C. C. Bowley, G. P. Crawford, and S. Zumer, Enhanced dynamic response of the in-plane switching liquid crystal display mode through polymer stabilization, *Appl. Phys. Lett.* **75**, 3264 (1999).

23 P. M. Alt, *Conference records of the 1997 Int'l Display Research Conference*, (Toronto, Canada, 1997), p. M-19.

24 R. L. Melcher, M. Ohhata, and K. Enami, *SID Tech. Digest* **29**, 25 (1998).

25 T. Uchida, K. Saitoh, T. Miyashita, and M. Suzuki, *Proc. the 17th Int'l Display Research Conf.* (Toronto, Canada, Sept. 15–19, 1997), p. 37.

26 S. Saito and T. Takahashi, Highly multiplexable achromatic LCD utilizing double layered homogeneously oriented nematic LC cells without twist, *Mol. Cryst. Liq. Cryst.* **207**, 173 (1991).

27 K. Lu and B. E. A. Saleh, A single-polarizer reflective LCD with wide-viewing-angle range for gray scales, *SID Tech. Digest Application Papers*, **27**, 63 (1996).

28 S. T. Wu and C. S. Wu, Mixed-mode twisted nematic cells for reflective liquid crystal displays, *Appl. Phys. Lett.*, **68**, 1455 (1996).

29 S. T. Wu and C. S. Wu, A biaxial film-compensated thin homogeneous cells for reflective liquid crystal displays, *J. Appl. Phys.* **83**, 4096 (1998).

30 D. J. Schott, Reflective LCOS light valves and their application to virtual displays, *Conference Records of the 1999 Int'l Display Research Conference* (Berlin, Germany, 1999), p. 485.

31 R. Chang, Application of polarimetry and interferometry to liquid crystal film research, Material Research Bulletin 7, 267 (1972).

32 A. Lien and H. Takano, Cell gap measurements of filled twisted nematic liquid crystal displays by a phase compensation method, *J. Appl. Phys.* **69**, 1304 (1991).

33 A. Lien, The general and simplified Jones matrix representations for the high pretilt twisted nematic cell, *J. Appl. Phys.* **67**, 2853 (1990).

34 K. H. Yang and H. Takano, Measurements of twisted nematic cell gap by spectral and split-beam interferometric methods, *J. Appl. Phys.* **67**, 5 (1990).

35 S. T. Tang and H. S. Kwok, A new method to measure the twist angle and cell gap of liquid crystal cells, *SID Tech. Digest* **29**, 552 (1998).

36 S. T. Wu and C. S. Wu, Bisector effect on the twisted-nematic cells, Jpn. *J. Appl. Phys.* **37**, L1497 (1998).

37 J. Grinberg, A. Jacobson, W. P. Bleha, L. Miller, L. Fraas, D. Bosewell, and G. Meyer, A new real-time noncoherent to coherent image converter: the hybrid field effect liquid crystal light valve, *Opt. Eng.* **14**, 217 (1975).

38 S. T. Wu, C. S. Wu, and C. L. Kuo, Reflective direct-view and projection displays using twisted-nematic liquid crystal cells, Jpn. *J. Appl. Phys.* **36**, 2721 (1997).

39 T. Sonehara, Photo-addressed liquid crystal SLM with a twisted nematic ECB mode, *Jpn. J. Appl. Phys.* **29**, L1231 (1990).

40 S. T. Wu, C. S. Wu, and C. L. Kuo, Comparative studies of single-polarizer reflective liquid crystal displays, *J. SID* **7**, 119 (1999).

41 H. S. Kwok, Parameter space representation of liquid crystal display operating modes, *J. Appl. Phys.* **80**, 3687 (1996).

42 P. J. Bos, K. R. Koehler/Beran, The π-cell: a fast liquid crystal optical switching device, *Mol. Cryst. Liq. Cryst.* **113**, 329 (1984).

43 C. Z. van Doorn, Dynamic behavior of twisted nematic liquid crystal layers in switched fields, *J. Appl. Phys.* **46**, 3738 (1975).

44 D. W. Berreman, Liquid crystal twist cell dynamics with backflow, *J. Appl. Phys.* **46**, 3746 (1975).

45 T. Uchida, Field sequential full color LCD without color filter by using fast response LC cell, *The 5th Int'l Display Workshops*, p. 151 (1998).

46 S. T. Wu and A. M. Lackner, Film-compensated parallel-alignment cells for high speed LCDs, *SID Tech. Digest* **25**, 923 (1994).

47 P. Bos, T. Buzak, and R. Vatne, A full-color field-sequential color display, *Proceeding SID* **26**, 157 (1985).

48 N. Koma, T. Miyashita, T. Uchida, and K. Yoneda, Using an OCB-mode TFT-LCD for high speed transition from splay to bend alignment, *SID Tech. Digest* **30**, 28 (1999).

Low Operating Voltage

1 Introduction

The low operating voltage of a liquid crystal display (LCD) contributes to a reduced power consumption and lowers the cost of driving electronics. In the TFT-LCD there is a continuous need to lower the operating voltage from 5, 3.3 to 2.5 V_{rms}. For a normally white display such as a 90° twisted-nematic cell under crossed polarisers, the operation voltage is determined by how quickly the dark state is obtained. In contrast, for a normally black display such as a homeotropic cell, the dark state occurs at $V=0$ so that the bright state governs the required operating voltage.

Several approaches, such as self-phase compensation, external phase compensation, material effect, pre-tilt angle effect and different LC configurations, could be used to lower the operating voltage. In this chapter we describe the underlying operation principles of each effect. In practice, one or more such effects can be combined to lower the operating voltage.

To make a fair comparison between different LC operation modes, unless otherwise specified, the following LC parameters are used: $K_{11}=13.4$, $K_{22}=6.0$ and $K_{33}=9.0$ pN, $\varepsilon_{\parallel}=10.3$, $\varepsilon_{\perp}=3.8$ and $\Delta n=(0.083, 0.085, 0.087)$ for the three primary RGB wavelengths (650, 550, 450 nm), respectively. These parameters are based on the two-bottle mixture (MLC-9100-000 and -100) that Merck developed for TFT-LCD applications.

2 Self-phase Compensation Effect

In a homogeneous cell as discussed in Chapter 3, the molecular alignment directions in the front and rear surfaces are anti-parallel. Therefore, in a high-voltage state the residual phase retardation originating from these boundary layers is additive. These boundary layers are quite difficult to orient by the applied electric field. As a result, its dark state voltage is too high to be practically useful for TFT-LCD applications.

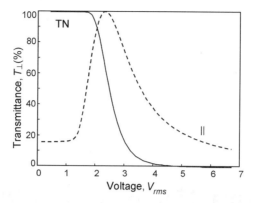

Figure 11.1 The voltage-dependent transmittance of a homogeneous cell (dashed lines) and a 90° TN cell (solid line). For both cells, $d\Delta n = 480\,\text{nm}$, $\Delta\varepsilon = 6.5$ and the pre-tilt angle $\alpha = 2°$. For the homogeneous cell, $\beta = 45°$ and for the TN cell, $\beta = 0°$

Twisting the rear LC alignment direction by 90°, the well-known twisted nematic (TN) cell is formed [1]. The normally white TN cell has been widely employed for display applications owing to its weak color dispersion, high contrast ratio and relatively low operation voltage. From the Gooch–Tarry first minimum [2], the $d\Delta n$ of the cell should be $(\sqrt{3}/2)\lambda$; where d is the cell gap, Δn is the birefringence and λ is the wavelength. For the green band centered at $\lambda = 550$ nm, $d\Delta n$ is equal to 480 nm. Figure 11.1 shows the voltage-dependent light transmittance (at $\lambda = 550$ nm) of a homogeneous cell (at $\beta = 45°$) and a 90° TN cell (at $\beta = 0°$) with the same $d\Delta n = 480$ nm, where β is the angle between the polariser axis and the front LC director. For simplicity, only the normalised transmittance $[T_\perp]$ of the LC cell sandwiched between two crossed polarisers is compared; the optical losses from polarisers and substrates are all ignored.

From Figure 11.1, the homogeneous cell has a total phase retardation of $\delta \simeq 1.75\pi$. As the voltage increases, $T_\perp$ reaches a maximum and gradually diminishes. However, the dark state voltage is relatively high. This is because the phase retardation of the boundary layers is additive, rather than subtractive. On the other hand, the TN cell has a good dark state with a contrast ratio greater than 300 : 1 at $V \simeq 4.5\,V_{\text{rms}}$. Such a low dark state voltage originates from the self-phase compensation of the boundary layers. When the applied voltage is about three times the threshold voltage, the bulk LC directors are reoriented along the electric field direction and the residual phase retardations of the orthogonal boundary layers cancel each other out, resulting in a low dark state voltage.

The TN result shown in Figure 11.1 serves as a benchmark for comparison. To further reduce the operation voltage of a 90° TN cell, the following effects can be taken separately or collectively: material parameters, the pre-tilt angle α, the twist angle θ, and the polarisation angle β effects.

3 Material Effects

The threshold voltage V_{th} of a general twisted nematic cell is related to the dielectric anisotropy $\Delta\varepsilon = \varepsilon_\parallel - \varepsilon_\perp$, the twist angle θ and elastic constants K_{ii} as [3]

$$V_{th} = \frac{\{K_{11}\pi^2 + \theta^2[K_{33} - 2K_{22}(1 - \alpha)]\}^{1/2}}{\varepsilon_0 \Delta\varepsilon} \tag{11.1}$$

Here $\alpha = 2\pi d/p\theta$, where p is the pitch length of the chiral dopant. For a homogeneous cell, $\theta = 0$ and Equation 11.1 is reduced to $V_{th} = \pi(K_{11}/\varepsilon_0\Delta\varepsilon)$. Thus, a low threshold voltage can be obtained by either enhancing the dielectric anisotropy, reducing the elastic constant, or a combination of both. From Chapter 1, two primary factors governing the splay elastic constant are molecular shape and molar volume [4]. Some compounds do possess a small K_{ii} [5]; however, decreasing the elastic constant would also slow down the response time of the LC device. Thus, to lower the threshold voltage through either increasing the dielectric anisotropy or decreasing the elastic constant depends on the specific need.

3.1 Δε effect

Figure 11.2 shows the $\Delta\varepsilon$ effect on the operation voltage of a 90° TN cell. For the comparison shown in Figure 11.2, the $d\Delta n$ value, polarisation angle $(\beta = 0°)$ and pre-tilt angle $(\alpha = 2°)$ are all kept the same. As $\Delta\varepsilon$ increases, both V_{th} and the dark state voltage (V_{dark}) decrease whereas the switching ratio remains unchanged $(V_{dark}/V_{th} \simeq 3)$.

From mean-field theory [6], the dielectric anisotropy of a LC compound is governed by the dipole moment and its angle of inclination with respect to the principal molecular axis. Many polar LC compounds have been developed and LC mixtures formulated with a wide range of $\Delta\varepsilon$. For TN and STN LCD applications, many cyano LC mixtures with $\Delta\varepsilon$ ranging as 10–15 are commercially available [7]. However, for TFT-LCDs, a high resistivity ($>10^{13}\,\Omega/cm$) is required in order to retain a high charge holding ratio and avoid image flickering. The fluoro compounds meet these requirements except that their dipole moment is smaller ($\mu = 1.4\,$Debye). The dielectric anisotropy of the axial fluoro compound is about 5. To enhance $\Delta\varepsilon$, difluoro and trifluoro compounds have been investigated extensively [8,9].

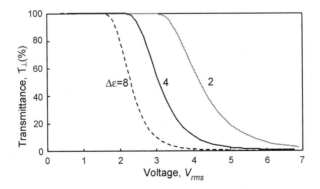

Figure 11.2 The voltage-dependent transmittance of a 90° TN cell with $\Delta\varepsilon = 8$, 4 and 2. Pre-tilt angle $\alpha = 2°$, $d\Delta n = 480\,$nm, $\lambda = 550\,$nm

3.2 Push-pull effect

One way to enlarge the dielectric anisotropy is to utilise the push-pull effect [10]. As discussed in Chapter 6, the nitro-amino tolane, with molecular structure shown below, exhibits a large dielectric anisotropy ($\Delta\varepsilon \sim 60$).

$$C_5H_{11}-N-\langle\!\!\!\bigcirc\!\!\!\rangle\equiv\langle\!\!\!\bigcirc\!\!\!\rangle-NO_2 \tag{I}$$

By adding 5–10% of such a high dielectric dopant to a host LC mixture, the threshold voltage could be reduced significantly. This effect is particularly manifested for the host LC mixtures with a small $\Delta\varepsilon$. For instance, the linearly conjugated diphenyl-diacetylene (PTTP) LCs described in Chapter 2 have a high birefringence, low viscosity, low melting point and wide nematic range, except that their dielectric anisotropy is too small ($\Delta\varepsilon \sim 1$) and their photo and thermal stability is inadequate. By doping 10% of compound (I) to a binary mixture PTTP-24/36, the threshold voltage is greatly reduced and the photo and thermal stability is improved.

Figure 11.3 depicts the concentration effect of the threshold voltage of the PTTP-24/36 mixture. A homogeneous cell with $d \sim 6\,\mu m$ and $\sim 2°$ pre-tilt angle was used for V_{th} measurements at $\lambda = 633\,nm$. The voltage-dependent phase change method was used for determining the threshold voltage [11]. In Figure 11.3, the circles represent experimental data of PTTP-24/36, the squares are for a 5% compound (I) dissolved in PTTP-24/36 and the triangles are for a 10% (I) in PTTP-24/36. Near to the threshold, the voltage-dependent phase change is linear [12]. From this linear extrapolation, the threshold voltage is obtained. From Figure 11.3, the threshold voltage of PTTP-24/36 at $f = 1\,kHz$ sine waves and room temperature is found to be $\sim 3.9\,V_{rms}$. Doping 5% and 10% of compound (I) lowers the threshold voltage to 1.95 and 1.5 V_{rms}, respectively.

Normally, increasing the dopant concentration would increase the mixture viscosity. However, in this case, the dopant and host compounds have comparable molecular sizes. As a result, the measured visco-elastic coefficient (γ_1/K_{11}) of the 10% doped mixture is only about 20% higher than that of the host.

Compound (I) has a red color. This implies that its absorption band extends to the visible spectral region. Figure 11.4 shows the wavelength-dependent absorption of a 1% compound (I) dissolved in a UV transparent ZLI-2359 host. A homogeneous cell with $d = 6\,\mu m$ was used for measurements. From Figure 11.4, the C5 nitro-amino tolane has an absorption peak located at $\lambda = 410\,nm$ and extends to the blue-green region. The dichroic ratio of this dye is around $8:1$ and is insufficient for the guest–host displays. On the other hand, to serve as a dielectric dopant, the inherent color is undesirable.

To remove color, the molecular conjugation needs to be reduced while preserving the push-pull effect. A simple way to achieve this goal is to replace the NO_2 polar group with a cyano. Indeed, its absorption peak is shifted to $\lambda = 370\,nm$, as shown in Figure 11.4. Beyond $\lambda = 420\,nm$, the cyano compound has little absorption.

Another interesting application of the tolane dye (Structure (I)) is in the holographic polymer-dispersed liquid crystal (HPDLC) devices. HPDLCs are formed with pre-polymers cured in the interference fringes of two coherent laser beams, e.g. argon-ion laser.

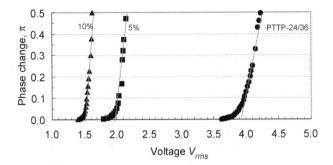

Figure 11.3 The voltage-dependent phase change of four homogeneous LC cells. Circles represent experimental data of the pure E63 cell, squares are a 5% compound (I) dissolved in E63, and triangles are a 10% (I) in E63. $\lambda = 633$ nm, $f = 1$ kHz sine waves; $T = 22°$C

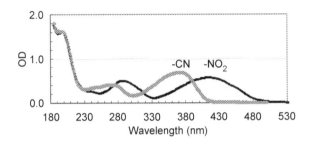

Figure 11.4 Optical density of a 1% nitro-amino and nitro-cyano tolane compound dissolved in a ZLI-2359 LC host. Homogeneous cell $d = 6$ µm, $T = 22°$C

Polymerisation occurs in the bright fringes, depleting the local monomer concentrations in these regions. Monomers diffuse from the dark fringes into the bright fringes where they are polymerised, and the liquid crystal is displaced into the dark fringe regions and forms droplets that are embedded in the polymer matrix. The droplet density is high in the dark fringes and low in the bright fringes. The alternating LC-rich and polymer-rich layers result in the switchable grating. Both transmissive [13] and reflective [14] gratings have been developed. Due to the tiny (about 100 nm) LC bubbles, the switching time is around 100 µs. However, also due to the small LC droplet sizes, the switching voltage is beyond 100 V$_{rms}$.

To lower the HPDLC operation voltage, doping with a high $\Delta\varepsilon$ tolane dye has been investigated [15]. Results show that by adding less than 1% of the 5NH-PTP-NO$_2$ dye, the threshold voltage of the HPDLC device is reduced by a factor of two. The increase in dielectric anisotropy alone is negligible and cannot account for the decreased voltage. Scanning electron microscope results reveal that a morphology change is the main mechanism for the observed voltage reduction.

Another type of high dielectric dopant, such as nitro-amino polyene compounds with the structure shown below, has been studied as well [16].

$$\text{(II)}$$

Two polyene homologues with $R_1=R_2=CH_3$ and C_6H_{13} have been synthesised. Their melting points are 239° and 88.5°C, respectively. The melting point is closely related to the side chain length. As the side chain length increases, the melting point drops due to increased molecular flexibility. The $\Delta\varepsilon$ of these polyene compounds is around 65.

There are several reasons for designing these linear polyene molecules: First, they exhibit a considerable charge transfer that is confined along the quasi one-dimensional π-conjugated bridge. Second, the incorporation of the polyene into a fused-ring bridge prevents isomerisation resulting in improved thermal and photochemical stability. Third, the ability to change alkyl groups allows one to control the melting temperature so as to tune the solubility to the LCs. Fourth, the two cyano and amino groups provide a huge push-pull effect for enhancing the dipole moment. Fifth, each cyano acceptor itself has a large dipole moment. Although the double cyano group enhances the dielectric anisotropy, it also substantially increases the viscosity.

3.3   Colorless high $\Delta\varepsilon$ compounds

For most display applications, a high $\Delta\varepsilon$ but colorless compound is preferred [17]. The following compound is colorless while possessing a relatively large $\Delta\varepsilon$ [18]:

$$\text{(III)}$$

The melting point (°C) and heat fusion enthalpy (ΔH, in kcal/mol) of the $n=1$ and 6 homologues are [84.0, 5.93] and [3.4, 5.32], respectively. Both homologues have a relatively low melting temperature and small ΔH that assure a good solubility with most LC hosts. The $n=6$ homologue, designated as III-6, is a clear liquid at room temperature. Its absorption spectrum is shown in Figure 11.5. From Figure 11.5, the absorption of compound III-6 occurs at $\lambda_{peak}\simeq 305$ nm and falls off quickly beyond 360 nm. In the visible spectral region, it is basically clear.

The dielectric anisotropy of compound III-6 is $\Delta\varepsilon\simeq 50$. Two factors contribute to the observed large dielectric anisotropy: the compound exhibits a considerable charge transfer that is confined along the quasi one-dimensional π-conjugated bridge [19], and the two cyano groups provide a large effective dipole moment.

Two Merck colorless and high $\Delta\varepsilon$ compounds are included here for comparison:

$$\text{(ZLI-3220)} \qquad \text{(IV)}$$

$$\text{(ZLI-1906)} \qquad \text{(V)}$$

The extrapolated $\Delta\varepsilon$ values of ZLI-3220 and ZLI-1906 are about 43 and 25, respectively. Due to their colorless characteristic, large dipole moment and relatively low viscosity, these

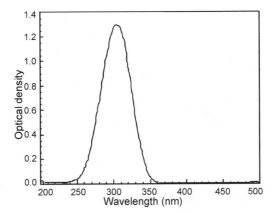

Figure 11.5 The absorption spectrum of compound III-6. During measurement, 1% of such a compound was dissolved in a LC host ZLI-2359. Cell gap = 6 μm

compounds are attractive for low-voltage TN and STN displays where a high resistivity is not required.

Other important applications of the high $\Delta\varepsilon$ compounds are to use as a dopant to reduce the operating voltage of some non-polar LCs or cholesteric display (see Chapter 8). For instance, dialkyl diphenyl-diacetylene (PTTP) LCs possess a high Δn, low melting temperature, wide nematic range, small heat fusion enthalpy and a low viscosity [20]. Thus, PTTP LCs are ideal host materials for IR and millimeter wave application where photostability is not a problem. However, the dielectric anisotropy of these non-polar PTTPs is too small ($\Delta\varepsilon \simeq 0.8$) resulting in a high threshold voltage. Doping 10% of the 5NH-PTP-NO$_2$ dye into a binary eutectic PTTP-24/36 mixture reduces the threshold voltage from 3.8 to 1.5 V$_{rms}$, and causes almost no change in viscosity.

3.4 Elastic constant effect

The elastic constant plays an equally important role in determining the threshold voltage as the dielectric anisotropy. The double bond is found to have a remarkable effect on the elastic constant [21]. Figure 11.6 shows the temperature-dependent splay elastic constant K_{11} of three fluorinated diphenyl-diacetylene liquid crystals containing a double bond at the first ($6d_1$), second ($6d_2$) and third ($6d_3$) positions. The parent compound PTTP-60F is also included here for comparison.

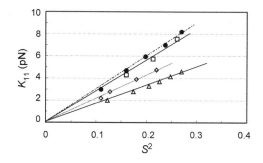

Figure 11.6 Double bond effect on K_{11} of PTTP-6OF homologues: PTTP-60F (●), $6d_1$ PTTP-60F (□), $6d_2$ PTTP-60F (△) and $6d_3$ PTTP-60F (◇). S is the order parameter

For each homologue, the splay elastic constant basically follows the mean-field theory (i.e. $K_{11} \simeq S^2$). In addition, the odd-even effect exists depending on the double bond position [22]. At a given S (i.e. the same T_r), K_{11} oscillates between the $6d_1$, $6d_2$, and $6d_3$ isomers. The physical origin of this odd-even effect is not completely understood. A computer model based on the interactive molecular graphics has been proposed [23]. This model is designed to determine the equilibrium structures and conformations of complex molecular ensembles based on the inter- and intra-molecular van der Waals interactions and conformational torsional potentials. From this model, the observed double bond effect on elastic constants could be attributed to a steric conformation of nematic ensembles.

From Figure 11.6, the K_{11} of the $6d_2$ homologue is nearly two times smaller than that of PTTP-60F at the same reduced temperature. A smaller K_{11} reduces the threshold voltage, nevertheless, it also lengthens the response time, which is proportional to γ_1/K_{11}; γ_1 is the rotational viscosity. Solely from the threshold voltage standpoint, the impact of a reduced elastic constant is equivalent to an enhanced dielectric anisotropy. However, to double $\Delta\varepsilon$ while preserving all the other desirable physical properties such as the viscosity and nematic phase of a LC compound is not an easy task. Substituting a carbon–carbon double bond in a proper position becomes a simple method to lower the threshold voltage.

4 Pre-tilt Angle Effect

A small pre-tilt angle is needed for the LC directors to relax in the same direction. The pre-tilt angle also affects the V_{th} of an LC device. Figure 11.7 shows the V–T curves of a 90° TN cell at three different pre-tilt angles, $\alpha = 2$, 5 and 10°. From Figure 11.7, a large pre-tilt angle tends to decrease the threshold of the V–T curve that, in turn, lowers the dark state voltage. The whole V–T curve shifts toward the lower voltage side as the pre-tilt angle increases. However, a large pre-tilt angle tends to slow down the response time because of weaker elastic restoring torque.

5 Twist Angle Effect

Figure 11.8 shows the V–T curves of three TN cells with twist angles $\theta = 110$, 90, and 70°. The pre-tilt angle of these cells is fixed at $\alpha = 3°$. To obtain the maximum $T_\perp$ for the green

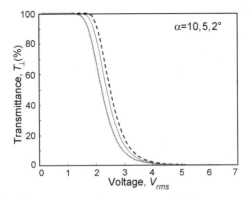

Figure 11.7 Pre-tilt angle effect on the VT curve of a 90° TN cell. From left to right, $\alpha = 10$, 5 and 2°. LC: MLC-9100-000

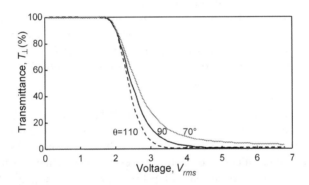

Figure 11.8 The voltage-dependent transmittance of three TN cells with twist angle $\theta = 110$ ($\beta = 10°$), 90 ($\beta = 0$), and 70° ($\beta = -4°$) and $d\Delta n = 550$, 480 and 420 nm, respectively, $\lambda = 550$ nm; $\alpha = 3°$.

wavelength, the $d\Delta n$ of each cell is adjusted to 550, 480 and 420 nm and the β angle set at $\beta = 10$, 0 and $-4°$, respectively. From Figure 11.8, the 90° TN cell exhibits the highest off-state transmission and its dark state occurs at about 4.5 V_{rms} and stays dark afterwards. For the 70° TN cell, its dark state transmission remains at about 5% at 5 V_{rms}. This is because the self-phase compensation effect gradually disappears as the twist angle deviates from 90°. By contrast, for the $\theta = 110°$ TN cell, its dark state voltage is reduced to about 3.5 V_{rms}. A small amount of light leakage is observed in the high voltage regime. Again, this results from imperfect self-phase compensation.

To accommodate the required larger $d\Delta n$ for the higher twisted cells one could select a higher Δn LC mixture so that the cell gap remains the same. Merck's multi-bottle mixtures are developed for such a purpose; they have a very similar $\Delta\varepsilon$, K_{ii} and viscosity, but different values of Δn.

6 Polarisation Angle Effect

Another self-phase compensation effect of a TN cell is realised at $\beta = \theta/2$; it is known as the bisector effect [24]. In these special circumstances, the boundary layers near the front and back substrates also compensate for each other. For example, if we set $\beta \simeq 45°$ for a 90° TN cell, the dark state would occur at a still lower voltage than that at $\beta \simeq 0°$. Care must be taken: although the bisector effect provides an effective way for reducing the operating voltage, its side effects on the light efficiency and color dispersion have to be taken into consideration as well.

6.1 Transmissive displays

Figure 11.9 depicts the $V\text{--}T$ curves of the 90° TN cell ($d\Delta n = 480$ nm) at $\beta = 45°$, instead of 0°. By comparing the results shown in Figures 11.1 and 11.9, we find that the dark state voltage is reduced by about 1 V. However, the color dispersion for the $\beta = 45°$ cell becomes more severe. This dispersion is more pronounced for the red and blue colors since the cell parameters are optimised at $\lambda = 550$ nm.

At $\beta = 0$, the polarisation rotation effect is dominant and the incoming linearly polarised light basically follows the molecular twist. The transmission of a TN cell is thus insensitive to λ. As β departs from 0°, the birefringence and polarisation rotation effects are both present. This is the mixed-mode twisted nematic (MTN) effect for transmissive cells [25]. The β angle determines the mixing ratio between these two effects. As β gets close to 45°, the birefringence effect becomes more apparent.

As the birefringence effect is quite sensitive to λ, the color dispersion for the MTN cell is larger than its corresponding TN cell at $\beta = 0°$. The larger the $d\Delta n$, the more pronounced the color dispersion. To suppress color dispersion, we could choose $0 < \beta < \theta/2$ and compromise for the slightly higher dark state voltage. The large dispersion shown in Figure 11.9 is not quite suitable for a direct-view display using a single LCD panel because its off-state is not a balanced white. For a projection display using three LCD panels, each panel can be filled with a different LC such that the $d\Delta n$ for each cell is optimised for each color. As a result, all three curves shown in Figure 11.9 should overlap and the color dispersion is eliminated.

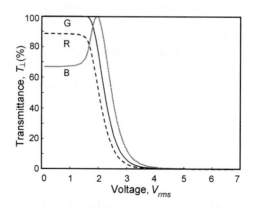

Figure 11.9   Voltage-dependent transmittance of a 90° TN cell. $d\Delta n = 480$ nm and $\beta = 45°$

6.2 Reflective displays

For transmissive displays, $T=1$ occurs for the 90° TN cell where the self-phase compensation effect is inherent. For reflective displays, although the 90° MTN cell [26] exhibits many outstanding features, e.g. weak dispersion, low operating voltage and fast response, its maximum reflectance $(R_\perp)$ drops to 88%. To boost $R_\perp$, the twist angle, β angle and $d\Delta n$ need to be optimised simultaneously, as discussed in Chapter 4. For instance, for the 70° MTN cell, $R_\perp \simeq 1$ occurs at $\beta = 0$–30°, depending on the detailed $d\Delta n$ value [27]. For the 60° MTN cell, $R_\perp = 1$ occurs at $\beta = 30°$ and $d\Delta n = 340$ nm. For these non-90° MTN cells, the β angle effect becomes important in order to achieve a low operating voltage. Please bear in mind that when $\beta \simeq \theta/2$, its dark state voltage is the lowest, but its color dispersion is the largest. A proper compromise between color dispersion and operating voltage is necessary.

75° MTN cell

The 75° MTN has been used for Sharp reflective direct-view displays, such as the color gameboy for handheld entertainment and other personal digital assistance [30]. From the lowest operating voltage standpoint, the β angle should be set at $\beta = 37.5°$. However, the normalised reflectance drops to 62%. For reflective direct-view displays, owing to the glare the display contrast is often limited to $\leq 50 : 1$. Thus, optimising the brightness is more important than the contrast ratio. Under such a design philosophy, how to obtain $R_\perp = 1$ is the primary concern.

Figure 11.10 plots the voltage-dependent reflectance of the 75° MTN with $d\Delta n = 270$ nm and $\beta = 15°$. The normalised reflectance of the green band indeed reaches 100%. The color dispersion is relatively weak. At 5 V, the contrast ratio exceeds 30 : 1.

45° TN cell

The 45° TN cell [31, 32] is frequently used for reflective projection displays because of its normally black mode. Figure 11.11 shows the voltage-dependent reflectance of the 45° TN cell at $\lambda = 650$ nm and $\beta = -10°$. Two $d\Delta n$ values are used for comparison: 450 nm (solid

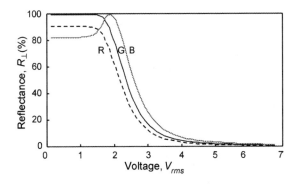

Figure 11.10 Voltage-dependent reflectance of a 75° MTN cell. $d\Delta n = 270$ nm and $\beta = 15°$

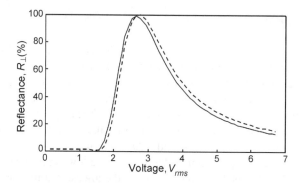

Figure 11.11 Voltage-dependent reflectance of a 45° TN cell at $\lambda=650\,\text{nm}$ and $\beta=-10°$ solid lines: $d\Delta n=450\,\text{nm}$; dashed lines: $d\Delta n=475\,\text{nm}$

lines) and 475 nm (dashed lines). The minimal $d\Delta n$ requirement for not seeing light leakage at this wavelength is 450 nm.

From Figure 11.11, the 45° TN cell exhibits two interesting features: it has a large $d\Delta n$, i.e. cell gap tolerance; and its on-state voltage is below $3\,\text{V}_{rms}$ and the voltage swing is merely 1.2 V. Thus, this mode is particularly attractive for silicon-wafer LCD applications where a low operating voltage is essential. However, this cell has two shortcomings. First, its dark state is dependent on the temperature and wavelength so that each light valve needs to be biased at a different voltage for each RGB color. As the temperature fluctuates, the dark state voltage could float and degrade the contrast ratio. Second, its viewing angle is narrow due to the large $d\Delta n$ value. Therefore, the 45° reflective TN cell is more suitable for projection than direct-view display.

6.3   Virtual displays

Most virtual displays employing liquid crystal on silicon use a light-emitting diode (LED) for illumination. LEDs have a low power consumption and high brightness. In virtual dis-

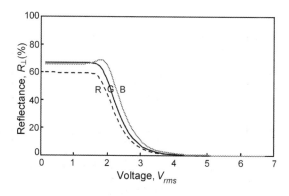

Figure 11.12 Voltage-dependent 90° MTN cell at $\beta=0°$. The $d\Delta n$ of the cell is 165 nm. R = 650, G = 550 and B = 450 nm

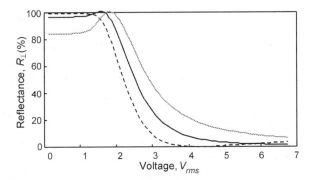

Figure 11.13 Phase compensation effect on a homogeneous cell. Cell $d\Delta n = 175$ nm (gray line), with -20 nm (solid line) and -40 nm (dashed lines) compensation films

plays, a fast response time for obtaining sequential colors and a low operating voltage for lengthening the battery lifetime are more critical than light efficiency. Under these circumstances, the 90° MTN at $\beta=0°$ can be considered.

Figure 11.12 shows the voltage-dependent reflectance of a 90° MTN cell with $\beta=0°$ and $d\Delta n = 165$ nm. The reflectivity for the specified RGB colors is between 60–68%. Thus, the color dispersion is weak. With merely 2.5 V voltage swing, a contrast ratio of about 100:1 can be achieved due to the built-in phase compensation of the 90° TN cell. The small $d\Delta n$ requirement enables a thin cell to be used for obtaining sequential colors. Sequential colors using a single light valve would not only save cost but also eliminate the pixel registration problem.

7 Compensation Films

In addition to the aforementioned self-phase compensation effects, compensation films provide another avenue to obtain a low operating voltage. For instance, the film-compensated thin homogeneous cell is demonstrated to exhibit a very wide viewing angle and fast response time for reflective displays [33, 34]. It is a potential approach for color sequential projection displays using silicon-wafer LCDs.

Figure 11.13 depicts the voltage-dependent reflectance of a phase-compensated homogeneous cell with $d_2\Delta n_2 = 175$ nm and $d_2\Delta n_2 = 0,-20$ and -40 nm, at $\lambda=550$ nm. For the uncompensated cell, the dark state voltage is much greater than 7 V. This is because the residual phase retardation from the boundary layers is additive, instead of subtractive. Adding a phase-matched uniaxial polycarbonate (PC) film with $d_2\Delta n_2$ to -20 nm, a good dark state occurs at $V=6$ V$_{rms}$. Increasing the compensation film $d_2\Delta n_2$ to -40 nm, the dark state voltage drops to 4 V$_{rms}$.

8 Homeotropic Cells

In a homeotropic cell [35], the LC directors are aligned nearly perpendicularly to the substrate surfaces. Thus, it exhibits an excellent dark state in the voltage-off state when sandwiched between two crossed polarisers. The operating voltage is thus determined by the bright state voltage.

Two methods have been commonly used to lower the on-state voltage: to reduce the threshold voltage and to increase the $d\Delta n$ value. To reduce V_{th}, one would need to find LC mixtures with large $\Delta \varepsilon$. For a homeotropic cell, the LC employed has a negative $\Delta \varepsilon$. A large but negative $\Delta \varepsilon$ means that the polar groups are in the lateral positions. The most commonly employed group is fluoro. Due to its modest dipole moment, the dielectric anisotropy of a difluoro compound is around -6.

Figure 11.14 depicts the $\Delta \varepsilon$ effect on the on-state voltage of a reflective homeotropic cell with $d\Delta n = 200$ nm at $\lambda = 550$ nm and $\beta = 45°$. As the dielectric anisotropy is increased from -2, -4 to -6, the on-state voltage is reduced from 6.6, 4.7 to 3.8 V_{rms}, respectively. For the case that $\Delta \varepsilon = -6$, the required voltage swing is less than 2 V. Thus, the homeotropic cell is also attractive for silicon-wafer-based reflective displays [36].

The $d\Delta n$ effect provides a simple way to lower the operating voltage. Figure 11.15 shows the V–R plots of three homeotropic cells with different $d\Delta n$ values: 175, 200 and 225 nm

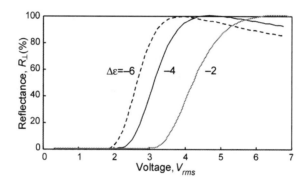

Figure 11.14 The $\Delta \varepsilon$ effect on the on-state voltage of a homeotropic LC cell. $d\Delta n = 200$ nm and $\beta = 45°$

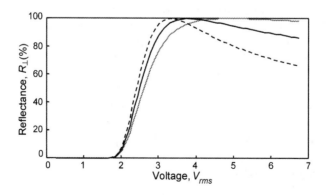

Figure 11.15 The voltage-dependent reflectance $R_{\perp}$ of three homeotropic cells. From right to left, $d\Delta n = 175$, 200 and 225 nm. LC mixture: $\Delta \varepsilon = -6$

(from right to left) at $\lambda = 550\,nm$. From Figure 11.15, the bright state shifts gradually toward the lower voltage side as $d\Delta n$ increases. In comparison to the self-phase compensated 90° TN cell and film-compensated homogeneous cell shown in Figures 11.1 and 11.13, the homeotropic cell shows the lowest operating voltage, highest contrast ratio, smallest $d\Delta n$ requirement and largest cell gap tolerance. Moreover, the K_{33} of many LCs is about two times larger than K_{11} and K_{22}. A larger K_{33}, although unfavorable from the threshold voltage viewpoint, would lead to a faster response time. Thus, the homeotropic cell will be favorable for both direct-view and projection displays if more negative LC materials become available.

References

1   M. Schadt and W. Helfrich, Voltage-dependent optical activity of a twisted nematic liquid crystal, *Appl. Phys. Lett.* **18**, 127 (1971).

2   C. H. Gooch and H. A. Tarry, The optical properties of twisted nematic liquid crystal structures with twist angles $\leq 90°$, *J. Phys. D.* **8**, 1575 (1975).

3   E. P. Raynes, The theory of super-twist transitions, *Mol. Cryst. Liq. Cryst. Lett.* **4**, 1 (1986).

4   H. Gruler, The elastic constants of a nematic liquid crystal, *Z. Naturforsch. Teil A* **30**, 230 (1975).

5   S. T. Wu, Q. T. Zhang, and S. Marder, High dielectric dopants for low voltage liquid crystal operation, *Jpn. J. Appl. Phys.* **37**, L1254 (1998)

6   W. Maier and G. Meier, A simple theory of the dielectric characteristics of homegeneous oriented crystalline-liquid phases of the nematic type, *Z. Naturforsch. Teil A* **16**, 262 (1961).

7   G. W. Gray, K.J. Harrison, and J.A. Nash, New family of nematic liquid crystals for displays, *Electron. Lett.* **9**, 130 (1973).

8   R. Tarao, H. Saito, S. Sawada, and Y. Goto, Advances in liquid crystals for TFT displays, *SID Tech. Digest* **25**, 233 (1994).

9   D. Klement and K. Tarumi, Liquid crystal material development of IPS TFT displays, *SID Tech. Digest* **29**, 393 (1998).

10 S. T. Wu, J. D. Margerum, M. S. Ho, and B. M. Fung, Liquid crystal dyes with high solubility and large dielectric anisotropy, *Appl. Phys. Lett.* **64**, 2047 (1994).

11 I. C. Khoo and S. T. Wu, Optics and Nonlinear Optics of Liquid Crystals, World Scientific Press, Singapore (1993).

12 H. J. Deuling, Elasticity of nematic liquid crystals, *Solid State Phys.* Suppl. 14: Liquid Crystals, edited by L. Libert, Academic Press, New York (1978).

13 R. Sutherland, L.V. Natarajan, V. P. Tondiglia, and T.J. Bunning, *Chem. Mater.* **5**, 1533 (1993).

14 K. Tanaka, K. Kato, S. Tsuru, and S. Sakai, Holographically formed liquid crystal/polymer device for reflective color display, *J. SID* **2**, 37 (1994).

15 J. Colegrove, H. Yuan, S. T. Wu, J. R. Kelly, C. Bowley, and G. P. Crawford, Drive-voltage reduction for HPDLC displays, *The 6th Int'l Display Workshops*, p.105 (1999).

16 S. T. Wu, Q. T. Zhang, and S. Marder, High dielectric dopants for low voltage liquid crystal operation, *Jpn. J. Appl. Phys.* **37**, L1254 (1998).

17 S. T. Wu, Q. T. Zhang, and S. Marder, Material Research Society Symposium Proceedings, "Liquid Crystal Materials and Devices" **559**, 235 (1999).

18 S. T. Wu, Q. T. Zhang, and S. Marder, U. S. patent pending (1999).

19 B. Kippelen, S. R. Marder, E. Hendrickx, J. L. Maldonado, G. Guillemet, B. L. Volodin, D. D. Steele, Y. Enami, Sandalphon, Y. J. Yao, J. F. Wang, H. Rockel, L. Erskine, and N. Peyghambarian, *Science* **279**, 54 (1998).

20 S. T. Wu, J. D. Margerum, B. Meng, L. R. Dalton, C. S. Hsu, and S. H. Lung, Room temperature asymmetric diphenyl-diacetylene liquid crystals, *Appl. Phys. Lett.* **61**, 630 (1992).

21 M. Schadt, M. Petrzilka, P. E. Gerber, and A. Villiger, Polar alkenyls: Physical properties and correlations with molecular structure of new nematic liquid crystals, *Mol. Cryst. Liq. Cryst.* **122**, 241 (1985).

22 S. T. Wu, C. S. Hsu, Y. N. Chen, and S. R. Wang, Fluorinated diphenyl-diacetylene and tolane liquid crystals with low threshold voltage, *Appl. Phys. Lett.* **61**, 2275 (1992).

23 M. Schadt, R. Buchecker, and K. Muller, Material properties, structural relations with molecular ensembles and electro-optical performance of new bicyclohexane liquid crystals in field effect LCDs, *Liq. Cryst.* **5**, 293 (1989).

24 S. T. Wu and C. S. Wu, Bisector effect on the twisted-nematic cells, *Jpn. J. Appl. Phys.* **37**, L1497 (1998).

25 S. T. Wu and C. S. Wu, Mixed-mode twisted nematic cell for transmissive liquid crystal display, *Displays* **20**, 231 (1999).

26 S. T. Wu and C. S. Wu, Mixed-mode twisted nematic cells for reflective liquid crystal displays, *Appl. Phys. Lett.* **68**, 1455 (1996).

27 C. L. Kuo, C. K. Wei, S. T. Wu, and C. S. Wu, Reflective display using mixed-mode twisted nematic liquid crystal cell, *Jpn. J. Appl. Phys.* **36**, 1077 (1997).

28 H. A. Van Sprang, Reflective liquid crystal display device with twist angle between 50 and 68° and the polarizer at the bisectrix, U.S. Patent 5,490,003 (1996).

29 K. H. Yang, A self-compensated twisted-nematic liquid crystal mode for reflective light valves, *Euro Display'96*, p. 449 (1996).

30 M. D. Tillin, M. J. Towler, K. A. Saynor, and E. J. Beynon, Reflective single-polarizer low and high twist LCDs, *SID Tech. Digest* **29**, 311 (1998).

31 I. A. Shanks, Electro-optical color effects by twisted nematic liquid crystal, *Electron. Lett.* **10**, 90 (1974).

32 J. Grinberg, A. Jacobson, W.P. Bleha, L. Miller, L. Fraas, D. Bosewell, and G. Meyer, A new real-time noncoherent to coherent light image converter: the hybrid field effect liquid crystal light valve, *Opt. Eng.* **14**, 217 (1975).

33 K. Lu and B. E. A. Saleh, A single polarizer reflective LCD with wide viewing angle range for gray scales, *SID Tech. Digest for Application Papers* **27**, 63 (1996).

34 S. T. Wu and C. S. Wu, A biaxial film-compensated thin homogeneous cell for reflective liquid crystal displays, J. Appl. Phys. **83**, 4096 (1998). *Liq. Cryst.* **24**, 811 (1998).

35 M. F. Schiekel and K. Fahrenschon, Deformation of nematic liquid crystals with vertical orientation in electric field, *Appl. Phys. Lett.* **19**, 391 (1971).

36 R. D. Sterling and W. P. Bleha, D-ILA^{TM} technology for electronic cinema, *SID Tech. Digest* **31**, 310 (2000).

12

Wide Viewing Angle

1 Introduction

The viewing angle is a critical problem for large panel liquid crystal displays [1]. For a handheld display, its panel size is usually no bigger than 10 cm and the viewing direction can be adjusted freely. Moreover, benefiting from the mirror-image effect discussed in Chapter 1, its viewing angle is equivalent to a two-domain transmissive display. Thus, the viewing angle of a palm-sized reflective display is not too big a concern. However, if a reflective display is going to compete with a transmissive one in the large-panel notebook and desktop computer markets, then its viewing angle is apparently inadequate.

Several technologies have been developed to widen the viewing angle of transmissive LCDs. We can roughly categorise them as the external, internal and combination methods. In the external method, the LCD structure is a conventional twisted nematic cell and a phase compensation film or diffusive optics is laminated outside the LCD panel. In the internal method, the LCD structure is modified internally. For example, in the in-plane switching method, the electrodes are arranged in the same substrate such that the longitudinal, rather than transversal, electric field is used to drive the LC molecules. The combination method integrates internal structure change and also adds compensation films.

The external approach has the advantage of the simple manufacturing process. It puts a burden on the compensation film. The disadvantage is that its viewing angle is limited to about $\pm 60°$. This is adequate for notebook applications. On the other hand, the internal method involves more complicated fabrication processes; nevertheless, its payoff is in the wider viewing angle. The state-of-the-art approaches involve both new electrode designs while adding at least one compensation film. The impressive results are a viewing angle wider than $\pm 80°$, little color shift, a contrast ratio higher than $300:1$, and a response time less than 30 ms. This approach is suitable for desktop monitors and LCD TVs where a wide viewing angle is critical.

In this chapter we first introduce two external approaches: compensation film and SpectraVue™ methods, and then advance to the transverse field and multiple domain methods.

2 Film-compensated TN Cells

Twisted nematic cells exhibit some attractive features, such as a low operating voltage, high contrast ratio and weak color dispersion, and have been used extensively for displays. However, the viewing angle of a TN cell is insufficient for large panel displays. Figure 12.1 plots the computer-simulated iso-contrast ratio of a 90° TN cell with $d\Delta n = 480$ nm. In the horizontal direction, the viewing angle is more or less symmetric. For the CR = 10 contour, the horizontal viewing angle is about $\pm 50°$. However, in the vertical direction, the viewing characteristic is asymmetric and limited to 15° and 40°, for the up-down and bottom-up directions, respectively. Beyond this region, reversed contrast would occur. For the personal use of small-sized displays, such a viewing angle is sufficient. As the display size increases, and more people are viewing the same display, a wider viewing angle is necessary.

To improve the viewing angle, the film-compensated TN cell has been studied extensively. This is a simple process that does not require modification of the well-defined process flow of the TN-LCD. A film can be laminated onto the LCD modules as the last step. From the manufacturing standpoint, this is the most cost-effective way to improve the viewing angle of TN LCDs.

A major challenge for the film-compensated TN cell is in the need for an asymmetric phase-compensation film. The asymmetric viewing characteristic of a TN cell is the direct consequence of the rod-like, optically anisotropic LC molecules. Figure 12.2 illustrates the director orientations of a LC cell in a voltage-on state. The refractive index ellipsoid is different when viewed from the right-hand side rather than from the left-hand side. As a result, the phase retardation that governs the light leakage and color shift depends on the viewing angle. Thus, an asymmetric phase compensation film is needed.

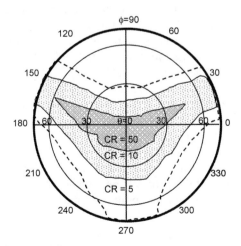

Figure 12.1   Iso-contrast contour of a 90° TN cell with $d\Delta n = 480$ nm

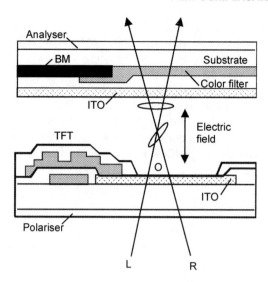

Figure 12.2 Structure of a TFT-TN cell. The electric field is in the longitudinal direction. The tilted middle LC directors cause different phase retardation as viewed from right or left oblique angles. BM = black matrix

2.1 Principles of phase compensation

When a phase-compensation film is used in conjunction with a LC cell, its total phase retardation is expressed as

$$\delta = 2\pi(d_1\Delta n_1 - d_2\Delta n_2)/\lambda \tag{12.1}$$

Here, $d_1\Delta n_1$ represents the phase retardation of the LC cell and $d_2\Delta n_2$ is for the film. As the birefringence of the LC is dependent on the angle, wavelength and temperature, ideally a phase-compensation film should have a similar angular, wavelength and temperature-dependent phase retardation as the LC material employed. To achieve these ultimate goals, several types of film, namely A, C, O and biaxial plates [2], have been developed to satisfy as many of the above-mentioned stringent requirements as possible. Some films are aimed at reducing the operating voltage as described in Chapter 3, and some are targeted to widen the viewing angle.

Compensation films

If the z axis represents the film thickness direction, then an A-plate is a polymer film which is uniformly stretched in one axis, say the x axis. Both positive ($n_x > n_y = n_z$) and negative ($n_x < n_y = n_z$) birefringence A-plates can be made [3]. If the film is stretched in the x and z axes such that $n_x > n_z > n_y$, or in the x and y axes such that $n_x > n_y > n_z$, then a biaxial film is formed. A biaxial film is useful to widen the viewing angle and reduce the operating voltage simultaneously [3]. By contrast, a C-plate is isotropic in the x–y plane and could also have a negative ($n_x = n_y > n_z$) or positive ($n_x = n_y < n_z$) birefringence along the z axis. Two methods for the preparation of a negative birefringence C-plate have been reported: spin-coating of a polymer solution [4, 5] and alternating depositions of thin SiO_2 and TiO_2 layers [6]. An O-plate has its molecular axis uniformly tilted at an oblique angle [7]. A twisted film [8, 9] has a similar but reversed structure to a twisted cell, and in principle is an ideal candidate for compensating twisted nematic LCDs. Discotic film [10] gives rise to a natural splay structure and has been widely used for wide viewing angle technology in transmissive displays. Hybrid aligned rod-like liquid crystal polymer is another series that offers asymmetric phase retardation [11, 12]. For many normally white reflective displays, a broadband quarter-wave film is used in conjunction with a polariser. For super-twisted nematic (STN) displays, a more complicated biaxial film or multiple retardation films are required [13, 14].

Theoretical analysis

The viewing angle is tied to the image contrast ratio and the contrast ratio is determined by the light leakage of the dark state. In a normally white display, the dark state occurs when $V \gg V_{th}$ in which the LC directors are aligned nearly perpendicularly to the substrates. The voltage-on state of a homogeneous cell is equivalent to a homeotropic cell, except for the boundary layers. Figure 12.3 shows a simplified structure of a homeotropic LC cell (or more generally, a birefringence film) under a bias voltage $V > V_{th}$. The tilt angle ϕ is assumed to be uniform throughout the cell. For a homeotropic cell at $V = 0$, the directors tilt angle is $\phi \simeq 0$. The light with wavelength λ is incident on the air–LC cell interface at an angle θ_i. For simplicity, let us assume that the glass, transparent electrode (indium-tin-oxide) and LC molecules all have the same refractive index n so that the light travels inside the medium at an angle θ which satisfies Snell's law:

$$n_{air} \sin \theta_i = n \sin \theta \qquad (12.2)$$

Here, the average refractive index n is equal to $(n_e + 2n_o)/3$. As observed from Equation 12.2, the light refraction effect has already led to a big improvement in the viewing angle for an LCD. For example, let us assume $n = 1.50$ for a glass and LC mixture. A 45° incident angle in air corresponds to only $\theta = 28°$ in the LC medium.

At an incident angle θ_i, the incoming linearly polarised light experiences a phase retardation after traversing the LC layer. The angular dependent phase retardation of a birefringence film is expressed as [15]

$$\delta(\theta, \lambda) = 2\pi (d\Delta n)_{eff}/\lambda \qquad (12.3)$$

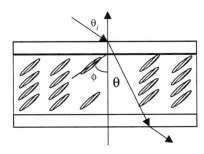

Figure 12.3 Light path of a uniformly tilted cell at an oblique angle

where

$$(d\Delta n)_{eff} = \frac{d}{\cos\theta}\left[\frac{n_e n_0}{[n_e^2\cos^2(\theta\pm\phi) + n_0^2\sin^2(\theta\pm\phi)]^{1/2}} - n_0\right] \qquad (12.4)$$

The first factor $1/\cos\theta$ in Equation 12.4 originates from the increased optical path length as shown in Figure 12.3, where d is the LC layer thickness at $\theta=0$. The $\pm$ signs in the second term represent the light coming from the left and right sides, respectively. It is clear that the phase retardation is quite asymmetric. Therefore, a compensation film with an asymmetric angular-dependent phase retardation is required.

Owing to the complicated angular dependence its physical meaning is not clear at first glance. To reveal the physical significance of Equation 12.4, let us look into the small angle region. Let us take a homeotropic cell with $\phi\simeq 0$ as an example. When θ is small the cosine and sine terms can be expanded as $\cos\theta\simeq 1-\theta^2/2$ and $\sin\theta\simeq\theta$. Retaining up to the θ^2 term, Equation 12.4 is simplified to

$$(d\Delta n)_{eff} \simeq d\Delta n\left[1 - \frac{3}{2}\frac{\Delta n}{n_e}\right]\theta^2 \qquad (12.5)$$

Equation 12.5 gives a clear account on the angular dependent $(d\Delta n)_{eff}$ of a homeotropic cell at $V=0$ or a TN cell at a high-voltage state. In the small angle region, the phase retardation is proportional to the product of $d\Delta n[1-3\Delta n/2n_e]$ and θ^2. Thus, the LC thickness, magnitude and sign of Δn and n_e all play important roles in determining the phase retardation at oblique angles. The major LC materials developed for display applications all exhibit a positive birefringence ($\Delta n>0$). Therefore, in order to compensate for the residual phase retardation of an LC cell at oblique angles, a compensation film with a negative Δn is needed. Ideally, this phase cancellation should take place at all angles and temperatures for all the wavelengths employed in a full-color display device.

2.2 Discotic compensation films

A discotic film exhibits a unique splayed structure, asymmetric angular-dependent phase retardation and negative birefringence, and it is particularly suitable for compensating the viewing angle of a TN cell [10].

Molecular structure

The molecular structure of the discotic compound commonly used is shown below:

The triphenylene-based disk-shaped compound was first found to exhibit a nematic phase by a French group [12] and was later devised by a Fuji group [13, 16] as a phase compensation film to improve the viewing angle of a LCD. The thermal behavior of this discotic compound is described below. In the heating scan it exhibits a columnar to discotic nematic transition at 50.6°C and a discotic nematic to isotropic transition at 164.2°C. While in the cooling scan it reveals an isotropic to discotic nematic at 162.3°C and a discotic nematic to columnar phase transition at 8.7°C.

Angular dependence

To prepare a discotic film, a substrate with a thin rubbed polyimide (PI) was first prepared. Then, discotic monomers were spin-coated or rolled over the PI surface. At the air–film interface, the discotic layer is splayed naturally at an angle ($\phi \simeq 40$–$68°$). The molecular tilt angle continuously evolves within the film. These disc orientations can be fixed by the UV polymerisation process. Since the disc orientation is similar to the boundary layer of a TN cell in its voltage-on state, it provides a natural phase compensation effect. Unlike a spin-coated negative birefringence film that has symmetric angular phase retardation, the angular phase retardation of the discotic film is asymmetric, as depicted in Figure 12.4. Thus, a discotic film is particularly attractive for display applications. Placing two such films on each side of a transmissive TN cell, the viewing cone is expanded to about $\pm 60°$ [17].

Wavelength dependence

Phase matching is another important parameter for a compensation film. If the $d\Delta n$ value of the film only matches with an LC cell at one wavelength, say green, then the light leakage for the red and blue will degrade the display contrast ratio. Therefore, the wavelength-dependent birefringence [15] of a phase retardation film should be similar to that of the LC mixture employed. In such circumstances, good phase compensation can be simultaneously satisfied for all the three primary colors and a high device contrast ratio can be obtained.

Based on the birefringence dispersion model described in Equation 2.1, in order to ensure a good phase matching, the film selected should have a similar molecular structure to the LC compounds employed. Similar electron conjugation leads to a similar λ^*. Therefore, the wavelength-dependent birefringence of the film will be similar to that of the LC employed.

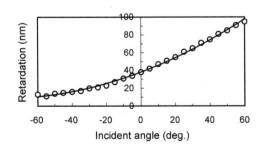

Figure 12.4 Light path of a uniformly tilted cell at an oblique angle

The Δn value of a compensation film is usually much smaller than that of a LC. Nevertheless, the total phase retardation value $d\Delta n$ can be adjusted by the film thickness.

To characterise the phase matching behavior, the measured birefringence of the discotic film is normalised to that of $\lambda = 550$ nm and compare their trend. The results are plotted in Figure 12.5. Also included in Figure 12.5 are the ratios $\Delta n/(\Delta n)_{\lambda=550\,nm}$ of two LC mixtures: ZLI-4792 (TFT mixture) and ZLI-5600-100 (STN mixture). The birefringences of ZLI-4792 and ZLI-5600-100 are 0.097 and 0.15, respectively. The mixture ZLI-4792 consists mainly of a fluorinated cyclohexane-phenyl ring compound, and ZLI-5600-100 is a biphenyl mixture. On the other hand, the discotic compound used in this study has a molecular conjugation larger than that of ZLI-4792 and similar to that of ZLI-5600-100. Thus, its wavelength-dependent birefringence is somewhat stronger than that of ZLI-4792, but very similar to that of ZLI-5610-100.

Temperature dependence

As described in Equation 2.1, the birefringence of a LC mixture decreases as the temperature increases. A decisive factor is the order parameter that is mainly determined by the clearing point of the LC mixture. For a stretched polymer phase compensation film, such as poly-carbonate, its birefringence is quite insensitive to temperature. As a result, a good phase match can only be preserved within a limited temperature range.

To illustrate the essence of such temperature matching, a phase compensation film made of twisted liquid crystal polymer (LCP) showing similar temperature dependence as LC mixtures is used as an example. Figure 12.6 shows the contrast ratio of a 240° STN cell with a temperature-matched and $-240°$ twisted LCP film and a non-temperature matched phase-compensation film. The clearing temperature of the LCP is 110°C. With a mismatched film, the contrast ratio drops rapidly as the temperature increases. On the other hand, the temperature-matched LCP film preserves a good contrast ratio up to 90°C.

Although the experimental results of the temperature-dependent birefringence of a dis-cotic film are not found in the literature, based on its phase transition temperature, we can predict that its temperature-dependency should be similar to that of the LC employed. The discotic film-compensated LCD should preserve a high contrast ratio over a relatively wide temperature range.

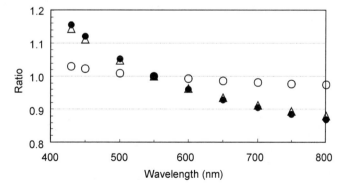

Figure 12.5 The birefringence dispersion of a discotic film (filled circles) and two LC mixtures, ZLI-4792 (open circles) and ZLI-5400-100 (triangles). The ratio here represents $\Delta n(\lambda)/\Delta n(\lambda = 550\,nm)$

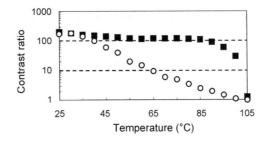

Figure 12.6 Temperature-dependent contrast ratio of a 240° STN display compensated with a temperature-matched twisted film (squares) and a non-temperature matched twisted film (circles)

2.3 Twisted compensation films

The purpose of using phase compensation is to minimise the light leakage of a LCD panel over a wide range of viewing angle. If the display employs a homogeneous or homeotropic LC cell, then a simple uniaxial film with equal but opposite sign of $d\Delta n$ should provide sufficient phase compensation. For TN [18] and STN [19] cells, their LC directors are twisted from the front to the back surface. To achieve a good compensation at the voltage-off state, the film should possess a similar twisted structure as the LC cell [20, 21]. If the compensation is designed to take place at a voltage-on state where the LC directors are reoriented nearly perpendicularly to the substrates, the compensation film should have a similar profile as the LC directors near the boundaries.

Figures 12.7(a) and (b) illustrate the molecular arrangement for compensating the phase retardation of normally black and normally white TN LC cells, respectively. In Figure 12.7(a) the LC directors twist 90° from the first to the Nth layer. By contrast, the compensation film has a reversed twist compared with the LC cell. The molecular axis in the N'th layer is perpendicular to that of the Nth layer of the LC cell, and so on. The ordinary ray in the N'th layer becomes the extraordinary ray in the Nth layer, and vice versa. The phase retardation incurred in the N'th layer cancels with that of the Nth LC layer. As a result, a dark voltage-off state is expected over a wide range of incident angle. A similar explanation holds for the normally white mode as shown in Figure 12.7(b). In a high-voltage state the LC directors are reoriented by the electric field and the TN cell behaves like a homeotropic cell except for the boundary layers.

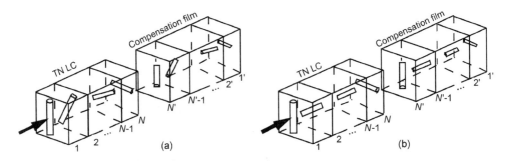

Figure 12.7 Twisted films for compensating (a) normally white and (b) normally black mode TN cells

Twisted discotic films have been theoretically analysed to have good compensation properties for a TN cell [22]. This enables a single- or two-domain TN cell to have a higher contrast ratio, wider viewing angle, lower operating voltage, faster response time and lower color shift. The fabrication processes of twisted discotic film are being developed [23]. They involve a two-component mixture of discotic and chiral discotic compounds. For a given film thickness, the twist angle of the film is determined by the concentration and twisting power of the chiral dopant. Discotic films with thickness up to 8 μm and twist angles between 45 and 90° have been fabricated. However, how to keep good uniformity over a large area is a challenging task.

Twisted phase-compensation films made from UV curable LC polymers are not only scientifically interesting but also technologically important. There are two major advantages of such twisted films: they exhibit the same twist structure as the LC layer, and the core molecule of the LC polymer can be the same as the LC compounds employed to assure the same Δn and temperature dispersion.

3 SpectraVue^{TM}

As analysed in Section 2 above, the obliquely incident light causes leakage through crossed polarisers and degrades the contrast ratio. This is the origin of the limited viewing angle. Usually, the backlight used in the LCD panel has a near-Lambertian distribution. If the incident light from the backlight can be collimated, then the TN cell will preserve a high contrast ratio. If the light can be diffused to the observer on the exit side, then a wide viewing angle, high contrast and weak color dispersion can be achieved simultaneously.

Allied Signals came up with this elegant idea and demonstrated a wide view TN LCD panel in 1995 [24]. The device configuration is plotted in Figure 12.8. A micro-prism and lens array is used to collimate the backlight to enter the LCD panel at a nearly normal (within ± 15°) angle. On the output side, a diffusive screen is used to diffuse the image to the viewer. As a result, avionics grade TFT LCD panels with a contrast higher than 100 : 1 at ± 60° viewing cone have been demonstrated [25]. These displays exhibit essentially no color shift with angle, and gray scale linearity is maintained.

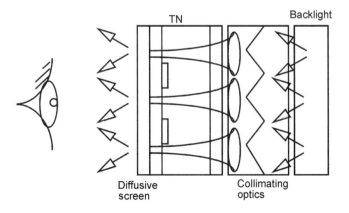

Figure 12.8 Device configuration of a SpectraVue^{TM} for a wide viewing angle

Two major drawbacks prevented SpectraVue™ becoming the dominant wide view approach: light throughput and cost. After adding the collimating lenses and diffusive screen, the light efficiency is reduced by nearly 50%. In addition, as the panel diagonal keeps on lengthening, the cost and weight of the complementary optics may soon become unaffordable.

4 Transverse Field Effect

In the TN cell shown in Figure 12.2, the applied electric field is in the longitudinal direction. The tilted LC directors in the bulk cause a different phase retardation viewed from either the right or from the left direction. This is the origin of the narrow and asymmetric viewing angle in the vertical directions.

Soref proposed an elegant driving scheme using a transverse electric field [26, 27]. The inter-digital electrodes are arranged in the same substrate such that the generated fringing field is in the transverse plane. The LC directors are rotated in the plane. Thus, this driving scheme is often referred to as the transverse field effect or in-plane switching (IPS) [28]. Because the LC molecules are rotated in-plane, its viewing angle is wider and more symmetric than that of the TN cell.

4.1 In-plane switching

As a transverse field is used to switch the LC directors in the IPS mode, both homogeneous and homeotropic alignments can be used [29]. In the homeotropic cell, the LC employed can have either a positive or negative $\Delta\varepsilon$, although the negative $\Delta\varepsilon$ LC is found to have more uniform transitions near the edges of electrodes.

Device configuration

Figure 12.9 depicts an early version of the Hitachi IPS cell using a negative $\Delta\varepsilon$ LC in homogeneous alignment [30, 31]. The front polariser is parallel to the LC directors and the rear analyser is crossed. In the voltage-off state, the incident light experiences no phase retardation so that the outgoing beam remains linearly polarised and is absorbed by the crossed analyser. In a voltage-on state, the fringing field reorients the LC directors, causes phase retardation to the incoming light and modulates its transmittance through the analyser.

The Hyundai [32] and Samsung [33] groups independently proposed another IPS mode using a positive $\Delta\varepsilon$ LC material in vertical alignment. The device configuration is shown in Figure 12.10. In the voltage-off state, the LC molecules are anchored homeotropically on both alignment layers, resulting in an excellent dark state between crossed polarisers. To compensate for the light leakage at oblique angles, a negative birefringence phase-compensation film, as described in Section 1 above, needs to be used. In the voltage-on state, the fringing field patterns are shown in Figure 12.10. Near the edges of the electrodes, the fields are in opposite directions. Thus, the LC alignment is equivalent to a two-domain vertical aligned cell. In the middle layer between the electrodes, the electric fields are symmetric. As a result, the LC molecules are not reoriented, leading to a dark line.

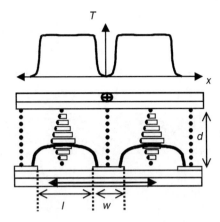

Figure 12.9 An IPS mode using a fringing field on a homogeneously aligned cell

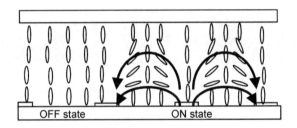

Figure 12.10 An IPS mode using a vertically aligned positive $\Delta\varepsilon$ LC cell

The Hyundai group later extended their pixel design from two to four domains [34]. In a pixel the patterned electrodes consists of two parts: the top part has electrodes in the horizontal direction, whereas the bottom part has electrodes in the vertical direction. As a result, the fringe field drives the top part of the LC directors to tilt down in the upper and lower directions, whereas it drives the bottom part of the LCs to tilt down to the left and right directions. The Samsung group also proposed a chevron shaped electrode to form four domains. The viewing angle is further improved.

Director profile

Figure 12.11 depicts the computer-simulated LC director profile and the corresponding light transmission through crossed polarisers [35]. In Figure 12.11 the cell parameters used for the calculations are a LC cell gap $d = 5\,\mu m$, electrode width $w = 5\,\mu m$, electrode gap $l = 10\,\mu m$, and the LC directors are at $10°$ with respect to the long edges of the inter-digital electrodes. The LC materials used for simulation have $\Delta\varepsilon = +5$ and homogeneous alignment. The applied voltage is $V = 7\,V_{rms}$ and the wavelength used to calculate light transmittance is $\lambda = 550\,nm$.

In general, the fringing field strength within the LC layers is dependent on the electrode gap and width, and the LC cell gap. In an IPS mode, l is usually greater than w and d so that the

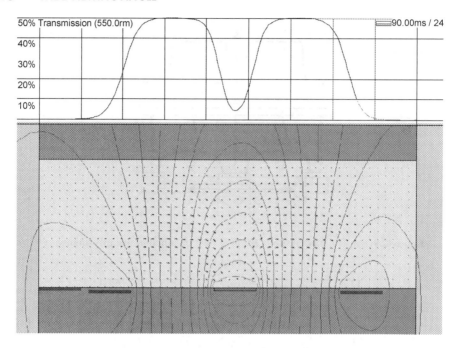

Figure 12.11 Computer-simulated electric field and LC director distribution (bottom) and corresponding transmittance (top) through crossed polarisers of the IPS mode (courtesy of Dr M. E. Becker of Autronic-Melchers GmbH)

fringing field between the gap is basically in the horizontal direction. Near the edges of the electrodes the fields are curved, and at the center the fields are flat. The reorientation angle results from the balance between the dielectric and the elastic torque. However, right above the center of the electrodes, the fringing field is nearly symmetric and the LC directors remain untwisted. As a result, the transmittance is reduced drastically right above the electrodes. In some IPS modes, opaque metals are used for the electrodes. Thus, the overall transmittance of the IPS mode is smaller than that of the TN mode. Widening the electrode gap would improve the aperture ratio, however, at the expense of increased operating voltage.

The threshold field and voltage of the IPS mode can easily be derived by solving the balance of dielectric and elastic torque. The results are expressed as follows [30]:

$$E_{th} = \frac{\pi}{d} \left(\frac{K_{22}}{\Delta \varepsilon} \right)^{1/2} \tag{12.6}$$

$$V_{th} \simeq E_{th} l = \frac{\pi l}{d} \left(\frac{K_{22}}{\Delta \varepsilon} \right)^{1/2} \tag{12.7}$$

If the electrode gap is increased for the purpose of enhancing the aperture ratio, the required voltage is increased proportionally. The threshold voltage of the IPS mode is different from TN by the ratio l/d. A typical electrode gap is $l \simeq 6$–$8\,\mu m$ and $d \simeq 4$–$5\,\mu m$. Thus, the IPS mode generally has a higher operating voltage than TN. To lower the operating voltage, LC mixtures with higher dielectric anisotropy need to be considered.

In the bulk area, the molecular reorientation angle is not uniform. As the distance from the bottom electrodes increases, the field strength gets weaker. Therefore, the LC directors are twisted from the bottom to the top substrates, as shown in Figure 12.11. Similarly to a TN cell, this twisting nature helps reduce the wavelength dependency of the transmitted light.

Viewing angle

The major advantages of the IPS over the TN mode are the much wider viewing angle, a uniform gray scale, and reduced color shift. Figure 12.12 shows the iso-contrast contour of a Hitachi IPS panel [36]. The viewing angle at the $CR=5$ contour exceeds $\pm 70°$. With the advances of new liquid crystal materials and 15 V driver ICs, an 18″ IPS panel with 25 ms response time has been demonstrated [37]. To further improve the IPS response time, a polymer stabilisation method has been developed. This improvement is achieved by introducing a low-density, stabilising polymer network that causes the LC directors to favor the zero-field orientation [38]. The trade-offs are in the reduced transmission and slightly higher drive voltage.

Color shift

Color shift is another important issue for liquid crystal displays. In an IPS mode, a yellowish color shift occurs at the $\phi = 45°$ azimuthal angle and a bluish color shift occurs at $\phi = -45°$, due to the phase retardation difference. To suppress the color shift, a chevron-shaped electrode similar to a two-domain structure has been proposed [39, 40]. Each pixel is divided into two domains where the LC directors face in opposite directions and the color shift is compensated for effectively.

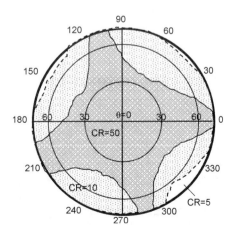

Figure 12.12   The iso-contrast contour of a Hitachi IPS LCD panel

4.2   Figure-on-plane mode

The conventional IPS LCDs utilise opaque metals for interdigital electrodes. Thus, the aperture ratio is reduced to nearly half that of the TN cell. To improve light efficiency, the ERSO group has developed a finger-on-plane (FOP) structure using ITO as electrodes [41]. Figure 12.13 shows a FOP structure where the finger electrodes are placed on top of the plane electrode. The vertical aligned LC is considered. The fringing field directions are shown as parabolic lines in Figure 12.13. Between adjacent electrodes, two fringing fields are generated. Above the electrodes, two fringing fields are twisting the LC molecules in a complementary way. So, the IPS drawback of a small aperture ratio is improved.

It is shown experimentally and theoretically that with proper finger electrode design the light modulation efficiency of an FOP LCD could reach more than 80% of TN LCD [42, 43]. Moreover, since a storage capacitor is automatically formed in the overlapping area of finger and plane electrodes, the overall light efficiency could be further improved. This property is not seen in the other LC mode, and is a unique feature of the FOP LCD.

4.3   Fringing field switching (FFS)

Another way to improve the aperture ratio while retaining a wide viewing angle is the fringe field switching (FFS) mode developed by Hyundai [44]. The basic structure of the FFS is similar to that of the IPS except for the much smaller electrode gap ($l \simeq 0\text{–}1 \, \mu\text{m}$). In the IPS mode, the gap l between the electrodes is larger than the cell gap d. The horizontal component of the electric field is dominant between the electrodes. However, in the FFS mode where $l<d$, the fringing field exists above the electrodes, similar to the FOP mode discussed above. The fringing fields are able to reorient the LC directors above the electrodes. Therefore, high transmittance is expected.

In a FFS mode, both positive and negative $\Delta\varepsilon$ liquid crystals can be used. The negative $\Delta\varepsilon$ LC tends to have a higher on-state transmittance than the positive LC. This is because the directors of the positive $\Delta\varepsilon$ LC tend to align along the field, so that it does not contribute to the phase retardation. Figure 12.14 shows a FFS structure with a homogeneous alignment and positive $\Delta\varepsilon$ LC mixture. The fringing field covers both the electrodes and the gaps.

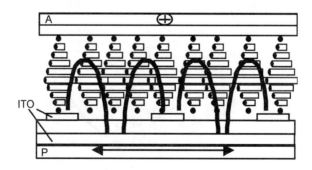

Figure 12.13   The device structure of an ERSO figure-on-plane mode

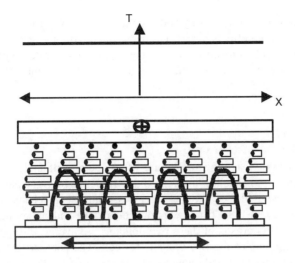

Figure 12.14 Device structure and LC orientation of the fringing field switching mode in a voltage-on state

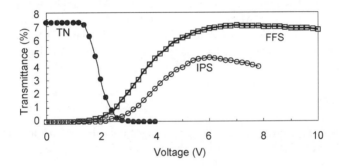

Figure 12.15 Voltage-dependent transmittance of the TN, IPS and FFS modes

Unlike the IPS mode, there is no dead zone prohibiting transmittance. The transmittance is therefore improved.

Figure 12.15 compares the panel transmittance of a TFT LCD operating at TN, IPS and FFS modes. The TN LCD has the normally white display, high transmittance and low operating voltage. On the other hand, the IPS and FFS modes are normally black under the crossed-polariser condition. The transmittance of the FFS mode is about 95% of that of the TN cell. By contrast, the IPS mode transmittance is about 60% of that of the TN cell. The viewing characteristic of the FFS is very similar to that of the IPS; both are much wider than that of the TN.

5 Multi-domain TN

To overcome the narrow viewing angle problem of a single-domain LCD, multiple domains is a logical extension [44]. Chronologically, two-domain TN and VA (vertical alignment)

cells were proposed first, followed by four-domain ones. To fabricate multi-domains, the rubbing and fringing field effects have been investigated. Among these various approaches, the four-domain VA with compensation film has been demonstrated with a viewing angle wider than $\pm 80°$, a contrast ratio higher than $300:1$, and a response (frame) time shorter than 25 ms.

5.1 Two-domain TN

In a two-domain approach, as shown in Figure 12.16, the pre-tilt angles of the domains face in opposite directions, so that the incident light from left or right passes both domains. To control the opposite tilts of the LC directors, multiple mechanical buffing [46, 47], multiple alignment layers [48], fringe electric fields [49], and photo-alignment [50] have been developed. The EO effects of these two domains are averaged. Figure 12.17 plots the iso-contrast ratio contours of the two-domain TN cell. The contour line at $CR = 10$ shows that both upper and lower vertical viewing angles are almost equal (around 25°). It is the averaged value of the upper (about 15°) and lower (about 35°) viewing angles of the single-domain TN shown in Figure 12.1.

The two-domain TN improves the upper vertical viewing angle at the sacrifice of the lower vertical viewing angle. Moreover, the disclination lines (around 5 μm wide) between two domains cause light transmission loss. Thus, the two-domain TN does not provide a sufficiently large viewing angle for large-sized displays.

5.2 Four-domain TN

To further expand the viewing angle, four domains are considered. The domain configuration is plotted in Figure 12.18. Each pixel is divided into four sub-pixels. The LC alignment direction for each sub-pixel is complementary.

Several methods have been developed to generate four domains. These include reverse rubbing, double evaporation, photo-alignment and domain control electrodes. The SiO_x

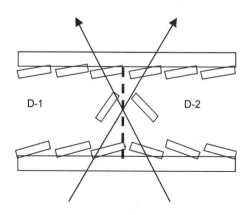

Figure 12.16 Director orientations of a two-domain TN cell

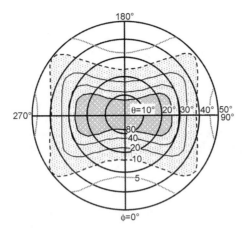

Figure 12.17 The iso-contrast contour of a two-domain TN display

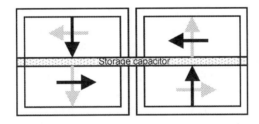

Figure 12.18 Rubbing directions of a four-domain TN cell. Solid and gray arrows represent the rubbing directions for top and bottom substrates, respectively

evaporation method is more suitable for small rather than large panel fabrication because of the need for a vacuum. The reverse rubbing processes [51] contain four steps: the substrate coated with the polyimide alignment layer is rubbed once; half of the domain is then protected with photoresist; the uncovered part is rubbed in the opposite direction; and the photoresist is removed. In the photo-alignment method [52], photosensitive polymer was coated over the substrate. To generate the pre-tilt angle, each domain was twice exposed to UV light using a photomask. By repeating these processes four times, a four-domain LCD is generated. In the domain control electrode method [53], four domains are formed through a controlled electric field on the TFT substrate and then stabilised by a polymer network.

The iso-contrast curve of the four-domain LCD shows symmetric up-down and left-right viewing characteristics. The CR greater than 10 is found over the $\pm 50°$ viewing cone without gray scale inversion. Besides the four-domain structure, adding compensation films is helpful to further widen the viewing angle [54]. Reducing the $d\Delta n$ value of the TN cell helps to increase the viewing angle [55, 56]. However, the trade-off is in the lower transmittance, i.e. higher power consumption.

5.3 *Amorphous-TN*

Continuing to divide the domain into a finer scale (8, 16 and 32 domains) would improve the viewing angle further, except that the fabrication processes would become too tedious. To overcome this problem, the Stanley group proposed an interesting concept on the amorphous domain TN (a-TN) [57]. The a-TN is fabricated by injecting a chiral molecule doped nematic LC into a cell at isotropic phase and slowly (at about $0.1°C/min$) cooled down to room temperature. The inner surfaces of the substrates are not rubbed. To obtain a 90° TN cell, the cell gap to chiral pitch ratio (d/p) is chosen to be 1/4.

As the temperature is reduced, the nucleation of the non-rubbing cell does not start from the surface but from the bulk region [58]. If the temperature is kept at least 30° below the clearing point, the LC molecules touching the polyimide surface are immobilised and anchored to the surface. After the cooling process is completed, the amorphous domain structure is formed.

The domain boundary causes negligible light scattering based on the result that the transmittance of the a-TN cell is only 3% lower than the conventional TN cell. The voltage-dependent light transmittance of the a-TN basically follows that of a nominal TN cell. The viewing angle of the a-TN cell with a phase-compensation film is greater than $\pm 50°$ without gray scale inversion [59].

In the a-TN cell, the aligned LC does not have any pre-tilt angle. Thus, when a finite voltage is applied to the cell, the reverse tilt disclination lines appear. To stabilise the disclinations, a LC polymer-stabilised a-TN is proposed [60]. A 0.5–1.0 wt% LC diacrylate monomer, a photo-initiator, and a material preventing thermal polymerisation are mixed in the chiral TN LC mixture and injected into a cell at the isotropic phase. The amorphous alignment is achieved by cooling down the cell to room temperature. The cell is biased at a small voltage ($V \simeq 2\,V$) while exposed to UV for polymerisation. As a result, the tilt angle in the bulk region is generated and fixed. The reverse tilt disclinations in the a-TN cell are frozen and immobilised.

The polymer-stabilised a-TN eliminates the reverse tilt, preserves the wide viewing angle as in the original a-TN, and improves the response time. The trade-off is the reduced brightness. In the case of 1% polymer concentration, the transmittance is reduced by about 15%.

6 *Multi-domain Vertical Alignment*

As discussed in Chapter 3, vertical alignment (VA) offers excellent contrast ratio and low operation voltage. A major problem is that its viewing angle is limited to $\pm 20°$. Thus, the single-domain VA is good for projection, but not suitable for direct-view displays. To improve the viewing angle, several approaches to achieve multi-domain vertical alignment have been developed [61]. A major difference between multi-domain VA and TN is in the dark and bright disclination lines. In normally white multi-domain TN cells, the disclination lines occur at a high voltage state, i.e. the domain boundaries are bright in a dark background. Therefore, black matrices to cover these disclinations lines are necessary. In most multi-domain VA (MVA) cells, the disclination lines occur at a null voltage state, which is black. Thus, no black matrix is needed.

Owing to their promising features, various MVA cells have been investigated extensively. We will begin with the Fujitsu protrusion method, patterned VA or surrounding electrode method, polymer-stabilised VA, backside exposure MVA, and then enhanced VA approaches.

6.1 Protrusions method

A Fujitsu group demonstrated a multi-domain VA (MVA) TFT-LCD by implementing protrusions on both TFT and color filter substrates, as shown in Figure 12.19 [62]. The protrusions were formed before coating the alignment layer. No rubbing on the alignment layer is needed. The domains are automatically controlled by the slopes of the protrusions.

In the voltage-off state, the LC directors align nearly vertically to the substrates. The light leakage phenomenon is analysed in Section 1 above. To obtain a wide viewing angle, a biaxial film could be added. A MVA panel with a viewing angle greater than 80° at CR>10 has been demonstrated [63]. In the voltage-on state, the LC ($\Delta\varepsilon<0$) directors are reoriented by the fringing electric field and they form multiple domains. For a given cell thickness, the transmittance depends on the protrusion gap, width and height. A wider protrusion gap would lead to a larger aperture and higher light throughput at the expense of a higher operating voltage. An early version of the Fujitsu panel has an aperture ratio of about 58%.

Real-virtual protrusions

To improve the aperture ratio and simplify the fabrication process, the Fujitsu group proposed a modified design using protrusions only on the color filter side. On the TFT side, the protrusions are replaced by a patterned ITO slit, as shown in Figure 12.20 [64]. The fringe fields generated near the edges of ITO electrodes serve as virtual protrusions. Combining the real and virtual protrusions, the wide viewing angle characteristic is preserved.

The replacement of bottom protrusions by fringing fields offers two advantages: a higher contrast ratio and a larger aperture ratio. The contrast ratio of a VA cell is determined by the light leakage in the voltage-off state. In the case of real protrusions, LC molecules are tilted at an angle, depending on the slope of the protrusions. In the ITO slit design, no such

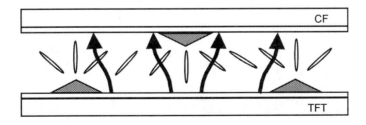

Figure 12.19 A four-domain vertical alignment cell using protrusions on both TFT and color filter substrates

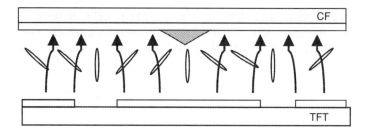

Figure 12.20 Four-domain vertical alignment cell using protrusions on color filter substrate. On the TFT side, patterned ITO is used to generate a fringing field

alignment perturbation occurs. Thus, the replacement of protrusions with ITO cuts out almost half of the light leakage. As a result, the contrast ratio is improved from 300 to 500 : 1. Moreover, the VA mode is operated at normally black mode. No black matrix is needed to block the light leakage around disclination lines. The real-virtual protrusion approach boosts the aperture ratio to 68% and reduces fabrication by one step.

The slit width has an important effect on the transmittance. When the slit is too narrow, the edge field effect is too weak to reorient the LC directors, leading to low transmittance. On the other hand, if the slit is too wide, the effective area of the pixel is decreased. A large portion of LC molecules is not activated. As a result, the transmittance is reduced. The optimal slit width is about 10 μm.

Enhanced vertical alignment (EVA)

A Samsung group proposed another real-virtual protrusion concept, called enhanced VA [65]. In the EVA mode, the protrusion and fringing field effects are integrated on the TFT side, as shown in Figure 12.21. The ITO covered protrusion at the center region of the sub-pixel is formed on gate lines by using metal/SiN$_x$/passivation/ITO layers. The operation mechanism is similar to that discussed above except that the disclinations are localised on the protrusion regions, as indicated by the thick lines in Figure 12.21.

The major advantage of the EVA mode is the simple fabrication process and large margin in cell assembly. However, etching ITO patterns take extra time. An XGA panel with

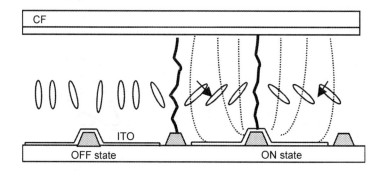

Figure 12.21 An enhanced VA cell with a protrusion and fringing field on the TFT side

viewing cone greater than 80°, 300:1 contrast ratio and 26 ms response time has been demonstrated.

6.2 Patterned vertical alignment (PVA)

Another way to generate MVA is to use patterned electrodes, as shown in Figure 12.22. The discrete electrodes on the top and bottom substrates only partially overlap [66]. This creates electric fields in the opposite directions, as indicated by arrows. The multiple domains are thus formed. In the middle plane of the electrode gap, the dielectric torque induced by these opposite electric fields is cancelled and the LC molecules remain untwisted.

In such a surrounding electrode [67], also called the patterned vertical alignment (PVA) mode [68], ITO patterning is needed only on the color filter side. No rubbing is required. Thus, the fabrication process is relatively simple and the yield is high. The transmittance of the PVA mode is governed by two factors: disclination lines and LC alignment in each sub-domain. The width of domain boundaries formed in the pixel is narrow. Because the PVA mode is normally black, it is not necessary to have a black matrix covering the disclination lines. Therefore, the PVA mode could, in principle, lead to a higher light throughput.

6.3 Polymer-stabilised vertical alignment (PSVA)

Similarly to the Fujitsu protrusion method, the ERSO group used polymer to stabilise the liquid crystals in the gibbous lattices. The device structure of the polymer-stabilised vertical alignment (PSVA) display is depicted in Figure 12.23 for a voltage-on state. Both top and bottom substrates have ITO and alignment layers. However, no rubbing is required. Because of the alignment layers, the negative $\Delta\varepsilon$ LC molecules are anchored perpendicularly on the boundaries. Symmetric and small pre-tilt around the pixel is introduced by the gibbous square boundaries without rubbing. In the voltage-on state, the LC reorientation angles depend on those near the gibbous boundaries.

Although the gibbous boundaries produce pre-tilt angles around the pixels, they are not strong enough to stabilise the director field as the voltage is abruptly switched on. The evolution of the texture takes a long time to stabilise resulting in a very slow response time.

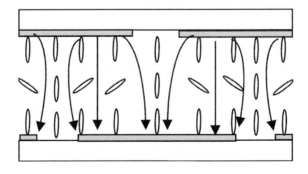

Figure 12.22 Device structure of a patterned electrode vertical alignment cell

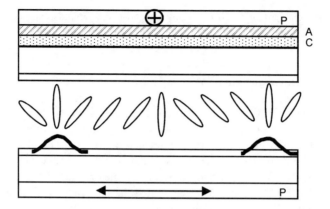

Figure 12.23 Device structure and fringing field directions of a polymer-stabilised vertical aligned cell. A = positive birefringence film, C = negative birefringence film, and P = polariser

To stabilise the director field, a small percent (less than 5%) of diacrylate monomer is dissolved in the LC mixture and injected into the cell. After UV polymerisation, the polymer network stabilises the director field in a symmetric form. In addition, this polymer network helps to accelerate the relaxation process. The rise time (0–90%) of a 4 μm cell ($d\Delta n = 0.3\,\mu$m) is about 16 ms and the decay time (100–10%) is about 8 ms. So the total response time is about 24 ms.

The viewing angle of the film-compensated PSVA cell is similar to that of the Fujitsu MVA cell. In all directions, the contour line with CR = 10 is over ± 70°. No gray scale inversion within 70° of the viewing cone was observed.

6.4 Reverse exposure MVA

In the PSVA cell the gibbous lattices are fabricated by a photolithographic method. To simplify the fabrication process, a reverse exposure method was developed by the ERSO group, as shown in Figure 12.24. The matrix of the ridge field is formed only on the TFT plate. Taking the bus lines of the TFT array as a mask, the pattern of ridge matrix is determined by the reverse exposure. Since the ridge surface is curved, it generates a different

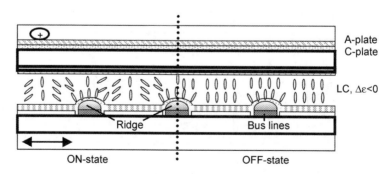

Figure 12.24 Device structure of the ERSO reverse exposed multi-domain vertical alignment cell

pre-tilt angle on the LC directors. It functions like protrusions or gibbous lattices as mentioned above. In the voltage-on state, the ridge and fringe field effects altogether generate a multi-domain in a pixel. The merits of this mode are that only a single layer for a ridge matrix is necessary for a multi-domain, and a pattern of ridge matrix is formed by and self-aligned with the TFT metal layers through the cost-effective reverse exposure method.

7 Axially Symmetric-aligned Microcell (ASM) Mode

Another wide view mode developed by Sharp is called the axially symmetric-aligned microcell (ASM) [69]. In the ASM mode, the LC alignment has spiral distribution, as shown in Figure 12.25, and is symmetric with respect to the cell normal. The ASM mode is obtained through the phase separation process from the mixture of LC and polymerisable resins, and the aligning force of the liquid crystal itself. The polymer walls are formed by UV exposure to construct microcell pixels. The axially symmetric LC alignment is stabilised by the polymer walls. Polymer walls play a crucial role in preventing the destruction of the axially symmetric orientation of a liquid crystal. No rubbing is required. The 90° TN is achieved by doping a chiral compound in the LC mixture employed with $d/p = 1/4$. The luminance of the ASM mode is about 81%, as compared to the TN cell.

Both positive and negative $\Delta\varepsilon$ LC mixtures, designated as P-type and N-type ASM modes, have been investigated. The P-type ASM is obtained by phase separation. However, for the N-type ASM, the LC alignment and tilt angles are controlled by polymeric monomers. After setting at the desirable direction and tilt angle, the monomers are polymerised by UV exposure. As a result, the alignment direction and tilt angle are memorised. Figure 12.26 shows a model of the polymer-controlled tilt appearing element [70]. The appearing elements induce the LC to align in a radical sense.

The P-type has a viewing angle of $\pm 60°$ at CR>10:1 in the right-left and up-down directions, and $\pm 40°$ for the 45 and 135° viewing directions. By contrast, the N-type has a viewing angle of $\pm 80°$ at CR>10:1 in the right-left and up-down directions. However, for the 45 and 135° viewing directions, the viewing cone remains at $\pm 40°$, as for the P-type. With a special phase retardation film, the viewing angle of the N-type ASM mode is wider than $\pm 80°$ in all directions.

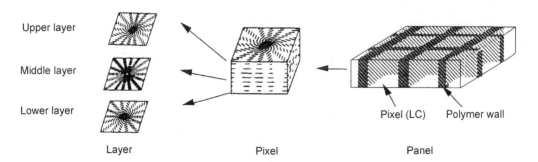

Upper layer

Middle layer

Lower layer

Pixel (LC)   Polymer wall

Layer   Pixel   Panel

Figure 12.25 Cell, pixel, and LC directors on the upper, middle and lower layers of the Sharp axially symmetric-aligned microcell [69]

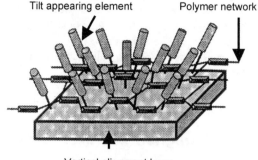

Tilt appearing element   Polymer network

Vertical alignment layer

Figure 12.26 Model of the polymer-controlled tilt appearing elements in the N-type axially symmetric-aligned microcell [70]

The ASM mode has been applied to a 42 inch plasma-addressed liquid crystal display [71]. The contrast ratio of the P-type is about 200 : 1 and of the N-type it is 300 : 1. The response time of a 4.5 μm N-type cell is about 30 ms, similar to that of a TN cell.

8 Film-compensated π-cell

In the π-cell approach [72], the pre-tilt angles in the top and bottom substrates are in opposite directions, i.e. anti-parallel, as shown in Figure 12.27. The opposite pre-tilt angles make two important contributions to the electro-optic properties of the π-cell. First, it gives rise to the optical self-phase compensation effect on the LC directors. As a result, the viewing angle is wide and symmetric. Second, the bend directors in a voltage on-state eliminate the back-flow effect, resulting in a fast relaxation time.

As the applied voltage increases, the LC directors go through splay and then bend deformation, as illustrated in Figure 12.28. Based on the Gibbs free energy calculation, below the critical voltage ($V_c \simeq 1.8$ V) splay is more stable than bend, and above V_c the bend state is more stable [73]. Thus, with proper phase compensation, the optically compensated bend (OCB) cell exhibits a wide and nearly symmetric viewing angle ($\pm 60°$ without gray scale inversion) and fast response time, and is a potential candidate for both transmissive and reflective display applications [74].

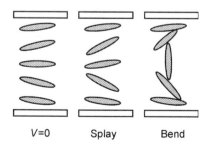

V=0   Splay   Bend

Figure 12.27 The splay and bend states of a π-cell

For a single domain device, the π-cell has been used as a high-speed light shutter without problem. However, for flat panel displays the π-cell exhibits some technical difficulties. It is found that the transition from splay to bend mode is heavily dependent on the applied voltage, pulse width and electrode size. In a display panel experiment, the Tohoku group reported the following detailed observations [75]. The transition starts from one or a few points in each pixel electrode and spreads to the whole area. When there are no starting points in a pixel electrode, the transition from splay to bend does not take place. The number of starting points increases as the applied voltage increases. Areas with no electrodes impede the spread of the bend alignment transition. The speed of the bend alignment transition increases as the applied voltage increases. Based on these analyses, how to initiate uniform splay-to-bend transition and reduce defects are challenging tasks before the OCB cell can be practically implemented for display panels.

References

1 A. Lien, H. Takano, S. Suzuki, and H. Uchida, The symmetry property of a 90° nematic liquid crystal cell, *Mol. Cryst. Liq. Cryst.* **198**, 37 (1991).
2 H. L. Ong, *Japan Display'92*, p. 247 (1992).
3 Y. Fujimura, T. Nagatsuka, H. Yoshimi, and T. Shimomura, Optical properties of retardation films for STN-LCDs, *SID Tech. Digest* **22**, 739 (1991).
4 S. T. Wu, Film-compensated homeotropic liquid crystal cell for direct-view display, *J. Appl. Phys.* **76**, 5975 (1994).
5 S. Z. D. Cheng, F. Li, E. P. Savitski, and F. W. Harris, Uniaxial negative birefringent aromatic polyimide thin films as optical compensators in liquid crystalline displays, *Trends in Polymer Science* **5**, 51 (1997).
6 J. P. Eblen, Jr. W. J. Gunning, J. Beedy, D. Taber, L. Hale, P. Yeh, and M. Khoshnevisan, Birefringent compensators for normally white TN-LCDs, *SID Tech. Digest* **25**, 245 (1994).
7 L. H. Wu, Shr-Jie Luo, C. S. Hsu, and S. T. Wu, Obliquely tilted discotic phase compensation films, *Jpn. J. Appl. Phys.* **39**, L869 (2000).
8 H. Hasebe, K. Takeuchi, and H. Takatsu, Properties of novel UV-curable liquid crystals and their retardation films, *J. SID* **3**, 139 (1995).
9 S. T. Wu, Design and fabrication of phase-matched compensation films for LCDs, *SID Application Digest* **26**, 21 (1996).
10 H. Mori, Y. Itoh, Y. Nishlura, T. Nakamura, and Y. Shinagawa, Novel optical compensation film for AMLCDs, *SID Tech. Digest* **27**, 941 (1997).
11 K. Inoue, T. Kurita, E. Yoda, T. Kaminade, T. Toyooka, and Y. Kobori, Novel viewing angle compensation film for TN-LCDs using a hybrid aligned rod-like liquid crystalline polymer, *The 5th Int'l Display Workshops*, p. 255 (1998).
12 T. Toyooka, E. Yoda, T. Yamanashi, and Y. Kobori, Viewing angle performance of TN-LCD with hybrid aligned nematic film, *Displays* **20**, 221 (1999).
13 T. Miyashita, Y. Miyazawa, Z. Kikuchi, H. Aoki, and A. Mawatari, A multicolor wide-viewing-angle STN-LCD with multiple retardation films, *SID Tech. Digest* **22**, 743 (1991).
14 S. Kondo, T. Yamamoto, A. Murayama, H. Hatoh, and S. Matsumoto, A fast-response black and white STN-LCD with a retardation film, *SID Tech. Digest* **22**, 747 (1991).
15 S. T. Wu, Phase-matched biaxial compensation film for CLDs, *SID Tech. Digest* **26**, 555 (1995).
16 T. Yamada, M. Okazaki, and Y. Shinagawa, *The 3rd Int'l Display Workshops*, p. 349 (1996).
17 H. Mori, Y. Itoh, Y. Nishiura, T. Nakamura, and Y. Shinagawa, *Jpn. J. Appl. Phys.* **36**, 143 (1997).

18 M. Schadt and W. Helfrich, Voltage-dependent optical activity of a twisted nematic liquid crystal cell, *Appl. Phys. Lett.* **18**, 127 (1971).

19 T. J. Scheffer and J. Nehring, A new highly multiplexable liquid crystal display, *Appl. Phys. Lett.* **45**, 1021 (1984).

20 Y. Takiguchi, A. Kanemoto, H. Iimura, T. Enomoto, S. Iida, T. Toyooka, H. Ito, and H. Hara, *Euro Display'90*, p. 96 (1990).

21 D. J. Broer, Molecular architectures in thin plastic films by in-situ photopolymerization of reactive liquid crystals, *SID Tech. Digest* **26**, 165 (1995).

22 M. Lu and K. H. Yang, Twisted discotic film compensated single-domain and two-domain TN LCDs, *SID Tech. Digest* **31**, 338 (2000).

23 T. Sergan, M. Sonpatki, J. Kelly, and L. C. Chien, Photo-polymerized discotic films for viewing quality improvement of liquid crystal displays, *SID Tech. Digest* **31**, 1091 (2000).

24 S. Zimmerman, K. Beeson, M. McFarland, J. Wilson, T. J. Creedle, K. Bingaman, P. Ferm, and J. T. Yardley, Viewing angle enhancement system for LCDs, *SID Tech. Digest* **26**, 793 (1995).

25 M. McFarland, S. Zimmerman, K. Beeson, J. Wilson, T. J. Creedle, K. Bingaman, P. Ferm, and J. T. Yardley, SpectraVue™ viewing angle enhancement system for LCDs, *Asia Display'95*, p. 739 (1995).

26 R. A. Soref, Transverse field effect in nematic liquid crystals, *Appl. Phys. Lett.* **22**, 165 (1973).

27 R. A. Soref, Field effects in nematic liquid crystals obtained with interdigital electrodes, *J. Appl. Phys.* **45**, 5466 (1974).

28 R. Kiefer, B. Weber, F. Windscheid, and G. Baur, In-Plane switching of nematic liquid crystals, *Japan Displays'92*, p. 547 (1992).

29 M. Oh-e, M. Yoneya, and K. Kondo, Switching of a negative and positive dielectric-anisotropic liquid crystals by in-plane electric fields, *J. Appl. Phys.* **82**, 528 (1997).

30 M. Oh-e, M. Ohta, S. Arantani, and K. Kondo, Principles and characteristics of electro-optical behavior with in-plane switching mode, *Asia Display'95*, p. 577 (1995).

31 M. Ohta, M. Oh-e, and K. Kondo, Development of super-TFT-LCDs with in-plane switching display mode, *Asia Display'95*, p. 707 (1995).

32 S. H. Lee, H. Y. Kim, I. C. Park, B. G. Rho, J. S. Park, H. S. Park, and C. H. Lee, Rubbing-free, vertically aligned nematic liquid crystal display controlled by in-plane field, *Appl. Phys. Lett.* **71**, 2851 (1997).

33 K. H. Kim, S. B. Park, J. U. Shim, J. H. Souk, and J. Chen, New LCD modes for wide-viewing-angle applications, *SID Tech. Digest* **29**, 1085 (1998).

34 S. H. Hong, Y. H. Jeong, H. Y. Kim, H. M. Cho, W. G. Lee, and S. H. Lee, Electro-optic characteristics of 4-domain vertical alignment nematic liquid crystal display with interdigital electrode, *J. Appl. Phys.* **87**, 8259 (2000).

35 M. E. Becker, H. Wohler, M. Kamm, and J. Kreis, Numerical modeling of IPS effect, *SID Tech. Digest* **27**, 596 (1996).

36 M. Ohta, M. Oh-e, and K. Kondo, Development of super TFT-LCDs with in-plane switching display mode, *Asia Display'95*, p. 707 (1995).

37 M. Ohta, M. Oh-e, S. Matsuyama, N. Konishi, H. Kagawa, and K. Kondo, 18″-diagonal super-TFTs with a fast response speed of 25 ms, *SID Tech. Digest* **30**, 86 (1999).

38 M. J. Escuti, C. C. Bowley, and G. P. Crawford, Enhanced dynamic response of the in-plane switching liquid crystal display mode through polymer stabilization, *Appl. Phys. Lett.* **75**, 3264 (1999).

39 W. S. Asada, N. Kato, Y. Yamamoto, M. Tsukane, T. Tsurugi, K. Tsuda, and Y. Takubo, An advanced in-plane-switching mode TFT-LCD, *SID Tech. Digest* **28**, 929 (1997).

40 Y. Mishima, T. Nakayama, N. Suzuki, M. Ohta, S. Endoh, and Y. Iwakabe, Development of a 19″ diagonal UXGA super TFT-LCM applied with super-IPS technology, *SID Tech. Digest* **31**, 260 (2000).

41 I. W. Wu, D. L. Ting, and C. C. Chang, Advancement in wide-viewing-angle LCDs, *The 6th Int'l Display Workshops*, p. 383 (1999).

42 S. H. Lee, S. L. Lee, H. Y. Kim, and T. Y. Eom, A novel wide-viewing-angle technology, *SID Tech. Digest* **30**, 202 (1999).

43 Z. Meng, H. S. Kwok, and M. Wong, Comb-on-plane switching electrodes for liquid crystal displays, *J. SID* **8**, 139 (2000).

44 S. H. Lee, S. L. Lee, and H. Y. Kim, Electro-optic characteristics and switching principle of a nematic liquid crystal cell controlled by fringe-field switching, *Appl. Phys. Lett.* **73**, 2881 (1998).

45 S. Tanuma, Y. Koike, and H. Yoshida, Japanese patent, JP 63-106624 (1988).

46 K. H. Yang, Two-domain twist nematic and tilted homeotropic liquid crystal displays for active matrix applications, *Int'l Display Research Conference*, p. 68 (1991).

47 K. Takatori, K. Sumiyoshi, Y. Hirai, and S. Kaneko, A complementary TN LCD with wide viewing angle gray scale, *Japan Display'92*, p. 591 (1992).

48 Y. Koike, T. Kamada, K. Okamoto, M. Ohashi, I. Tomita, and M. Okabe, A full-color TFT-LCD with a domain-divided twisted-nematic structure, *SID Tech. Digest* **23**, 796 (1992).

49 A. Lien and R. A. John, Two-domain twisted nematic liquid crystal displays fabricated by parallel fringe field method, *SID Tech. Digest* **24**, 269 (1993).

50 A. Lien, R. A. John, M. Angelopoulos, K. W. Lee, H. Takano, K. Tajima, A. Takenaka, K. Nagayama, Y. Momoi, and Y. Saitoh, UV-type two-domain wide viewing angle TFT-LCD panels, *Asia Display'95*, p. 593.

51 J. Chen, P. J. Bos, D. R. Bryant, D. L. Johnson, S. H. Jamal, and J. R. Kelly, Four-domain TN-LCD fabricated by reverse rubbing or double evaporation, *SID Tech. Digest* **26**, 865 (1995).

52 M. S. Nam, J. W. Wu, Y. J. Choi, K. H. Yoon, J. H. Jung, J. Y. Kim, K. J. Kim, J. H. Kim, and S. B. Kwon, Wide-viewing-angle TFT-LCD with photo-aligned four-domain TN mode, *SID Tech. Digest* **28**, 933 (1997).

53 T. Ishii, H. Murai, M. Suzuki, S. Kaneko, H. Matsuyama, K. Kobayashi, and Y. Hirai, *The 5th Int'l Display Workshop*, p. 73 (1998).

54 J. Chen, K. H. Kim, J. J. Jyu, J. H. Souk, J. R. Kelly, and P. J. Bos, Optimum film compensation modes for TN and VA LCDs, *SID Tech. Digest* **29**, 315 (1998).

55 L. Pohl, G. Weber, R. Eidenschink, G. Baur, and W. Fehrenbach, *Appl. Phys. Lett.* **38**, 497 (1981).

56 J. Chen, P. J. Bos, J. R. Kelly, N. D. Kim, and S. T. Shin, Optical simulation of electro-optical performance of low-nd multi-domain TN displays, *SID Tech. Digest* **28**, 937 (1997).

57 Y. Toko, T. Sugiyama, K. Katoh, Y. Iimura, and S. Kobayashi, TN-LCDs fabricated by non-rubbing showing wide and homogeneous viewing angular characteristics and excellent voltage holding ratio, *SID Tech. Digest* **24**, 622 (1993).

58 Y. Iimura, S. Kobayashi, T. Sugiyama, Y. Toko, T. Hashimoto, and K. Kato, Electro-optic characteristics of amorphous and super-multidomain TN-LCDs prepared by a non-rubbing method, *SID Tech. Digest* **25**, 915 (1994).

59 T. Sugiyama, Y. Toko, T. Hashimoto, K. Katoh, Y. Iilura, and S. Kobayashi, Analytical simulation of electro-optical performance of amorphous and multidomain TN-LCDs, *SID Tech. Digest* **25**, 919 (1994).

60 T. Hashimoto, K. Katoh, H. Hasebe, H. Takatsu, Y. Iwamoto, Y. Iimura, and S. Kobayashi, *Int'l Display Research Conference*, p. 484 (1994).

61 H. Vithana, Y. K. Fung, S. H. Jamal, R. Herke, P. J. Bos, and D. L. Johnson, A well-controlled tilted-homeotropic alignment method and a vertically aligned four-domain LCD fabricated by this technique, *SID Tech. Digest* **26**, 873 (1995).

62 K. Ohmuro, S. Kataoka, T. Sasaki, and Y. Koike, Development of super-high-image-quality vertical alignment-mode LCD, *SID Tech. Digest* **28**, 845 (1997).

63 A. Takeda, S. Kataoka, T. Sasaki, H. Chida, H. Tsuda, K. Ohmuro, Y. Koike, T. Sasabayashi, and K. Okamoto, A super high image quality multi-domain vertical alignment LCD by new rubbing-less technology, *SID Tech. Digest* **29**, 1077 (1998).

64 Y. Tanaka, Y. Taniguchi, T. Sasaki, A. Takeda, Y. Koibe, and K. Okamoto, A new design to improve performance and simplify the manufacturing process of high quality MVA TFT-LCD panels, *SID Tech. Digest* **30**, 206 (1999).

65 J. O. Kwag, K. C. Shin, J. S. Kim, S. G. Kim, and S. S. Kim, Implementation of new wide viewing angle mode for TFT-LCDs, *SID Tech. Digest* **31**, 256 (2000).

66 A. Lien and R. A. John, Multi-domain homeotropic liquid crystal display for active matrix application, *Euro Display'93*, p. 21 (1993).

67 N. Koma, Y. Baba, and K. Matsuoka, No-rub multi-domain TFT-LCD using surrounding-electrode method, *SID Tech. Digest* **26**, 869 (1995).

68 K. H. Kim, K. H. Lee, S. B. Park, J. K. Song, S. N. Kim, and J. H. Souk, Domain divided vertical alignment mode with optimized fringe field effect, *Asia Display'98*, p. 383 (1998).

69 N. Yamada, S. Kohzaki, F. Funada, and K. Awane, Axially symmetric aligned microcell mode, *SID Tech. Digest* **26**, 575 (1995).

70 Y. Kume, N. Yamada, S. Kozaki, H. Kisishita, F. Funada, and M. Hijikigawa, Advanced ASM mode, *SID Tech. Digest* **29**, 1089 (1998).

71 N. Yamada, K. Himejima, T. Akai, H. Uede, M. Imai, K. Endo, T. Kakizaki, R. van Gelder, and C. van Winsum, A 42-in PALC display with wide viewing angle using ASM mode, *SID Tech. Digest*, **29**, 203 (1998).

72 P. J. Bos and K. R. Koehler/Beran, The □-cell: a fast liquid crystal optical switching device, *Mol. Cryst. Liq. Cryst.* **113**, 329 (1984).

73 T. Miyashita, P. Vetter, M. Suzuki, Y. Yamaguchi, and T. Uchida, *Proc. of the 13th Int. Display Research Conference* (Strasbourg, France, August 31, 1993), p. 149.

74 C. L. Kuo, T. Miyashita, M. Suzuki, and T. Uchida, Improvement of gray-scale performance of OCB display mode for AMLCDs, *SID Tech. Digest* **25**, 927 (1994).

75 N. Koma, T. Miyashita, T. Uchida, and K. Yoneda, Using an OCB-mode TFT-LCD for high speed transition from splay to bend alignment, *SID Tech. Digest* **30**, 28 (1999).

Index